The Emil and Kathleen Sick Lecture-Book Series
in Western History and Biography

MILLS AND MARKETS

A History of the Pacific Coast
Lumber Industry to 1900

By Thomas R. Cox

UNIVERSITY OF WASHINGTON PRESS

Seattle and London

Library of Congress Cataloging in Publication Data

Cox, Thomas R   1933–
    Mills and markets.

    (The Emil and Kathleen Sick lecture-book series in
western history and biography)
    Bibliography: p.
    1. Lumber trade—Pacific Coast—History. 2. Saw-
mills—Pacific Coast—History. I. Title. II. Series.
HD9757.A5C69   338.4′7′6740979   74-9505
ISBN 0-295-95349-7

61683

For MARY, *who made it all possible*

The Emil and Kathleen Sick
Lecture-Book Series
in Western History and Biography

UNDER the provisions of a Fund established by the children of Mr. and Mrs. Emil Sick, whose deep interest in the history and culture of the American West was inspired by their own experience in the region, distinguished scholars are brought to the University of Washington to deliver public lectures based on original research in the fields of Western history and biography. The terms of the gift also provide for the publication by the University of Washington Press of the books resulting from the research upon which the lectures are based. This book is the second volume in the series.

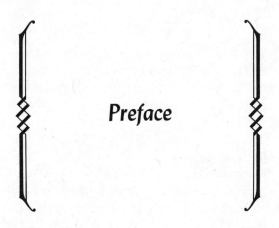

Preface

Long before permanent title to the area had been established, enterprising Europeans and Americans had commenced harvesting and sawing into lumber the forest resources that stretch along the Pacific coast of North America. Shipping their cut by sea, they found widely scattered outlets for the product of their labors. From these embryonic beginnings a major industry evolved. By the 1880s it had assumed vast dimensions and exerted widespread influence. However, as a writer for the San Francisco *Journal of Commerce* noted, few realized its importance. Even in San Francisco, gateway to one of the coastal lumber industry's main markets and the center from which it was controlled, the industry presented an unimpressive front; "but these piles of lumber, these modest little coops of offices represent thousands upon thousands of acres of land, millions upon millions of capital . . . [and] employment to . . . hands all the way from Santa Cruz on the south to Alaska on the north."[1] This anonymous observer was correct in his assessment of both the importance of the industry and the public's failure to recognize it. It is a failure that has been widely shared since.

Throughout their history the cargo mills—as the sawmills that produced lumber for this sea-based trade were known—served as more than just a source of employment and of building materials for the Far

---

1. Reprinted in *Canada Lumberman and Millers', Manufacturers', and Miners' Gazette* 3 (15 Oct. 1883):307. See also San Francisco *Alta California*, 1 Jan. 1886.

West. Collectively, they encouraged the permanent settlement of sizable areas bypassed by miners and agriculturists, produced one of the West Coast's leading exports, and, through this, drew money into a region that, for all its wealth in precious metals, was short on investment capital.

Like any industry, the coastal lumber industry was influenced by what transpired in the scattered realms that were its major markets: San Francisco and its hinterlands, southern California, Latin America, Hawaii, Australia, and China. Indeed, the cargo mills were often more affected by what transpired in the lands of the Pacific Basin than by what was taking place in other parts of the United States. The discovery of mineral deposits in Australia and their subsequent exploitation, the building of railroads in North China, and the rise of plantation agriculture in Hawaii had a great impact on them; the Civil War, relatively little. Not until long after the completion of the transcontinental railroads did the industry turn to the area east of the Cascades and Sierra Nevada as the major outlet for its cut; only then did mills of the Pacific Coast come to be more a part of the economy of the United States than that of the Pacific Basin.

The coastal lumber industry grew to maturity in the age of enterprise, a period that saw many another industry become both larger and increasingly more oligopolistic. But no John D. Rockefeller, no Standard Oil emerged to bring order into the chaos that was the uncertain, highly competitive world of the cargo mills. Then, and in years to come, the lumber industry remained fragmented, perhaps the most purely competitive of any major industry in the United States. To be sure, lumbermen repeatedly took steps that in other fields led to trusts and other forms of combination, but their efforts brought neither order nor lasting combination to the cargo trade. Too many factors were working in the opposite direction: the scattered resource base upon which the mills drew; the ease of entry into the business; the body of federal land laws that, for all its failures, made the erection of an effective timber monopoly nearly impossible; and, perhaps too, the absence of leaders with that peculiar combination of characteristics possessed by Rockefeller and his ilk.

Though the failure of centralization set it apart, the coastal lumber industry underwent many changes that closely paralleled developments in other areas of industrial enterprise. Firms integrated their holdings both vertically and horizontally, expanded and modernized their plants to remain competitive, and sought new markets and new products that would give them an advantage over rivals. Since they marketed their cut by sea, the cargo mills adjusted not only to technological changes in logging and sawmilling, but to those in marine transportation as well. Entrepreneurial leadership played a vital role in determining which firms prospered. Those who rose to prominence were not always those with the greatest knowledge of lumbering. Sometimes, as in the case of Asa Mead Simpson, a leader emerged whose expertise lay more at sea than in the woods or mills. But, whether advantage was gained ashore or at sea, it seems clear that those firms that prospered did so not just because they were the largest and financially soundest (in their early years they often were not), but because of the quality of leadership they enjoyed.

The coastal lumber industry was never made up solely of large and successful firms. There was a host of small, marginal operations; numerous mills failed for every one that grew large and prospered. Sawmills appeared on tiny, inhospitable ports, among sparse stands, and on cutover lands as well as in the more promising areas around Puget Sound, Coos and Humboldt bays, and Grays Harbor. What the smaller, less successful firms did had its impact on the larger. The history of the Pacific Coast lumber industry is the history of both.

Much has been written on the lumber industry of the Far West. But, although there have been studies of individual companies and areas of production, as yet no history of the cargo trade in its entirety has appeared. This book is designed to fill that void: it both synthesizes the earlier, more narrowly focused studies and goes beyond them. It investigates important aspects of the cargo trade that have not been analyzed previously, such as the market forces operating in the various outlets where the cut of the cargo mills was sold, the tiny mills on the more remote and inhospitable anchorages of the Pacific coast, and the maritime aspects of this seaborne trade.

At the same time, there are certain things this work is not. It is not a study in labor history. Some attention is necessarily given to the labor force of the lumber industry, but those wishing to pursue that subject should consult Vernon Jensen's *Lumber and Labor* and works by Harold Hyman, Robert Tyler, and Robert Pike. All give more attention to labor per se than does the present volume.

Nor is this an econometric study. The statistical record of the lumber industry of the Pacific Coast is notoriously incomplete. The records of few firms have survived; many relevant government records have been destroyed. Moreover, much of the available data is of questionable accuracy. In 1866 the Portland *Oregonian* confessed that its correspondents had "utterly failed to get anything like correct statistics"; some years later the *Northwest Lumberman* complained that the industry was "in the hands of powerful monopolists, who have the notion that a dark policy is the better one of them." [2] Both publications went ahead and published statistics on the lumber industry anyway.

To be sure, more could have been done on labor and with the economic data that do exist, but to have done so would have necessitated writing an essentially different book. The present volume is a study in institutional business history, a work which seeks to trace the evolving structure, the changing patterns of marketing and management, that marked the Pacific Coast lumber industry during its formative years. The result, it is hoped, is a work that will provide insight not only into the development of the industry and the Far West, but also into the forces at work during the last half of the nineteenth century forging the modern, industrial United States of the twentieth century. If it also provides a background useful to econometricians and students of labor history when they turn their attention to the industry, so much the better.

Parts of two chapters in this book were originally published elsewhere, in slightly altered forms. A portion of chapter 9 appeared as "Lower Columbia Lumber Industry, 1880–93," in the *Oregon Historical Quarterly* 47 (June 1966):160–78. Part of chapter 8 appeared as "Lumber and Ships: The Business Empire of Asa Mead Simpson" in *Forest History* 14 (July 1970):16–26. Both are used here by permission of the original publishers.

It would be impossible to name here all of the people who have been of help to me, but there are some to whom a special debt is owed. Professor Earl Pomeroy of the University of Oregon provided invaluable guidance at the beginning of this study. His high standards of scholarship served as both an inspiration and example; his vast knowledge, generously shared, eased the problems of research; and his skilled editorial touch smoothed the prose that resulted. Edwin R. Bingham, also of the University of Oregon, gave generously of his time even though he was officially on leave to pursue work of his own. More

2. *Oregonian*, 23 Feb. 1866; *Northwest Lumberman*, reprinted in *West Shore* 8 (July 1882):126.

than the usual debt is also owed to Professor Robert Peterson of the University of Montana, who first suggested that there was a topic here worthy of study and guided me in my initial steps; to Richard Berner of the Henry Suzzallo Library, University of Washington, for sharing knowledge gained in his own research; to Martin Schmitt of the University of Oregon library, who gave me free access to the then unopened records of William Kyle and Sons; to Adolphus Andrews, Jr., of Pope and Talbot, Incorporated, who turned over an office to my use and gave me unfettered freedom in using the old records of his firm; and to Marion R. Leitner, whose careful editorial eye made her far more than just a typist. Finally, thanks of a very special sort go to my wife Mary for her immeasurable tolerance, her quiet encouragement, and her many sacrifices.

Of course, whatever faults remain are the product of the author, not of these or the many others who have helped during the long course of this study's preparation.

THOMAS R. COX

San Diego, California

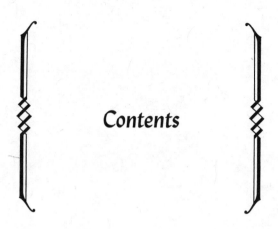

Contents

Illustrations

❧ Illustrations ❧

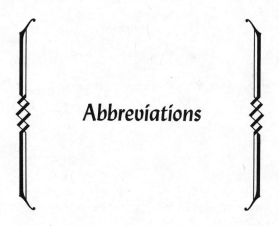

Abbreviations

AC	E. G. Ames Collection, University of Washington library, Seattle
APSA	John C. Ainsworth Papers, Ships' Accounts, University of Oregon library, Eugene
CHSQ	*California Historical Society Quarterly*
EL	George H. Emerson Letter Books, University of Washington library, Seattle
FNP	Fort Nisqually Papers (Hudson's Bay Company), Huntington Library, San Marino, Calif.
JKCP	John Kentfield and Company Papers, Bancroft Library, Berkeley, Calif.
LP	William A. Leidesdorff Papers, Huntington Library, San Marino, Calif.
MP	John Minto Papers, Oregon Historical Society library, Portland
OHQ	*Oregon Historical Quarterly*
OICP	Oregon Improvement Company Papers, University of Washington library, Seattle
PBMCP	Port Blakely Mill Company Papers, University of Washington library, Seattle
PHR	*Pacific Historical Review*
PNQ	*Pacific Northwest Quarterly*
PTP	Pope and Talbot, Inc., Papers, San Francisco
SP	Abel Stearns Papers, Huntington Library, San Marino, Calif.
USBFC	United States Bureau of Foreign Commerce
WHQ	*Washington Historical Quarterly*
WKSP	William Kyle and Sons Papers, University of Oregon library, Eugene

∽ *Abbreviations* ∽

WMCP   Washington Mill Company Papers, University of Washington library, Seattle

WP   Elkanah Walker Papers, Huntington Library, San Marino, Calif.

Mills and Markets

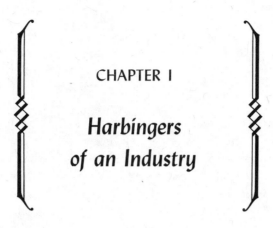

CHAPTER I

Harbingers
of an Industry

As it has been for years, the production of lumber is today one of the most important economic undertakings in Oregon, Washington, and northern California. It is also one of the oldest. Before the arrival of the first whites, Indians along the Northwest coast had been felling Western red cedars and splitting them into planks to use in the construction of their dwellings or fashioning them into giant dugout canoes. When Sir Francis Drake put in for repairs to the *Golden Hind* in 1579 somewhere in the vicinity of present-day San Francisco, forests of the area supplied material that went into refurbishing his vessel. Timber stands along the Pacific coast of North America have been utilized, to one extent or another, ever since.

Although the maritime lumber trade of the Pacific can claim no such antiquity, it does have a long history. The trade was first developed to serve a chain of missions established in Alta California by the Franciscans. Father Junipero Serra founded Mission San Diego de Alcalá, the oldest of these, in 1769. The settlement soon was in need of construction timbers and lumber in quantities too sizable to obtain from local sources. But there was timber to the north—reports of Spanish explorers made that clear. As a result, in 1776 Don Diego Choquet sailed northward in the *San Antonio* to obtain a cargo of construction materials. He arrived at Monterey on 21 May, took on a cargo of pit-sawn timbers, and weighed anchor for the return south on 30 June.[1]

1. Prior to the mechanization of sawmill operations, most lumber was cut by hand at sawpits. On sawpits, see Alan K. Brown, *Sawpits in the Spanish Redwoods* (San Mateo, Calif.: San Mateo

Thus, the year that saw Thomas Jefferson and Adam Smith draft documents that were to make 1776 long remembered was also the one during which, casually and without fanfare, the Pacific lumber trade began.

During the next seven decades awareness of the commercial value of the forests of the Far West slowly grew. By the end of the period, lumber production was firmly established along the Pacific coast. So, too, was a pattern of business that was to last into the twentieth century. Production was centered in sparsely populated areas where forest and sea met. Without significant demand near to hand, producers had to export much of their cut. At a time when land transportation was primitive at best, the Pacific provided routes by which markets in California, Latin America, and Hawaii could be reached. Over the years many changes occurred in logging, milling, and shipping, but the basic pattern remained. Not until the twentieth century did a new sort of lumber industry come to predominate on the Pacific coast of North America. This new industry looked inland rather than to sea for its primary markets, and was served by rails rather than ships. By the time this new industry emerged, the economy of the eastern United States was approaching maturity. But the older industry of the cargo mills (those engaged in the maritime lumber trade) was tied not to a developed economy, but to one whose foundations were just being laid. Its history is thus more than a chapter in the growth of commerce and industry: it is also an important chapter in the development of the Far West.

No established pattern of mills and markets existed in 1776 when Don Diego Choquet sailed with the Pacific lumber trade's first cargo. Only after decades of hesitant probing, of trial shipments, sporadic production, and pioneering efforts by a motley group of Spanish, British, and Americans did anything that could be called a lumber industry emerge.

Lumber continued to be cut in the vicinity of the northern missions

County Historical Association, 1966); C. Raymond Clar, *Harvesting and Use of Lumber in Hispanic California* (Sacramento, Calif.: Sacramento Corral of Westerners, 1971), pp. 9–10; San Jose *Pioneer*, 16 June 1877, 11, 17 Dec. 1880. On Choquet's voyage see Francisco Palou, *Fray Francisco Palou's Historical Memoirs of New California*, ed. Herbert E. Bolton, 4 vols. (Berkeley: University of California Press, 1926), 4:113–15, 141, 390. See also Eugène Duflot de Mofras, *Travels on the Pacific Coast*, trans. Marguerite Eyer Wilbur, 2 vols. (Santa Ana, Calif.: Fine Arts Press, 1937), 1:170; Winifred Davidson, *Where California Began* (San Diego, Calif.: McIntyre Publishing Co., 1929), pp. 86–96.

in the years following that in which the first cargo went to San Diego, and, though much of it was used locally, some was shipped to the timber-short southern missions intermittently. In 1782, for example, pine roof timbers were sent to Santa Barbara for use in the construction of new mission buildings.[2]

In spite of this early activity along the coast of California, the first cargo of West Coast forest products destined for a foreign port was not shipped from that area. Nor was it a Spanish enterprise. In 1788 Captain John Meares, a retired lieutenant of the British Royal Navy, arrived off the coast of the Pacific Northwest from Macao with two merchantmen, the *Felice Adventurer* and *Iphigenia Nubiana*, flying the flag of Portugal. Meares had been to the Northwest before, during the season of 1786/87. Low on supplies and with his crew racked by scurvy, Meares had received assistance from captains of the South Sea and East India companies. But because he was then sailing under British colors, and the companies of his rescuers had been granted monopolies to trading rights in the area by the British monarch, Meares's benefactors considered him an interloper and insisted that he promise to leave the area and never again encroach upon their monopoly rights.[3] However, Meares had seen enough of the Northwest to realize that great profits might be won there and so he returned in 1788 to tap the resources of the region. Since his vessels now flew Portuguese colors, Meares was not legally bound to recognize the English monopolies; he conveniently ignored his sworn bond.

Furs were the main attraction for Meares in the Northwest, but he also had plans for exploiting the forests of the area. He brought with him on his return a number of Chinese laborers whom he set to work felling trees and shaping them into spars and ship's timbers; "the woods of this part of America are capable of supplying, with these valuable materials, all the navies of Europe," he reported.[4] Meares instructed the captain of the *Iphigenia* that during "the time you remain in port your carpenters shall be employed in cutting down spars, and sawing plank; particularly boats knees and timbers,—all which bear a good price in China." [5] Meares was here carrying out the orders of his financial

2. Hubert Howe Bancroft, *History of California*, 7 vols. (San Francisco: The History Co., 1884–90), 1:463; Clar, *Harvesting and Use of Lumber in Hispanic California*, p. 7.
3. John Meares, *Voyages Made in the Years 1788 and 1789 from China to the North West Coast of America* . . . (London, 1790), pp. i–xl.
4. Ibid., p. 224.
5. Meares to W. Douglas, 21 Feb. 1788, ibid., appendix 2. James Strange of the East India Co., who visited Nootka Sound in 1786, also saw value in the forests of the area. "There is no doubt,"

backers, who in 1787 had informed him that spars of every size were constantly in demand in China and had told him to procure as many as he could.[6]

In September Meares set sail for China in the *Felice* with a cargo of furs and ship's spars;[7] he left behind the *Iphigenia* and a third vessel, the *North West America*, which had been put together by his laborers after their arrival in the Northwest.[8] Meares's hopes for developing a burgeoning trade were to go for naught, however. En route across the Pacific he ran into rough weather and had to jettison his cargo. To make matters worse, the Spanish, moving to head off foreign encroachment on lands claimed by them, seized the vessels Meares had left behind. As the Nootka Controversy came to a head and threatened to plunge Spain and England into war, the Northwest's nascent international timber trade, as well as the more lucrative fur trade, was held in temporary abeyance.[9]

The shipments of forest products remained infrequent and the quantities small until 1793. Then, just as would happen time and again in the decades to come, events outside the region made their influence felt. Uneasy about increasing foreign interest in California and the lands to the north of it—lands long claimed, though ineffectively controlled, by Spain—Governor Joaquín Arrillaga ordered the strengthening of the presidios against possible aggression. The building necessitated by the order led to a greatly increased demand for forest products. The *Princesa*, under Don Diego Fidalgo, and the *Arunzuza*,

Strange wrote, "that the timber with which this coast is covered (and which in its size and fine grain is nowhere to be excelled) would compose a very valuable addition to our trading, as this article carries a very advanced price in China and is always in demand there, especially such as is fit for masts and spars." Quoted in W. A. Carrothers, "The Forest Industries of British Columbia," in A. R. M. Lower, *The North American Assault on the Canadian Forest: A History of the Lumber Trade between Canada and the United States* (Toronto: Ryerson Press, 1938), p. 254.

6. Merchant proprietors to Meares, 24 Dec. 1787, Meares, *Voyages*, appendix 1.

7. Ibid., pp. 223–24.

8. The *North West America*, a schooner of twenty tons, is often referred to as the first vessel built by Europeans in the Pacific Northwest; but craft were launched even earlier by Russians in Alaska. The *North West America* was brought from China in a disassembled state aboard the *Iphigenia* and put together at Nootka Sound in September 1788. See ibid., pp. 221–22 and *passim*.

9. Ibid., pp. 272–73, 360–61, appendix. See also William R. Manning, "The Nootka Sound Controversy," American Historical Association *Annual Report* (1904) (Washington, D.C.: U.S. Government Printing Office, 1905), 1:279–478; F. W. Howay, ed., *The Dixon-Meares Controversy* (Toronto: Ryerson Press, 1929). As Howay notes, Meares had a "reckless disregard of truth" and his account, "a brief to support his exorbitant claims for damages against Spain," must be used with caution. However, there is no apparent reason for distorting his account of cutting spars and timbers and even so hostile a critic as Captain George Dixon did not dispute this portion of Meares's record. It is assumed, therefore, to be accurate.

under José Tobar, were pushed into service to carry the lumber, a capacity in which they continued to operate for some time.[10] Not all the timber used came from the ports of northern California. Logging had commenced in the hills behind Santa Barbara, and shipments of lumber were leaving that port for points farther down the coast.[11] Monterey was the main source of supply, however; from it lumber was being shipped in quantities that were not to be dispatched again until the 1830s.[12]

It is not clear when the first successful exports of timber left the West Coast. In 1816 Bryant, Sturgis and Company instructed the master of their ship, *Mentor*, that while he was in Northwestern waters "it would be well to take a large number of spars between decks, which will find a good market in China."[13] It cannot be determined whether the instructions were carried out. If they were not, the first shipment may then have been to timber-short Peru. In 1818 the *Two Catherines*, an American vessel out of Providence, Rhode Island, was commissioned by the Peruvian viceroy to carry construction timber from Alta California to the naval station at Callao. The vessel reached San Blas, Mexico, in July and, for reasons that are not clear, ran afoul of local authorities there. Again, it is not known whether the mission was completed, but the viceroy's action indicates that Monterey lumber was already known in Peru.[14] In 1819 the *Arab* of Boston left the Hawaiian Islands for the West Coast to pick up spars for the Chilean

10. Sherwood D. Burgess, "Lumbering in Hispanic California," *CHSQ* 41 (1962): 238; Clar, *Harvesting and Use of Lumber in Hispanic California*, p. 7. Fidalgo is more often remembered as the captain of one of the Spanish vessels which, by seizing the *Iphigenia* and the *North West America*, had triggered the Nootka Controversy.

11. Bancroft, *History of California*, 1:652. No lasting export trade in lumber developed at Santa Barbara, however. In the 1840s Eugène Duflot de Mofras was writing that there "is no wood suitable for building close to Santa Barbara." He went on to explain that since "it would be too troublesome and costly to bring wood down from the mountains seven or eight leagues away, timber is shipped in via Monterey from the sawmills belonging to Mission Santa Cruz." See Duflot de Mofras, *Travels*, 1:193, 196.

12. Burgess, "Lumbering in Hispanic California," pp. 238–39.

13. Quoted in F. W. Howay, "Brig Owyhee in the Columbia, 1827," *OHQ* 34 (1933): 328.

14. William Manning, ed., *Diplomatic Correspondence of the United States Concerning the Independence of Latin American Nations*, 3 vols. (New York: Oxford University Press, 1925), 2:943; Abraham P. Nasitir, *French Activities in California* (Stanford, Calif.: Stanford University Press, 1945), p. 507. A possible explanation of the difficulties of the *Two Catherines* can be found in the letters of Sir George Simpson. In 1824 he recommended against trying to trade on the west coast of Latin America, for "the Spaniards on that Coast are not to be depended on or trusted and ready to take any advantage and the Guarda Costas little better than Licensed Pirates at least it was so under the old Regime." Simpson to A. Colville, 9 Aug. 1824, in Simpson, *Fur Trade and Empire*, ed. Frederick Merk, rev. ed. (Cambridge, Mass.: Belknap Press, 1968), p. 245.

market. In 1827 the American brig *Owyhee* took a mixed cargo of furs and spars from the Columbia River to Whampoa, China.[15]

When Sir George Simpson visited the Hawaiian Islands in behalf of the Hudson's Bay Company in 1829, he noted that the demand for lumber in "the Islands has hitherto been met by shipments to fill up from Boston and by small quantities taken from the Coast by the Fur Traders." Simpson doubted that there was much potential in the Hawaiian lumber market, but he did state: "Timber, however, we know to be in great demand at all the Southern Ports, from St. Francisco [Bay] even to the southward of Valparaiso. . . . That it is an article of Trade in great demand, there is no manner of doubt; we have learnt so, from every Ship that has visited the Columbia for several years. . . ."[16] The evidence seems to indicate that exportation of forest products from the West Coast, while not regular, was not uncommon during the days of the maritime fur trade.

Just when regular shipments began to be made is also not clear. In the late 1820s the Frenchman Auguste Duhaut-Cilly visited the Russian establishment at Fort Ross, some one hundred miles north of present-day San Francisco. "We went with M. Shelikof to see his felling of wood," reported Duhaut-Cilly. "Independently of the needs of the establishment, he cuts a great quantity of boards, small beams, thick planks, etc., which he sells in California, the Sandwich Islands and elsewhere: he has entire houses built which may then be transported taken apart."[17] If the report is true, Russians must be given the credit for developing the first regular foreign trade in West Coast timber.

There seems to be ample ground for questioning the claim, however. Eugène Duflot de Mofras visited Fort Ross in 1841 and, though he described his stay in detail, he made almost no mention of lumber.[18]

15. Howay, "Brig Owyhee," pp. 324–29.
16. Simpson, *Part of Dispatch from George Simpson, Esqr., Governor of Ruperts Land, to the Governor & Committee of the Hudson's Bay Company, London, March 1, 1829 . . .* , ed. E. E. Rich (Toronto: Champlain Society, 1947), pp. 109–10, 87. Cf. Duflot de Mofras, *Travels,* 2:99; Simpson to John McLoughlin, 10 July 1830, Simpson, *Fur Trade and Empire,* p. 327.
17. Auguste Duhaut-Cilly, "Account of California in the Years 1827 and 1828," trans. Charles Franklin Carter, *CHSQ* 8 (1929): 320, 326.
18. Duflot de Mofras, *Travels,* 2:1–10, 163–64. His statement that the Russian-American Co. "could derive substantial profits from shipping its timber to the Sandwich Islands and certain points along the coast of California, Mexico, and even Chile where there is a dearth of this commodity" (2:166) clearly implies that the Russians had developed no significant trade of this sort. Duflot de Mofras was attaché of the French legation in Mexico. For an analysis of his tour of the West Coast see Rufus Kay Wyllys, "French Imperialists in California," *CHSQ* 8 (1929): 116–29.

Moreover, the voluminous Russian-American Company papers fail to substantiate Duhaut-Cilly's remarks. The Russians appear to have sent only one shipment to Hawaii and not to have sold in California until they departed from the area permanently in 1841. The Russians had no sawmill with which to produce large quantities of lumber for export; their beams and planks had to be hewn or pit-sawn. Duhaut-Cilly appears to have been misled by the overly sanguine reports of his Russian hosts.[19]

Thus it remained for the Hudson's Bay Company to develop the first regular foreign trade in West Coast forest products that can be verified. Governor George Simpson and Dr. John McLoughlin, the company's chief factor in the Northwest, were eager to protect the position of the Hudson's Bay Company in the region. As a means to this end, they advocated broadening the scope of the company's activities. "We must avail ourselves of all the resources of this Country if we have to Compete for the trade of it with the Americans," McLoughlin explained, "as we may depend they will turn everything they possibly can to account." [20] The erection of a sawmill seemed a logical step in this direction. McLoughlin requested the necessary machinery from London, the equipment was sent, and in late October 1827, at a site some six miles up the Columbia River from Fort Vancouver, the mill went into operation. Exports of lumber began not long thereafter.[21]

In the winter of 1827/28 McLoughlin sent Captain Aemilius Simpson to Monterey in the schooner *Cadboro* to obtain information on potential markets. The captain not only secured provisions that were

19. Patricia M. Bauer, "The History of Lumbering and Tanning in Sonoma County, California, since 1812" (master's thesis, University of California, Berkeley, 1950), p. 61; *The Timberman* 35 (1934): 95; C. Raymond Clar, *California Government and Forestry from Spanish Times to the Creation of the Department of Natural Resources in 1927* (Sacramento, Calif.: State of California, Department of Natural Resources, Division of Forestry, 1959), pp. 15–16; Ernest Ingersoll, "In a Redwood Logging Camp," *Harper's New Monthly Magazine* 66 (1883): 198.

20. McLoughlin to the Governor, Deputy Governor, & Committee of the Hudson's Bay Company (hereafter McLoughlin to Governor), [5] Aug. 1829, *The Letters of John McLoughlin from Fort Vancouver to the Governor and Committee*, ed. E. E. Rich, 3 vols. (Toronto: Champlain Society, 1941–44), 1:77. Cf. Simpson to McLoughlin, 15 Mar. 1829, Simpson, *Fur Trade and Empire*, pp. 309–10.

21. G. Simpson to A. Simpson [ca. Oct. 1828], Simpson, *Fur Trade and Empire*, p. 298; Simpson, *Part of Dispatch*, pp. 84–85; Donald H. Clark, "Sawmill on the Columbia," *The Beaver* 281 (1950): 42–44; Donald Hathaway Clark, "An Analysis of Forest Utilization as a Factor in Colonizing the Pacific Northwest and in Subsequent Population Transitions" (Ph.D. diss., University of Washington, 1952), pp. 21–22; Edmond S. Meany, Jr., "History of the Lumber Industry of the Pacific Northwest to 1917" (Ph.D. diss., Harvard University, 1935), pp. 59–63.

sorely needed at Fort Vancouver, but also returned with favorable reports regarding the market for lumber in California and Hawaii.[22] McLoughlin quickly dispatched a cargo to the islands, where it was found that Captain Simpson's information had indeed been correct: the cargo sold in Honolulu for prices as high as $100 per thousand board feet.[23] Other shipments soon followed, and, in addition, markets were found in California, along the west coast of South America, and according to some commentators, even in China.[24] A trial shipment was also sent to England.[25] Encouraged by the profits earned from the first shipments, Sir George Simpson directed McLoughlin to build additions to the company's fleet of vessels, to keep the sawmill running constantly, and to move the mill to the falls of the Willamette River if enough timber to meet the demand could not be obtained at the original location.[26]

But Simpson was being overly optimistic. Demand was limited in the islands and prices were unstable. Moreover, lumber from other sources was soon competing with the Hudson's Bay Company product. Because of the shortage of money in California and South America,

22. E. E. Rich, *Hudson's Bay Company, 1670–1870,* 3 vols. (New York: Macmillan, 1960), 3:614; McLoughlin, *Letters,* ed. Rich, l:lxxiii; Simpson, *Part of Dispatch,* p. 75; McLoughlin to Governor, 14 Nov. 1827, McLoughlin, *Letters,* ed. Rich, 1:53–54; Alfred L. Lomax, "Dr McLoughlin's Tropical Trade Route," *The Beaver* 295 (1964):12. For biographical sketches of Captain Simpson, see R. Harvey Fleming, ed., *Minutes of Council Northern Department of Ruperts Land, 1821–31,* Publications of the Champlain Society, Hudson's Bay Company Series, 3 (Toronto: Champlain Society, 1940), pp. 454–55; Hubert H. Bancroft, *History of the Northwest Coast,* 2 vols. (San Francisco, 1884), 2:441, 447.

23. McLoughlin to Governor, 5 Aug. 1829, 11 Oct. 1830, McLoughlin, *Letters,* ed. Rich, 1:76–77, 93; Simpson, *Part of Dispatch,* pp. 86–87, 109–10.

24. McLoughlin to Governor, 24 Nov. 1830, 20 Oct. 1831, McLoughlin, *Letters of Dr. John McLoughlin Written at Fort Vancouver, 1829–1832,* ed. Burt Brown Barker (Portland: Binfords & Mort, 1948), pp. 163, 217; McLoughlin to R. Charlton, 18 Nov. 1830, ibid., p. 157; McLoughlin to A. Simpson, 24 Nov. 1830, 13 Aug. 1831, ibid., pp. 159–61, 207; McLoughlin to Arnold & Woollett, 24 Nov. 1830, ibid., p. 170; McLoughlin to A. McDonald, 9 Dec. 1830, 4 June 1831, ibid., pp. 175, 196; McLoughlin to G. Simpson, 20 Mar. 1831, ibid., pp. 183–84; McLoughlin to Governor, 31 Aug. 1833, McLoughlin, *Letters,* ed. Rich, 1:110; J. Douglas to G. Simpson, 18 Mar. 1838, ibid., 1:277–78; Carrothers, "Forest Industries of British Columbia," p. 255. Not all the lumber and shingles dispatched from the Northwest came from Fort Vancouver. See McLoughlin to A. McDonald [4 June 1831], McLoughlin, *Letters,* ed. Barker, p. 197; Joseph Collins Lawrence, "Markets and Capital: A History of the Lumber Industry of British Columbia" (master's thesis, University of British Columbia, 1960), p. 3.

25. Governor, Deputy Governor & Committee of the Hudson's Bay Co. to McLoughlin, 27 Sept. 1843, McLoughlin, *Letters,* ed. Rich, 2:305.

26. Simpson to McLoughlin, 15 Mar. 1829, Simpson, *Fur Trade and Empire,* pp. 309–10. See also McLoughlin to Governor [5 Aug. 1829], 11 Oct. 1830, McLoughlin, *Letters,* ed. Barker, pp. 24, 144; McLoughlin to R. Charlton, 4 Aug. 1829, ibid., p. 27; McLoughlin to G. Simpson, 20 Mar. 1830, ibid., pp. 94–95.

sales could be made only by resorting to barter. The little mill on the Columbia proved fully capable of supplying all the lumber for which markets could be found, and McLoughlin suspended work on the project on the Willamette shortly after it was begun.[27] Epidemics of "intermittent fever" further complicated matters by so reducing the available manpower that the mill on the Columbia had to close down from time to time.[28]

If the lumber business proved to have limited potential and to be fraught with problems, it nonetheless served as a valuable adjunct to the company's other business. The annual supply ships, on their return to London from the Northwest, normally carried cargoes of lumber to Hawaiian and South American ports. Occasionally, the first stop was at Monterey, where there was an intermittent demand for lumber, flour, and salt salmon from the Northwest. Moreover, vessels often made short runs down the coast or to Hawaii in the period between their arrival with supplies in the fall and their departure for London with the year's take of furs the following spring.[29] On such voyages lumber and other products from the Northwest could be sold at a profit while needed supplies were also obtained: sugar, molasses, rice, and salt were all purchased in Hawaii in large quantities. In addition, the operation of the company sawmill provided a way of keeping employees productively occupied during the slack season.[30]

The Hudson's Bay Company was to remain active in the island trade for years. At first Richard Charlton, the British consul in Hawaii, acted as the company's agent in the island kingdom, but, as the trade grew and Charlton came to have trading interests of his own, the

27. R. Charlton to McLoughlin, 7 Nov. 1829, Simpson, *Fur Trade and Empire*, p. 319; Rich, *Hudson's Bay Company*, pp. 704–5. The Fort Vancouver mill was rebuilt and improved in 1838. Thereafter there was even less need of a mill at Willamette Falls. See J. Douglas to Governor, 18 Mar., 18 Oct. 1838, McLoughlin, *Letters*, ed. Rich, 1:259–60, 285.
28. McLoughlin to F. Heron, 14 Oct. 1830, McLoughlin, *Letters*, ed. Barker, p. 149; McLoughlin to Governor, 24 Nov. 1830, ibid., pp. 162–63, 168; McLoughlin to A. McDonald, 9 Dec. 1830, ibid., p. 175.
29. Sometimes late arrival of the supply vessels or other problems made such runs impractical, however. See McLoughlin to F. Heron, 14 Oct. 1830, ibid., p. 149; McLoughlin to Governor, 20 Oct. 1831, ibid., pp. 215–16; McLoughlin to G. Simpson, 15 Mar. 1832, ibid., p. 259; McLoughlin to Governor, 28 May 1834, McLoughlin, *Letters*, ed. Rich, 1:120.
30. Governor to Simpson, 5 Mar. 1830, Simpson, *Fur Trade and Empire*, p. 321; Simpson, *Part of Dispatch*, pp. 84–85, 93–94; D. McTavish to J. Hargrave, 2 Apr. 1842, James Hargrave, *The Hargrave Correspondence, 1821–1843*, ed. G. P. de T. Glazebrook, Publications of the Champlain Society, 24 (Toronto: Champlain Society, 1938), p. 384; McLoughlin to A. McDonald, 9 Dec. 1830, McLoughlin, *Letters*, ed. Barker, p. 176; Lomax, "Dr McLoughlin's Tropical Trade Route," p. 14.

London office decided that the company should handle its own business in the islands. In 1833 George Pelly, cousin of the company's governor, John Henry Pelly, was sent out to assume the post.[31] The company's bark *Vancouver* was the first ship to make customs entry under the constitutional government instituted in the kingdom in 1840. Shortly after, James J. Jarves, an American resident of Honolulu, reported that the Hudson's Bay Company was trying to monopolize trade with the islands and seemed to have a good chance of being successful.[32] By 1846 business had risen to the point that the company found it necessary to erect a two-story office-warehouse building in Honolulu.

Uncut timbers were in demand in Hawaii for use as masts and yards. There was also demand for flour and salt salmon, but bills of lading show that lumber was the most important single item shipped to the islands. A typical cargo was that of the bark *Cowlitz* in 1845: 752 barrels of flour, 52,000 feet of lumber, 200 barrels of salt salmon, 53,000 shingles, and 45 spars. Delivery could often be made in twenty-one days from the Columbia River; the bark *Columbia* once made it in thirteen. Unlike the first shipments that had sold for as high as $100 per thousand board feet, most of the later sales appear to have been in the neighborhood of $40 per thousand. Occasionally the prices were much lower.[33]

In spite of Jarves' fears, the Hudson's Bay Company at no time had a

31. McLoughlin, *Letters*, ed. Rich, 1:xciii; G. Simpson to McLoughlin, 1 Mar. 1842, ibid., 2:269; McLoughlin to Governor, 19 July 1845, ibid., 3:91–92; Simpson, *Part of Dispatch*, pp. 109–10; D. Finlayson to J. Hargrave, 29 Feb. 1836, Hargrave, *Correspondence*, pp. 230–31; Thomas G. Thrum, "History of the Hudson's Bay Company Agency in Honolulu," Hawaiian Historical Society *Eighteenth Annual Report* (1911), pp. 35–49.

32. James J. Jarves, *History of the Hawaiian or Sandwich Islands* (Boston, 1843), p. 366. Another leading American resident of Hawaii, Gerrit P. Judd, felt company officials greatly overestimated the potential for trade in the islands. There was, he noted, little unoccupied land and even that had cattle on it. Laura Fish Judd, *Honolulu: Sketches of Life in the Hawaiian Islands from 1828 to 1861* (Honolulu: Star-Bulletin Publishing Co., 1928), p. 88.

33. J. Douglas to Governor, 18 Oct. 1838, McLoughlin, *Letters*, ed. Rich, 1:267; McLoughlin to Governor, 24 Oct. 1839, ibid., 2:4; J. Douglas to Governor, 14 Oct. 1839, ibid., 2:227; McLoughlin to G. Simpson, 1 Mar. 1842, ibid., 2:272; McLoughlin to Governor, 19 July, 20 Nov. 1845, ibid., 3:85, 144; Duflot de Mofras, *Travels*, 2:88–89; D. McTavish to J. Hargrave, 2 Apr. 1842, Hargrave, *Correspondence*, p. 384; Lomax, "Dr McLoughlin's Tropical Trade Route," pp. 13–15. It should be noted that Lomax misread the letter regarding the cargo on the *Cowlitz* and thought that the value listed on entry was the sale price. The letter indicates not how low prices had dropped, but rather to what lengths the company was willing to go in order to save on customs charges. The closure of the company's agency in Honolulu in 1860 did not mean that Northwest-Hawaiian trade was coming to an end, but that it was passing into the hands of others. See below, chaps. 4 and 5.

monopoly of the island trade.[34] Americans, New Zealanders, and Californians all sent cargoes of lumber to Hawaii. One of the first sources of this competition was Monterey.

During the late Spanish period in California, men who had jumped ship from foreign vessels calling in the waters of northern California found that the forests of the area provided excellent hideouts until their vessels sailed. The motley collections of men that resulted gradually coalesced into California's first foreign colony. Among the first to appear on the scene was an American named William Smith, more commonly known as "Bill the Sawyer." He built a shack in the San Mateo redwoods near present-day Woodside where he was joined two years later by an English mutineer, James Peace, and in 1821 by a demoted English naval officer, John Coppinger. These and others who drifted into the area supported themselves by working as whipsawyers, shingle-splitters, and at related jobs.[35] An even larger collection gradually grew up at Villa de Branciforte near Mission Santa Cruz. By 1835 the village's population stood at one hundred fifty; five years later it had risen to an estimated two to three hundred.[36]

Loggers were also working in the woods farther south along Monterey Bay. Michael Lodge had obtained a grant on Soquel Creek and hauled the lumber he cut there out to what is now known as Capitola Beach. Others were operating in the forests near Monterey.[37] Once deserting seamen reached the woods, one California pioneer noted, "they were perfectly safe, as no person dared to go after them, for if they did, they would never return alive." Under the circumstances, lumbering expanded rapidly.[38]

34. For example, see the discussion of American trading activity in Adele Ogden, "Boston Hide Droghers along California Shores," *CHSQ* 8 (1929):298–99.
35. Burgess, "Lumbering in Hispanic California," p. 239; Clar, *California Government and Forestry*, pp. 28–31, 34–36; B. F. Alley, *History of San Mateo County, California, . . .* (San Francisco, 1883), pp. 112, 114–20; Brown, *Sawpits in the Spanish Redwoods*, pp. 4–14; Charles Brown, "Statement of Recollections of Early Events in California" (MS, Bancroft Library, Berkeley, Calif., dated 1878), pp. 15, 18–19.
36. Duflot de Mofras, *Travels*, 1:163, 216–18; Sir George Simpson, *Narrative of a Journey Round the World, during the Years 1841 and 1842*, 2 vols. (London, 1847), 1:364; Burgess, "Lumbering in Hispanic California," pp. 239–40; H. A. Van Coenen-Torchiana, *The Story of Mission Santa Cruz* (San Francisco: P. Elder & Co., 1933), pp. 217–31, 300–301; E. S. Harrison, *History of Santa Cruz County, California* (San Francisco, 1892), p. 44.
37. Van Coenen-Torchiana, *Mission Santa Cruz*, pp. 341–43; Leon Rowland, *The Story of Old Soquel* (Soquel, Calif.: Soquel Print Shop, 1940), p. 5. Henry J. Bee, "Recollections," San Jose *Pioneer*, 13 Jan. 1877.
38. John Henry Brown, *Reminiscences and Incidents of Early Days of San Francisco (1845–50)* (San Francisco: Grabhorn Press, 1933), p. 82.

Thomas O. Larkin, remembered as the first and only American consul in Alta California, provided the vision and organizational ability to build an export lumber trade on the base provided by this motley body of woods workers. Shortly after his arrival in Monterey in 1833, Larkin began to purchase the cut of whipsawyers operating in the Santa Cruz area. He was soon dispatching cargoes of lumber to John Temple and Abel Stearns in Los Angeles and to a "Don Antonio" in Santa Barbara. Other merchants in the area, including William A. Richardson, Nathan Spear, and Jacob P. Leese, made similar shipments, but Larkin's activity as a shipper of lumber quickly overshadowed theirs.[39] The bulk of Larkin's shipments was sent on the bark *Don Quixote*, a vessel that was destined to continue in the coasting trade to the end of the period of Mexican rule. Larkin also engaged in the retail lumber trade in the Monterey area.[40] In time, he expanded his operations to include shipments to Hawaii, Mexico, and the Pacific coast of South America in spite of efforts by Mexican authorities to halt the exportation of timber. By 1844 the ship *Fama* was making annual trips to the islands, often with lumber as a part of the cargo. Other vessels, including the *Don Quixote*, were making the passage on a less regular basis, carrying goods both for Larkin and for others.[41]

A pattern quickly emerged that was to last for more than a decade. Individual woods workers cut the redwoods and pines that grew in the hills behind Monterey Bay and sawed them into boards, planks, and timbers. Larkin purchased the cut of these sawpit operators and had it

39. N. Spear to Abel Stearns, 7, 30 May, 18 July, 11 Aug., 5 Dec. 1834, SP, box 60; J. P. Leese to Stearns, 29 May 1836, SP, box 41; W. A. Richardson to Stearns, 12 Dec. 1835, SP, box 51.
40. Burgess, "Lumbering in Hispanic California," p. 240; Larkin to J. Temple, 22 July 1839, Thomas Oliver Larkin, *The Larkin Papers*, ed. George P. Hammond, 10 vols. (Berkeley: University of California Press, 1951–64), 1:18; Stearns to Larkin, 7 May 1834, SP, box 60. "Don Antonio" was probably Don José Antonio Aguirre, described by Duflot de Mofras as "the most prosperous merchant in the country [around Santa Barbara], a man with whom nearly all the local residents have mutual interests." See Duflot de Mofras, *Travels*, 1:194.
41. Larkin, *Papers*, 1:xi; Consular Returns, 1844, ibid., 3:6–7; A. Thompson to Larkin, 2 July 1839, 18 June 1840, 14 Oct. 1841, ibid., 1:15, 45, 125; W. Garner to Larkin, 7 July 1839, 2 Aug. 1841, ibid., 1:16–17, 104; L. Chamberlain to Larkin, 25, 27 Apr. 1840, ibid., 1:37–39; J. Jones to Larkin, 16, 21 July, 10 Sept., 7 Nov. 1841, ibid., 1:97–98, 118, 134; J. Weed to Larkin, 16 July 1841, ibid., 1:103; N. Don to Larkin, 16 Jan., 21 Mar. 1842, ibid., 1:156, 178; J. Paty to Larkin, 29 Feb., 26 Mar. 1842, ibid., 1:169–70; Marshall & Johnson to Larkin, 16 Mar. 1842, ibid., 1:176; A. Stearns to Larkin, 3 May 1842, ibid., 1:217; Larkin to Mott, Talbott & Co., 8 July 1845, ibid., 3:262–63; John Paty, "Journal of Captain John Paty," *CHSQ* 14 (1935):320 and *passim*; Clar, *California Government and Forestry*, pp. 28–29; Adele Ogden, "New England Traders in Spanish and Mexican California," in *Greater America: Essays in Honor of Herbert Eugene Bolton*, ed. Adele Ogden and Engel Sluiter (Berkeley: University of California Press, 1945), pp. 401–2. Cf. Meany, "History of the Lumber Industry," p. 70.

transported to Monterey by cart or ship, or shipped directly to coastal or foreign markets. Larkin generally paid $40 per thousand board feet for one-inch redwood boards and sold them for $50 or, less often, $55. Since transporting the lumber from Santa Cruz to Monterey usually cost $10 per thousand, Larkin would appear, at first glance, to have been operating on the narrowest of margins. The facts of the matter were quite the reverse.

The sale of lumber was but a part of Larkin's business, and not the most important part. His investments included a store, wharf, ships, flour mill and bakery, and brickyard. The store was his key holding. Through it he did business with the numerous visiting whalers and with people scattered over a wide area. Business transactions of such a scale in currency-short California necessitated accepting much payment in kind. In the Santa Cruz–Monterey area, at least, lumber served as a medium of exchange that helped keep the wheels of business turning—much as hides did throughout California. The economic realities of California, together with Larkin's willingness to engage in nearly any enterprise that would yield a profit, had driven him into the lumber trade. Even when his prices seemed no higher than his costs, the trade yielded him profits, for Larkin paid his loggers in credits at his store where the markup was from 50 to 150 percent. The indirect profits from his sales were clearly high.[42]

Still, Larkin apparently considered the $40 per thousand he had to pay for boards exorbitant. In the spring of 1834 he sent crews of his own to cut lumber to fill an order for nine thousand feet. He was to pay them $10 per thousand. These inexperienced men found that the work of the whipsawyer was not so easy as it appeared. By the end of the summer they had cut but three thousand feet and run up expenses of $375. To fill his order, Larkin found it necessary to call on Mark West and William Trevathan, two independent whipsawyers of the Santa Cruz area. They soon got out the lumber, but Larkin was in a poor position to bargain and had to pay a premium price for their cut. As Larkin's own crews gained experience, or felt the goad of

42. J. Belden to Larkin, 28 Mar., 5, 13, 20 Apr., 29 May, 23 June, 20, 30, 31 July, 7, 15, 22 Aug., 20 Sept. 1842, Larkin, *Papers*, 1:182–84, 187–88, 191–92, 204–5, 227–28, 243, 250–51, 254–56, 262–63, 268–70, 277, 289–90; W. Garner to Larkin, 5 Aug., 7 Sept. 1842, ibid., 1:260–61, 283; Larkin to J. Belden, 18 Aug. 1842, ibid., 1:270–72; J. Weekes to Larkin, [May], 30 June, 8 Sept. 1844, ibid., 2:133–34, 156–57, 221; J. Majors to Larkin, 11 July 1844, ibid., 2:165–66; Burgess, "Lumbering in Hispanic California," pp. 240–41; Robert J. Parker, "Larkin, Anglo-American Businessman in Mexican California," in *Greater America*, pp. 421–22.

competition, they too began to produce in quantities. By October they had turned out approximately forty-one thousand feet of lumber, which brought Larkin a profit of some $800. Encouraged, he sent others into the woods, including crews that cut shingles for $3.50 per thousand.[43]

In 1835 Larkin's half-brother, J. B. R. Cooper, was also entering the field of lumber production. He built the first commercial sawmill in California at El Molino Rancho in present-day Sonoma County. Larkin and Cooper combined their forces in marketing the production of the mill.[44]

Larkin's attempts at expanding his operations to include production as well as marketing were unsuccessful. To keep his crews active through the winter of 1834/35, Larkin had to move them to an area where he had to pay $4 per tree for stumpage. By 1835 he was having trouble with his shingle foreman. Moreover, Larkin was continually plagued by a high turnover in laborers. Once they had gained a minimum of skill, few workers were willing to cut for Larkin at $10 per thousand when, as independent operators, they could receive $40 per thousand. Sherwood Burgess has aptly described Larkin's labor policy as "hardly enlightened." In the end, Larkin withdrew his crews from the woods and reverted to dependence upon independent whipsawyers for the lumber he marketed.[45]

Cooper's sawmill was no more successful than Larkin's excursion into timber cutting. With the mill, Cooper had thought he would "be able to supply this coast with wood at a much better rate than has ever been done before"; but though the mill operated until the winter of 1840/41, when it was washed away by floods, it failed to live up to Cooper's expectations. By 1838 he reported that, after investing $12,000 and three years' hard work in the enterprise, he was "head and ears in debt." His millwright had "prooved himself incompetent for

43. Burgess, "Lumbering in Hispanic California," p. 241; Clar, *California Government and Forestry*, pp. 21–22; Larkin to A. Stearns, 22 July 1834, Thomas O. Larkin, *First and Last Consul: Thomas Oliver Larkin and the Americanizing of California: A Selection of Letters*, ed. John A. Hawgood, 2nd ed. (Palo Alto, Calif.: Pacific Books, 1970), pp. 6–7. This was one of the earliest shingle operations in California. Regarding the claim that George C. Yount introduced shingles into California in 1833, see Clar, *California Government and Forestry*, pp. 23–25; Brown, "Statement of Recollections," pp. 8–9.
44. Cooper to Stearns, [?] Feb., 15 Mar. 1835, SP, box 17; Clar, *California Government and Forestry*, pp. 25–27; Clar, *Harvesting and Use of Lumber in Hispanic California*, pp. 11–12. Clar incorrectly states that Cooper's mill was erected in 1834.
45. Burgess, "Lumbering in Hispanic California," pp. 241–42. In 1847 Larkin once again dispatched his own crews into the woods, this time to make shingles. See Larkin, contract with J. Sturzenegger and A. Cannifax, 23 June 1847, Larkin, *Papers*, 6:222.

any business, a compleat wretch and no mechanic of any kind. . . ." [46]

Marketing, as well as lumber production, posed problems for Larkin. Customers, he complained, were slow in paying their bills. To complicate matters further, lumber was not always available through his independent suppliers, either to pay the bills they had run up at his store or to meet the demands of the market.[47] When lumber became available during late fall or winter months, sea captains sometimes refused to brave the exposed roadstead off Santa Cruz in order to pick it up.[48] Moreover, Larkin was plagued by thieves and arsonists who stole or burned his forest products as they lay unprotected on the beach.[49]

In spite of all these problems, Larkin's lumber trade was prospering by the early 1840s. He was shipping to Peirce and Brewer in Honolulu, to Parrott and Company in Mazatlán, to John Sutter in New Helvetia, and perhaps even to buyers in Valparaiso and, on around Cape Horn, in the eastern United States. Moreover, the continued growth of Monterey itself increased the demand for construction materials. The expanding trade at the port created a need for spars and ship's timbers. In the meantime, Yerba Buena, as San Francisco was then known, was beginning to constitute an additional outlet.[50]

As cutting increased, local officials expressed concern. In November 1834 Governor José Figueroa published a decree prohibiting the exportation of lumber and holding the masters of vessels responsible for

46. Cooper to Stearns, 15 Mar. 1835, 6 Mar. 1838, SP, box 17; Clar, *California Government and Forestry*, pp. 25–27.

47. Burgess, "Lumbering in Hispanic California," p. 240; J. Weekes to Larkin, [May], 8 Sept. 1844, Larkin, *Papers*, 2:133–34, 221; J. P. Leese to Stearns, 7 Nov. 1835, SP, box 41.

48. A. Thompson to Larkin, 1 Nov. 1844, Larkin, *Papers*, 2:269. Maritime transportation sometimes caused problems even in the best of weather. See N. Spear to Stearns, 11 Aug., 25 Dec. 1834, SP, box 60; Cooper to Stearns, [1834?], SP, box 17; Leese to Stearns, 29 May 1836, SP, box 41.

49. J. Belden to Larkin, 7 Aug. 1842, Larkin, *Papers*, 1:262–63; Larkin, offer of reward and public notice, 1 June 1843, ibid., 2:17, 18; N. Dawson, statement regarding lumber fire, 19 June 1843, ibid., 2:21; J. Weekes to Larkin, 30 June 1844, ibid., 2:156; Parker, "Larkin, Anglo-American Businessman in Mexican California," p. 422.

50. J. Paty to Larkin, 3 Aug. 1842, Larkin, *Papers*, 1:259; E. Grimes to Larkin, 6, 20 Aug. 1844, ibid., 2:184, 210; Larkin to J. Bennett, 10 Feb. 1843, ibid., 2:6–8; Larkin to Parrott & Co., 18 Aug. 1844, ibid., 2:208; Reuben L. Underhill, *From Cowhides to Golden Fleece* (Stanford, Calif.: Stanford University Press, 1939), p. 53; Burgess, "Lumbering in Hispanic California," pp. 243–44; Duflot de Mofras, *Travels*, 1:225, 259; Hiram Bingham, *Residence of Twenty-one Years in the Sandwich Islands* (Canandaigua, N.Y., 1855), p. 572. Duflot de Mofras' statement that lumber was exported "only to the Sandwich Islands" is controverted by the other sources. See Duflot de Mofras, *Travels*, 1:259.

any breaches of the law.[51] In spite of this ban, shipments continued. "Some years since," the alcalde of San José reported, "a timber-cutting was established by foreigners in part of my jurisdiction [the Pulgas redwoods(?)], and it is well known that besides selling and dealing with the natives without paying a tax . . . they, not content with selling the lumber to those within our land, ship it abroad in trading vessels." [52]

It seems that little was actually done to halt the exportation of lumber. The scattered bodies of woods workers may have been beyond the reach of Mexican officials, but Thomas O. Larkin and the vessels he loaded at Monterey and Santa Cruz clearly were not. Yet there is no evidence that his trade suffered from official interference.

Production of lumber not only continued in established areas, it commenced in new ones. By the 1840s, William R. Garner, an Englishman who had deserted ship in Monterey in 1826, had crews of from one hundred to two hundred men whipsawing lumber on San Francisquito Rancho southeast of Carmel.[53] In 1840 George Patterson and John Parker, deserters from the British navy, began logging in the hills behind San Francisco Bay. According to Duflot de Mofras, Pierre Sicard and Joseph Leroy also were cutting in the East Bay hills by 1841. Although he did not indicate their method, he probably meant by sawpit rather than a plant powered by water or steam. In the south, Juan Bandini began the first substantial lumbering when, in 1841, he began cutting near Cajón Pass in the San Bernardino Mountains. Other areas, including the lower Napa Valley and the area north of the Golden Gate near present-day Corte Madera and Mill Valley, also were being brought into production.[54]

These scattered, small-time enterprises that appeared around 1840 combined with the established enterprises of Thomas O. Larkin and the Hudson's Bay Company to make the production of lumber a familiar, if not ubiquitous, sight in the forests of the Far West. But it

51. Clar, *California Government and Forestry*, pp. 28–29; Hubert Howe Bancroft, *California Pastoral, 1769–1848* (San Francisco, 1888), pp. 440–41.

52. Quoted in Brown, *Sawpits in the Spanish Redwoods*, p. 10.

53. Clar, *California Government and Forestry*, p. 22. See also Brown, "Statement of Recollections," pp. 15, 18–19.

54. Clar, *California Government and Forestry*, pp. 23–28, 31–33; Sherwood D. Burgess, "The Forgotten Redwoods of the East Bay," *CHSQ* 30 (1951):1–15; Duflot de Mofras, *Travels*, 1:224–25. There had apparently been some cutting in the latter area as early as 1815. See Brown, *Sawpits in the Spanish Redwoods*, p. 3; Clar, *Harvesting and Use of Lumber in Hispanic California*, p. 7.

would be claiming too much for them to say that they—or the earlier activities of Meares, the Russian-American Company, or the American fur traders who carried forest products to Hawaii—laid the foundations of a Pacific Coast lumber industry.

Records of the smaller operations are not extant. For most of them, little other than names and locations is known. Yet it seems clear that the felling of timber and sawing of logs in which they were engaged was more akin to the gathering of beaver pelts by mountain men or to placer mining than to industrial activity in the usual sense of the term. Their capital outlay was small, their fixed investments almost nil. These operations were run by men who were more laborers than capitalists, men who gathered forest resources and peddled them nearby when buyers could be found.[55] None of the important lumber firms on the Pacific Coast was to spring from such humble origins. The greatest contribution of these men lay in their demonstrating that logging could be carried out successfully on sites other than those immediately tributary to Fort Vancouver in the north and Monterey and Santa Cruz in the south. It was an almost self-evident truth that needed little demonstration.

The lumber operations of Larkin and the Hudson's Bay Company were more important, but even they can hardly be said to have provided the foundation on which the area's lumber industry was to be erected. Both marketed lumber as adjuncts of larger enterprises. It would be misleading to label either Larkin or John McLoughlin a lumberman, for their overall success was not dependent on the sale of forest products at a profit—though such sales were, of course, welcomed. Both were engaged in diverse, essentially preindustrial mercantile activities. Larkin and the Hudson's Bay Company had more in common with John Jacob Astor and his mercantile empire than with John D. Rockefeller and the Standard Oil Company of the coming industrial age. Larkin, especially, was a nonspecialized merchant-capitalist, a sort of Far Western equivalent of the sedentary urban merchants who were the key figures in America's economy down to 1815.[56]

55. The obvious exception to this generalization was the aristocratic Juan Bandini, one of southern California's leading figures, who seems to have entered lumbering in hopes of re-establishing his family's fortune. See Clar, *California Government and Forestry*, pp. 31–33; Bancroft, *History of California*, 2:709–10; R. H. Dana, Jr., *Two Years before the Mast: A Personal Narrative of Life at Sea* (New York, 1840), pp. 222–25; Larkin, notes on principal men, undated, *Papers*, 4:323.
56. For a revealing discussion of the sedentary merchant-capitalist in the United States and his gradual eclipse by manufacturers, see Glenn Porter and Harold C. Livesay, *Merchants and*

Larkin and the Hudson's Bay Company may not have been the architects of a Pacific Coast lumber industry, but they supplied those who were with a valuable body of information. They demonstrated that, though the market for lumber was limited in northern California and almost nonexistent in the Pacific Northwest, it was possible to sell significant quantities in southern California, Hawaii, and beyond. They made it clear that, even when the cost of production seemed to consume all profits, a satisfactory return could result from the associated sale of supplies to workers, from fuller use of vessels in which investment capital was already tied up, and from other indirect sources. Their activities also revealed some of the problems that were to plague the West Coast's lumber industry when it finally did emerge as a distinct entity. They had to cope with shortages of shipping, capital, and dependable labor. They found that the markets they reached by sea were easily glutted and remarkably unstable. They discovered that they were themselves dependent on sources of supply so distant that operations were sometimes slowed or halted as a result.

Thomas O. Larkin and the Hudson's Bay Company may not have built a new industry on the Pacific Coast, but they pointed the way toward one. They were harbingers of an industry to come.

Manufacturers: Studies in the Changing Structure of Nineteenth Century Marketing (Baltimore: Johns Hopkins Press, 1971). See also Kenneth Wiggens Porter, *John Jacob Astor, Business Man*, 2 vols. (Cambridge, Mass.: Harvard University Press, 1931).

CHAPTER II

Modest
Beginnings

QUICKENING commercial activity and an associated rise in demand for timber products resulted in efforts to increase lumber production during the 1840s. One consequence was the establishment of the first steam- and water-powered sawmills on the Pacific Coast that were intended not as adjuncts of other enterprises but as independent ventures. Their owners—Captain Stephen Smith and Henry H. Hunt and his associates—were the real founders of the Pacific Coast lumber industry.

Important though these men were as pathbreakers, they were overshadowed in the commercial life of the times by such familiar figures as John McLoughlin and Thomas O. Larkin and by newcomers such as Francis W. Pettygrove. McLoughlin and Larkin had long dealt in forest products, while Pettygrove—like other merchants and traders newly drawn to the area—soon found himself of necessity dealing in them too. All three men seem to have sold more lumber during the period than either Smith or Hunt. The old commercial pattern that the three represented was continuing at the same time that a new pattern was unobtrusively emerging.

Joseph Chapman, the first Anglo-Saxon resident of Los Angeles, apparently built the first mechanical sawmill on the Pacific Coast when he erected a saw- and gristmill near San Gabriel in 1822.[1] This was

1. Chapman's mill preceded by several years the mill constructed by George C. Yount, which is sometimes given the honor of being California's first sawmill. Neither mill had any connection with the lumber trade on the Pacific, however. See Clar, *California Government and Forestry*, pp.

followed by the Hudson's Bay Company mill at Fort Vancouver in 1827. Some time later, either in 1841 or 1842, Peter Lassen built a small water-powered mill on a tributary of the San Lorenzo River some six miles above Santa Cruz. After its completion, Isaac Graham paid Lassen one hundred mules, but the reason for the payment is not clear. It may have been for his labor, or perhaps Lassen was a partner whose interest in the mill Graham was buying. The cut of this mill, like the lumber Graham had been whipsawing earlier, undoubtedly found its way into Larkin's lumber trade.[2]

In spite of these precursors, the first mills of real importance did not appear until 1844. In that year Stephen Smith, a Baltimore sea captain who had traded at Yerba Buena in 1841, returned to the Bay Area. Aboard his ship, *George and Henry*, was the machinery for a steam sawmill. He proceeded to erect his mill, the first steam-powered sawmill on the Pacific Coast, at a site six miles from Bodega Bay, a small harbor some fifty miles north of San Francisco Bay.[3]

Steam mills represented a significant improvement over those driven by water, for they freed millmen from having to locate their plants near flowing streams. Still, Smith's mill was not an immediate success. By the time the captain had it running, California's economy was in the doldrums. Moreover, Smith at first lacked the means of shipping his cut to San Francisco—as Yerba Buena had by then come to be known—and had to sell in the limited market around Bodega Bay.[4]

But Smith persevered and by 1846 was able to announce proudly

17–19; C. A. Menefee, *Historical and Descriptive Sketch Book of Napa, Sonoma, Lake, and Mendocino Counties* . . . (Napa City, Calif., 1873), pp. 125–27.

2. Clar, *California Government and Forestry*, pp. 29–30, 35–37, 50–51; Burgess, "Lumbering in Hispanic California," p. 244; Clar, *Harvesting and Use of Lumber in Hispanic California*, p. 12. Lassen built a second mill in the Sacramento Valley in 1846. See J. Sutter to Larkin, 2 Mar. 1846, Larkin, *Papers*, 4:218.

3. Bancroft, *History of California*, 4:395–96, 565–66, 679; Theodore H. Hittell, *History of California*, 4 vols. (San Francisco, 1885–97), 2:376; Juan B. Alvarado, "Historia de California" (MS, Bancroft Library, Berkeley, Calif.), 5:5–7; Clar, *California Government and Forestry*, pp. 38–39; Everett R. Sanford, "A Short History of California Lumbering" (Master's thesis, University of California, Berkeley, 1924), p. 32; J. P. Munro-Fraser, *History of Sonoma County* . . . (San Francisco, 1880), pp. 52–56; Larkin, notes on principal men, undated, *Papers*, 4:334; Lyman L. Palmer, *History of Mendocino County, California* (San Francisco, 1880), pp. 133–36. It is interesting to note that, though Smith's mill is almost universally described as having been built in 1843, Larkin's consular returns show that Smith's vessel, the *George and Henry*, did not enter San Francisco Bay until 6 Feb. 1844. If this is in fact when the ship arrived, the equipment for the mill, which was aboard, could not have been installed until 1844. See Larkin, *Papers*, 3:6–7.

4. B. Reed to Larkin, 13 Oct. 1845, Larkin, *Papers*, 4:23; A. Thompson to Larkin, 13 Oct. 1845, ibid., 4:22; W. Leidesdorff to Larkin, 21 July 1846, ibid., 5:149–50; S. Smith to Leidesdorff, 26 Apr. 1846, LP, box 2; Smith to Stearns, 15 Nov. 1847, SP, box 60.

that his mill was cutting three thousand feet during a ten-hour workday.[5] By then the captain's business had begun to assume a pattern that he was to perfect in the years to come. Smith had his cut carted to Bodega Bay, stowed aboard merchantmen, and transported to San Francisco for sale through a lumberyard he and William A. Leidesdorff had opened there. Smith sold additional cargoes for shipment to Hawaii. When American control came to California, the economy responded favorably. The Bodega mill was an almost immediate beneficiary. Difficulties continued to plague Smith's operation, but the corner had been turned.[6]

At least three other mechanical sawmills were cutting lumber in California by 1845: the mill of Rousillion and Sainsavain on the San Lorenzo, the mill of John Hames on Soquel Creek, and the mill of William Blackburn on Branciforte Creek. All three were near Santa Cruz. James Peace had a mill in operation in the San Mateo redwoods soon thereafter. These mills were equipped with sash or muley saws; the first circular saw was not put in operation in California until 1847.[7] However, none of these mills had the production—or the importance —of Smith's. Rather than new departures, they were small, water-powered plants that were improvements on the old sawpit operations that supplied lumber to merchants such as Larkin.

Increased activity also came to the forests of the Pacific Northwest in the early 1840s. With the arrival of the Great Reinforcement at Jason Lee's mission in the Willamette Valley in 1840, the American population of the Oregon country reached about one hundred fifty. Thereafter the colony grew at an accelerating rate as news of the opportunities in the area spread. Growth in population led to an increase in trade, to the establishment of sawmills, and, coupling the two, to increased export of lumber from the region.

In 1842 the Island Milling Company, an American firm, erected a sawmill on John McLoughlin's claim by the falls of the Willamette.[8]

---

5. Smith to Leidesdorff, 26 Apr. 1846, LP, box 2.
6. Smith to W. Davis, 25 Jan. 1847, William Heath Davis Papers, Henry E. Huntington Library, San Marino, Calif., box 8; Smith to Stearns, 10 Feb., 27 June 1847, SP, box 60; Smith to Leidesdorff, 25 Jan., 10, 20 Feb. 1847, LP, box 3; 5, 29 May, 5, 15 July 1847, LP, box 4; Smith to Leidesdorff, 25 Dec. 1847, William A. Leidesdorff Papers, California Historical Society Library, San Francisco.
7. Sanford, "Short History of California Lumbering," p. 32; Burgess, "Lumbering in Hispanic California," p. 244. Monterey *Californian*, 27 Mar. 1847; Clar, *California Government and Forestry*, p. 42.
8. The enterprise was known after January 1845 as the Oregon Milling Co. It is unclear just who were its backers. See Oregon City *Oregon Spectator*, 28 Jan. 1845; J. Minto to F. G. Young, 22

As early as 1835 the governor and committee of the Hudson's Bay Company had written McLoughlin that whenever Americans "attempt to establish a Post on shore, we should have a party to oppose them, and to undersell them even at a loss." In 1841 Governor George Simpson had directed McLoughlin to take steps to maintain control of the import-export business of the Willamette colony and to utilize the power source at Willamette Falls so as to prevent its falling into the hands of others. Now, faced with the challenge of the Island Milling Company, McLoughlin moved at last. He revived plans for a sawmill at the falls; indeed, he had both a sawmill and a grist mill constructed there in 1843. He opened a branch store on the site a year later. Though he continued to reside in Fort Vancouver, McLoughlin managed the mills personally. In an attempt to protect the enterprise against seizure if the land south of the Columbia should become American, McLoughlin declared that the mills were a private venture of his own. In 1845 he purchased the property from the company, though McLoughlin may have acted more to protect the company's interests than to acquire title to the tract for himself.[9]

McLoughlin planned to export the lumber from the new mill, for "the Lumber here is of a Better quality than at Fort Vancouver and at present can be sawed cheaper." He contracted for the delivery of twelve hundred logs at the new mill "so as to be able to compete with the Americans at the Sandwich Islands."[10] Demand in the islands seemed to be increasing, but, though McLoughlin was eager to take advantage of the opportunity to sell lumber there, he had no vessel at his immediate disposal. He chartered a noncompany vessel to carry a cargo from the new mill to the islands; however, the shortage of shipping continued and, with the log contract still on his hands,

---

Nov. 1905, MP, box 2; Arthur L. Throckmorton, *Oregon Argonauts: Merchant Adventurers on the Western Frontier* (Portland: Oregon Historical Society, 1961), p. 28.

9. McLoughlin to Governor, 1 July 1846, McLoughlin, *Letters,* ed. Rich, 3:155–57; "McLoughlin's Willamette Falls Claim," ibid., 3:195–219; John McLoughlin, "Reminiscences" (John McLoughlin, Private Papers, microfilm copy, University of Washington library, Seattle), p. 14; J. McLoughlin, memorial to Congress, pp. 14, 44–53, 109–12, ibid.; McLoughlin, narrative concerning Willamette Falls claim, ibid.; Meany, "History of the Lumber Industry," pp. 73–76; Throckmorton, *Oregon Argonauts,* pp. 28, 37–38, 43–45, 50; Clark, "Analysis of Forest Utilization," pp. 22–23. The Hudson's Bay Co. had already been considering building a new mill, however, for accessible stumpage was becoming scarce near Fort Vancouver. See McLoughlin to Governor, 24 May 1841, McLoughlin, *Letters,* ed. Rich, 2:39; Governor to McLoughlin, 21 Dec. 1842, ibid., 2:297–98.

10. McLoughlin to J. Pelly, 12 July 1846, McLoughlin, *Letters,* ed. Rich, 3:165–66; McLoughlin to Governor, 1 July 1846, ibid., 3:155.

McLoughlin grew increasingly concerned. Faced with these circumstances, he decided to lease the mill to an American, Walter Pomeroy, for two years with the proviso that if Pomeroy shipped any lumber to Hawaii, he would offer it to the Hudson's Bay Company agency there before he made it available to others.[11] If John Minto was correct in his belief that Pomeroy was one of the partners in the Island Milling Company, the contract may have represented McLoughlin's coming to terms with the rival firm.[12]

The sawmills built at the falls of the Willamette by John McLoughlin and the Island Milling Company were not the only ones constructed at this time. In 1843 American settlers erected a sawmill on the Tualatin Plains and another on the Columbia River opposite Puget Island. In spite of the efforts of the Hudson's Bay Company to keep American settlers south of the Columbia River, a party led by Michael T. Simmons moved northward and took up claims near the southern end of Puget Sound. Not long thereafter, they erected the first sawmill in what was to become the state of Washington on a site by the lower falls of the Deschutes River. Much of the capital for erecting the mill apparently came from George W. Bush, Washington's first black resident. The cut of this mill, it has been claimed, was marketed through the Hudson's Bay Company at Fort Nisqually and found its way to Victoria and the Hawaiian Islands. The first shipment was supposedly made on the Hudson's Bay Company steamer *Beaver* in 1848.[13] The bulk of the mill's cut seems to have been consumed locally, however. Other mills also appear to have been erected at about this time to satisfy local demand in the Willamette Valley.[14]

11. McLoughlin to Governor, 1 July 1846, ibid., 3:158–59. Normal freight from the falls to the Hawaiian Islands was $16 per thousand board feet. See J. N. Couch to J. T. Cushing, 16 June 1844, WP, box 9.
12. Minto to Young, 22 Nov. 1905, MP, box 2.
13. Michael T. Simmons, brief of evidence in relation to the Puget Sound Agricultural Co.'s claims against the United States [n.d., ca. Sept. 1865], FNP, no. 1206; Ezra Meeker, *The Busy Life of Eighty-five Years* . . . (Seattle, Wash.: by the author, 1916), p. 111; Meany, "History of the Lumber Industry," pp. 97–98; Clark, "Analysis of Forest Utilization," pp. 23, 28–30; Iva L. Buchanan, "An Economic History of Kitsap County, Washington, to 1889" (Ph.D. diss., University of Washington, 1930), pp. 48–49, 55–56, 82–83; Edmond S. Meany, "First American Settlement on Puget Sound," *WHQ* 7 (1916):139–41; C. A. Snowden, *History of Washington: The Rise and Progress of an American State*, 4 vols. (New York: Century History Co., 1909), 2:438, 4:349; Ruby El Hult, "The Saga of George W. Bush," in Ellis Lucia, ed., *This Land around Us: A Treasury of Pacific Northwest Writing* (Garden City, N.Y.: Doubleday, 1969), p. 348. This Deschutes River should not be confused with the stream of the same name in present-day Oregon.
14. Duflot de Mofras, *Travels*, 2:111; John A. Hussey, *Champoeg, Place of Transition: A Disputed History* (Portland: Oregon Historical Society, 1967), p. 68.

In 1844 a different sort of sawmill appeared on the south bank of the Columbia River some thirty miles upstream from Astoria. There Henry H. Hunt and Tallmadge B. Wood built a mill to produce lumber for the export trade, for Wood believed that lumber "is and will be a great source of wealth to this country." [15]

However, it was difficult to secure vessels for carrying their cut from the treacherous, seldom-visited Columbia. Albert E. Wilson was taken on as a third partner to solve the problem.[16] Wilson was associated with J. P. Cushing and Company, a Massachusetts firm that had risen to power in the China trade and was now undertaking to challenge the Hudson's Bay Company for the trade of the Oregon country. With Wilson as a partner, the Cushing company's brig, *Chenamus*, was available to the enterprise. In March of 1844 the owners dispatched the first cargo of lumber from the mill to the Hawaiian Islands. Unfortunately, increasing competition for the market of the islands had driven prices down. The cargo of fifty thousand board feet brought only some $20 per thousand above the cost of freight. Apparently the enterprise still appeared promising, in spite of the low prices, for in the following year James Birnie, who had been in charge of Fort George, the Hudson's Bay Company post at Astoria, left the company to become a fourth partner in the milling operation.[17]

It is possible to reconstruct a rather detailed picture of Hunt's mill, for both contemporary descriptions and one of its account books have survived.[18] The millsite appears to have been well chosen. Deep waters came close to the river's bank at this point, making the loading of vessels safe and relatively easy. According to the captain of the *Chenamus*:

15. Wood to brother, 19 Feb. 1846, *OHQ* 4 (1903):83.
16. John Tod stated that in 1844 only six vessels entered the Columbia: four American, one Belgian, and one British (a sloop of war). See Tod to E. Ermatinger, 1845, Oregon Historical Society, Portland. See also Dorothy O. Johansen and Charles M. Gates, *Empire of the Columbia: A History of the Pacific Northwest*, rev. ed. (New York: Harper and Row, 1967), pp. 214–15; Throckmorton, *Oregon Argonauts*, p. 57; Hubert Howe Bancroft, *History of Oregon*, 2 vols. (San Francisco, 1886–88), 2:48. For an early description of the hazards of navigating the Columbia, see Duflot de Mofras, *Travels*, 2:55–60.
17. Throckmorton, *Oregon Argonauts*, pp. 38–40; Tallmadge B. Wood, "Letters of Tallmadge B. Wood," ed. Florence E. Baker, *OHQ* 3 (1902):395–98, 4 (1903):80–85; John Minto, "From Youth to Age as an American," *OHQ* 9 (1908):128; Avery Sylvester, "Voyages of the Pallas and Chenamus, 1843–45," *OHQ* 34 (1933): 263, 269–72, 364–71; McLoughlin to Governor, 24 June 1842, McLoughlin, *Letters*, ed. Rich, 2:56, 58; J. Douglas to G. Simpson, 5 Mar., 4 Apr. 1845, ibid., 3:182–83, 189–90.
18. The official designation of the plant appears to have been the Astoria Mill, but it was almost universally referred to as "Hunt's mill."

It is situated on the west bank of the Columbia . . . on a little stream that will carry it about 9 months in a year. This stream has a fall of about 60 ft. The mill is on the precipice or bank, and is carried by an overshot wheel 20 ft. in diameter. Thirty-six inches of water with this wheel will carry two run of saws,—Logs can be had at it, 2 or 3 years, with little more trouble than to roll them in, and this can be done on each side of the stream. There is a gradual ascent for ½ a mile back, well covered with timber.[19]

Another observer wrote that in "the vicinity of the mill there is some better timber than I have seen in any other part of the country." [20] The mill, which had a capacity of about three thousand feet per day, used spruce, fir, and hemlock logs. Captain John H. Couch, who took cargoes of lumber from the mill to the Hawaiian Islands, reported that the lumber was of high quality and sold readily in the islands.[21] At least some of the laborers at the mill were Hawaiians, or Kanakas as they were then known.[22] Wages were low compared with those being paid in the Willamette Valley at the time. One employee worked sixteen days for $6.15. Ninian Eberman, the "most constant employee," worked for $30.00 a month, and Tallmadge Wood received $25.55 for nineteen days' work as sawyer.[23] Wood and others also sold shingles to the mill.

As did Larkin, the partners in the Astoria Mill ran a store, but too few records have survived to tell if it was as great a source of profit as Larkin's was for him. It seems unlikely. What the records do make clear, however, is that, though a few lumber sales were made locally, over 95 percent of the mill's cut was exported. The *Chenamus* made three trips to the islands within eighteen months of the establishment of Hunt's mill; other vessels also visited the mill to take on cargoes.[24]

19. Sylvester, "Voyages of the Pallas and Chenamus," pp. 367–68. This description closely parallels that of John Minto, who worked at the mill in 1845. Minto to G. H. Himes, 2 Dec. 1910, Minto to M. W. Gaston, 23 Feb. 1911, MP, file 2; memorandum, Minto scrapbook, MP, box 2.
20. Joel Palmer, *Journal of Travels over the Rocky Mountains, to the Mouth of the Columbia River,* . . . (Cleveland: Arthur H. Clark Co., 1906), p. 204. See also Meany, "History of the Lumber Industry," pp. 79–80.
21. Couch to Cushing, 24 Mar. 1845, WP, box 9.
22. According to Minto, the work force numbered seventeen in the winter of 1845. He identified seventeen employees listed in the account book as whites, five as Hawaiians, two as Indians, and the cook, J. D. Sauls, as a Negro. See Minto to Himes, 2 Dec. 1910, MP, file 2.
23. Cf. Throckmorton, *Oregon Argonauts,* p. 57. Minto described Eberman as the "most constant employee" until he left for California during the Gold Rush. Wood, of course, was a partner in the mill, but he seems to have drawn additional wages as sawyer and to have split shingles independently as well. Unlike the account book, Minto identified L. L. M. "Indian" Cooper as the sawyer. Minto to Himes, 2 Dec. 1910, MP, file 2.
24. Astoria Mill, account book, Oregon Historical Society, Portland, *passim;* Sylvester, "Voyages of the Pallas and Chenamus," pp. 367–70; Palmer, *Journal,* pp. 204–5; Honolulu *Polynesian,* 31 Aug. 1844.

Hunt's mill, built to service the export trade, was prospering in its chosen role; the *Chenamus* and the company store were there as adjuncts of the mill, not the reverse. In Hunt's operation, like Smith's but unlike Larkin's or McLoughlin's, lumber production was the primary concern.

The close tie between mercantile and lumber interests, which the Astoria Mill represented by its connections with J. P. Cushing and Company, was equally evident in other enterprises in the region. New stores were being established to serve the growing population of the Willamette Valley, and ships that brought them merchandise were often employed in carrying exports from the Oregon country between trips or on their return voyages. The chronic shortage of cash that plagued the growing settlement contributed to this development. Storekeepers found that most of their customers could pay for their purchases only with grain or other products of the area, and the merchants turned to the Hawaiian Islands and California in an attempt to market the products they acquired. Lumber was among the many items they were soon exporting.

One such merchant was George Abernethy. Abernethy, having come to the Oregon country to serve as steward for Jason Lee's Methodist mission, persuaded Lee to locate the mission store at Willamette Falls in 1841, and managed the store until 1846, when he purchased it from the mission for $30,000. In the same year he acquired control of the Island Milling Company and, shortly thereafter, dispatched a cargo of lumber to Honolulu. He continued to ship occasional cargoes to the islands in the years that followed.

Abernethy was in a disadvantageous position. Most of the few vessels operating out of the Columbia River were controlled by competitors, such as Benjamin Stark and F. W. Pettygrove, who owned the barks *Toulon* and *Mariposa*. They agreed to carry a cargo of Abernethy's lumber to Hawaii in 1846, but refused to bring goods back from the islands except at exorbitant freight rates. Abernethy found rates to California were also high: $2 per barrel on flour and $20 per thousand board feet on lumber. Moreover, Abernethy was hampered by the annual payments he had to make to the Mission Society. Under the terms by which he purchased the store, he had to pay the mission $2,500 each year. Of this, $500 was to be in cash. Abernethy was hard pressed to keep up these cash payments, for wheat was the main medium of exchange in the valley and trade often took the form of barter. His uncertain financial position seems to have been what prevented Abernethy, who had been plagued by the lack of a vessel, from forming a partnership with Captain Roland Gelston of the *Whiton* when the captain was in Oregon in 1847.[25]

Also active in the Pacific trade was Captain John H. Couch, who, like Albert E. Wilson, was associated with J. P. Cushing and Company. He first came to the Columbia River in 1840 in charge of the brig *Maryland*. Because of unusually high water, Couch was able to take his vessel up the Willamette River as far as the falls, but Couch recognized that the practical head of navigation for ocean-going vessels was some twelve miles downstream and that there, not at the falls, the future metropolis of the region was likely to develop. At that point, on a heavily wooded tract on the steep west bank of the Willamette, Couch staked out a claim to 640 acres. In time the city of Portland was to grow on and around the site.

In spite of his perspicacity, Couch's first visit to the Willamette was a commercial failure. In 1841 he sailed the *Maryland* to Hawaii, sold her, and returned to Massachusetts on a whaler. Duflot de Mofras apparently had Couch's failure in mind when he wrote that he doubted whether "the sporadic efforts of American traders with limited facilities at their disposal" would generate serious competition for the Hudson's Bay Company.[26]

The Cushings, convinced there were profits to be made in the Northwest, were undeterred. In 1842 they again dispatched Couch to the Columbia, this time in charge of the *Chenamus*. Aboard was a large stock of merchandise with which the captain opened a store at Willamette Falls to compete with the Hudson's Bay Company and the Methodist mission for the trade of the American settlement. Couch put the *Chenamus* into the Oregon-Hawaii trade at first under his own command, then under his associate, Captain Avery Sylvester. Not long thereafter the vessel began hauling cargoes of lumber from the Astoria Mill.[27]

Like others, Couch found that the shortage of specie in the

25. McLoughlin to Governor, 20 Nov. 1845, McLoughlin, *Letters*, ed. Rich, 3:126; Throckmorton, *Oregon Argonauts*, pp. 23, 27, 34, 36, 50, 58, 65.

26. Duflot de Mofras, *Travels*, 2:131–32. For further information on this highly competitive trade, see Alfred L. Lomax, "Hawaii-Columbia River Trade in Early Days," *OHQ* 43 (1942):328–38; Meany, "History of the Lumber Industry," p. 68 and *passim*.

27. Cushing to Couch, 7 Oct. 1841, 15 Sept. 1843, WP, box 9; Couch to Cushing, 10 Apr. 1842, 16 June, 7 Nov. 1844, 24 Mar. 1845, WP, box 9.

Willamette Valley made profitable operation of a store difficult. Penetration of the Hawaiian market, where the Hudson's Bay Company was well entrenched, proved equally hard. By 1845 Couch and the Cushings were ready to abandon their venture. Couch offered to sell the merchandise and accounts of his store to the Hudson's Bay Company for $30,000; the Cushings withdrew the *Chenamus* from the area. The company turned down his offer, however, and Couch continued to operate his store for two more years. During that time he found, as Abernethy already had, that operating a mercantile business in the Oregon country without controlling vessels to supply it and market the produce it took in was a hazardous proposition. Stark and Pettygrove refused to haul goods to Couch except at the same ruinous rates they had offered Abernethy. When Couch refused to pay such high rates, his competitors allowed their vessels to return in ballast from the Hawaiian Islands rather than lower their charges and thus allow Couch to stock his store with goods he might market at prices competitive with those charged at Pettygrove's stores. In the face of this situation, Couch gave up and returned to Newburyport in 1847.[28]

By the mid-1840s Francis W. Pettygrove was becoming the leading American entrepreneur in the Northwest for reasons other than just his control of shipping. Pettygrove had been sent to the Oregon country in 1843 to establish a store for the New York firm of Benson and Brother. With Pettygrove aboard the bark *Fama* was fifteen thousand dollars' worth of goods with which to stock the store. A second cargo was sent out on the *Toulon* in 1845 and a third on the *Mariposa* in 1846. Only the Hudson's Bay Company stores could offer customers the variety Pettygrove did. Moreover, Pettygrove was not plagued by the lack of capital that so hampered Abernethy, especially after John McLoughlin bought his son David an interest in Pettygrove's Portland store in 1846 for $20,000.

Pettygrove's holdings expanded rapidly. He first opened a store in 1843 at Willamette Falls. Within two years he had acquired a granary at Champoeg and had opened a second store next to Couch's claim at

28. McLoughlin to Governor, 24 June 1842, McLoughlin, *Letters*, ed. Rich, 2:56, 58; McLoughlin to Governor, 19 July, 20 Nov. 1845, ibid., 3:88, 126–27; J. Douglas to G. Simpson, 5 Mar. 1845, ibid., 3:182; P. Ogden to J. Douglas, 16 Mar. 1845, ibid., 3:289; McLoughlin, "Reminiscences," McLoughlin private papers, part 2, pp. 9–10; Sylvester, "Voyages of the Pallas and Chenamus," pp. 265–70, 360–66, 368–69; Duflot de Mofras, *Travels*, 2:129; Throckmorton, *Oregon Argonauts*, pp. 29–30, 32, 39, 40, 58, 90; Lomax, "Dr McLoughlin's Tropical Trade Route," pp. 14–15.

the head of navigation on the Willamette. He and Asa Lovejoy platted a townsite there and then flipped a coin to decide who would have the honor of selecting the new town's name. Pettygrove, the winner, called it Portland, after the chief city of his home state of Maine. Pettygrove's brother-in-law, Philip Foster, also had a store. It was strategically located east of Oregon City where the Barlow Trail issued from the Cascade Mountains.[29] By 1846 Pettygrove's position was so strong that he sought to gain control of the area's salmon trade.

Like other businessmen in the Northwest, Pettygrove made trade with Hawaii and California an important adjunct of his mercantile operations. The *Toulon*, under Captain Nathaniel Crosby, Jr., arrived in the Northwest from New York via Hawaii in late 1845 with goods consigned to Pettygrove. Pettygrove then put the vessel in the island trade and, later, into direct trade with California. Similar voyages were being made soon thereafter by the *Mariposa*. Such voyages allowed Pettygrove not only to market products of the Oregon country that he had acquired through his business enterprises, but also to replenish the stocks in his stores with goods available in Hawaii or San Francisco.[30]

Pettygrove, like Abernethy and Couch, thus found himself engaged in the Pacific lumber trade not as a result of any plan or foresight, not because he was a lumberman, but simply because it was a necessity for anyone engaged in business in the region. He and the other American merchants in the Northwest had been driven into the trade by the same forces that had driven the Hudson's Bay Company and Larkin into it. These enterprises represented the mainstream, for other than Hunt's Astoria Mill, operations built specifically for the export lumber trade had not yet appeared. But the way had been pointed. Hunt, his associates, and Pettygrove, like McLoughlin before them, had found that doing business in the Northwest virtually necessitated dealing in lumber and that exporting it was both feasible and profitable. The *Oregon Spectator* summed it up succinctly. These shipments of lumber and produce, the paper stated, give "promise of what Oregon may yet do." Or, again, they at least show that "there is something to ship from Oregon." [31]

29. For revealing comments on Foster, see Sylvester, "Voyages of the Pallas and Chenamus," pp. 266, 363–64.
30. Portland *Oregonian*, 5 Aug. 1854; D. Lownsdale to S. Thurston, 10 Aug. 1849, *OHQ* 14 (1913):225; Nathaniel Crosby, Jr., accounts, Oregon Historical Society, Portland; Oregon City *Oregon Spectator*, 15 Apr. 1847.
31. Oregon City *Oregon Spectator*, 15 Apr., 13 May 1847; see also 14 May 1846.

The same could have been said of California's redwood coast. Stephen Smith's mill near Bodega·Bay demonstrated that lumber could be cut on the northern coast and shipped to San Francisco at a profit. Moreover, he and Hunt together made it clear that an entrepreneur could succeed on the Pacific Coast as a lumberman without also engaging in the host of other commercial activities that made up the business of a Pettygrove or Larkin.

The Pacific Coast lumber industry had begun, but it had not yet developed into a major undertaking. Merchants who were willing to buy and sell whatever the region produced so long as they could do so at a profit continued to overshadow lumber manufacturers like Smith and Hunt as the 1840s reached their midpoint.

CHAPTER III

The Winds
of Change

AMERICANS moving to the West Coast in the early forties were quickly disabused of their illusions about the economic opportunities there. Settlers in the Willamette Valley found that, though the soil was rich and the climate mild, their new home was in a valley far removed from markets for its products, a valley where the small quantities of money the settlers brought with them passed with startling rapidity into the coffers of the Hudson's Bay Company or the stores of Abernethy, Couch, or Pettygrove. Merchants, as well as farmers, sometimes became disillusioned. John Couch returned to Newburyport; the Cushings gave up their attempts to gain a foothold in the trade of the Northwest. Duflot de Mofras' prediction that the Hudson's Bay Company had little to fear from the American interlopers seemed in the process of being borne out.[1] Economic problems contributed to growing political discontent. During the wet months of 1846–47, conditions reached their nadir; it was, in Dorothy Johansen's trenchant phrase, Oregon's "winter of discontent." [2]

Conditions in California were, on the whole, somewhat better. The ports of Alta California were visited more frequently than those of the Northwest, thus making possible the exportation of larger quantities of

1. Duflot de Mofras, *Travels*, 2:131–32.
2. Johansen and Gates, *Empire of the Columbia*, pp. 211–21. See also Throckmorton, *Oregon Argonauts*, pp. 53–61, 90; James H. Gilbert, *Trade and Currency in Early Oregon: A Study in the Commercial and Monetary History of the Pacific Northwest* (New York: Columbia University Press, 1907), pp. 66–67, 71–72.

the area's produce and, in turn, the importation of more consumer goods. Moreover, if John Sutter knew whereof he spoke, California's overland immigrants were better off financially than Oregon's.[3]

There were economic problems in California, nevertheless. The wheat crop in the Santa Clara valley failed in 1845. Shipowners on the East Coast hesitated to dispatch vessels to California when a war with Mexico seemed imminent.[4] Increasing competition, primarily from the Northwest, cut into profits in the lumber trade with Hawaii. Even Larkin was not immune from the effects of these developments, and creditors were soon complaining that some of his bills were long overdue.[5]

Conditions continued to deteriorate in 1846. The demands for horses during the military campaigns of that year left ranchers without the necessary mounts to round up their cattle for slaughtering. In addition, these same ranchers, convinced that prices would drop even further, put off making purchases. William A. Leidesdorff, a leading merchant of Yerba Buena, summed it up well when he wrote Larkin, "We must call this year a lost one altogether." [6] Others were even harder hit than Larkin and Leidesdorff. Logging in the redwoods of the East Bay region was abandoned altogether during the years 1842–46.[7]

Once California came under the control of the United States, the slump came to an abrupt end. At the same time, speculation in land commenced. Leidesdorff advised Larkin, "If you want to speculate in lots here, now is your time, before the large party arrives that is exspected next month from the States." [8] Soon Larkin was involved in schemes to make Benicia the dominant city of the new California and

3. Sutter to Larkin, 2, 8 Oct. 1845, Larkin, *Papers*, 4:4, 10–11.

4. A. Thompson to Larkin, 13 Oct. 1845, ibid., 4:22; B. Reed to Larkin, 13 Oct. 1845, ibid., 4:23; Ogden, "Boston Hide Droghers," p. 300.

5. Underhill, *From Cowhides to Golden Fleece*, p. 53; J. Jarves to Larkin, 4 Oct. 1845, Larkin, *Papers*, 4:8. A correspondent, writing to Larkin from Valparaiso, noted the increased number of shipments from the Columbia River to the Hawaiian Islands and suggested that Larkin look to Chile for a less competitive outlet for his goods: "I believe it would be more profitable" than shipping anywhere else, he wrote. For reasons that are not clear, Larkin does not appear to have taken advantage of the advice. See F. Atherton to Larkin, 11 Feb. 1845, Larkin, *Papers*, 3:33–36.

6. W. Leidesdorff to Larkin, 21 July 1846, Larkin, *Papers*, 5:149–50. See also Brown, *Sawpits in the Spanish Redwoods*, p. 14.

7. Burgess, "Forgotten Redwoods of the East Bay," pp. 3–4. Lumbering did not stop in the Santa Cruz area, however, nor did Larkin cease purchasing lumber there. See Larkin to W. Swasey, 18 June 1846, and Swasey to Larkin, 23 June 1846, Larkin, *Papers*, 5:49–50, 68–69.

8. Leidesdorff to Larkin, 12 July 1846, Larkin, *Papers*, 5:130–31. John Sutter also saw great opportunities arising as a result of the change of sovereignty. Sutter to Leidesdorff, 27 July 1846, LP, box 2.

was dispatching cargoes of lumber to the site for use in building the would-be metropolis.[9]

Activity elsewhere had an even greater impact on lumbering. In San Francisco an estimated two hundred houses were erected in 1847.[10] One observer wrote that "almost everybody is dieing with the land feaver, the Water Lots have nearly all be[en] Sold, selling from ten to two thousand dollars the lot; . . . a person cannot anchor a launch in less than three fathoms of water for fear of being driven off by the owner. . . ." [11] Construction was also under way in Monterey. Larkin reported a great need for lumber there and offered captains William Phelps and John Paty $15 per thousand board feet freight to bring lumber to Monterey from Santa Cruz.[12]

New sources of lumber were developed in response to the rising demand, and established ones—such as Stephen Smith's mill at Bodega Bay—prospered. Charles Brown and Dennis Martin built sawmills, perhaps the first in present-day San Mateo County, and Salvador Vallejo let a contract for the construction of a mill on Napa Creek.[13] During the mid-decade slump, lumber prices had dropped as low as $20 per thousand, but in the face of renewed demand soon reached $50 to $60 once again.[14] The rising prices in California were noticed as far away as Australia, and cargoes of eucalyptus lumber were dis-

9. Larkin, *Papers*, 6:xi; Larkin to R. Semple, 28 July, 14, 27 Aug., 8 Sept. 1847, ibid., 6:240, 273–74, 297, 324–26; Larkin to H. Cambuston, 2 Aug. 1847, ibid., 6:249; Larkin to S. Smith, 2 Aug. 1847, ibid., 6:249–50; R. Semple to Larkin, 5 Aug. 1847, and other dates, ibid., 6:259 and *passim;* Smith to Larkin, 15 Aug. 1847, ibid., 6:278; Larkin to Leidesdorff, 12 Apr. 1847, LP, box 3; Monterey *Californian*, 20 Apr. 1847; Parker, "Larkin, Anglo-American Businessman in Mexican California," p. 423.

10. Larkin, *Papers*, 6:xi; Brown, *Reminiscences*, p. 43.

11. H. P. Richardson to Stearns, 23 July 1847, SP, box 51.

12. Larkin to W. Phelps and J. Paty, 6 Sept. 1847, Larkin, *Papers*, 6:318; Larkin to R. Semple, 8 Sept. 1847, ibid., 6:326.

13. Clar, *California Government and Forestry*, pp. 42, 52, 73–74; Frank M. Stanger, *Sawmills in the Redwoods: Logging on the San Francisco Peninsula, 1849–1967* (San Mateo, Calif.: San Mateo County Historical Association, 1967), pp. 27–29; Sutter to Leidesdorff, 14 Aug. 1846, LP, box 2. Brown, *Sawpits in the Spanish Redwoods*, p. 15, suggests that these mills were probably not constructed until 1849. Regarding production in the Napa Valley, see also E. Bale to Larkin, 5 July 1847, Larkin, *Papers*, 6:231; Larkin and H. Green, contract, 16 July 1847, ibid., 6:233; S. Brown to Larkin, 10 Sept. 1847, ibid., 6:332; Bale to Leidesdorff, 5 Mar. 1847, LP, box 3; N. Spear to Leidesdorff, 23 Dec. 1846, LP, box 3; G. Nye to Leidesdorff, 7 Feb. 1847, LP, box 5; R. Kilburn to Leidesdorff, n.d., LP, box 5.

14. J. A. Moerenhout to French minister of foreign affairs, 25 Sept. 1847, reprinted in Abraham P. Nasitir, ed., "The French Consulate in California, 1843–1856," *CHSQ* 12 (1933):350; S. Smith to Stearns, 10 Feb. 1847, SP, box 60. This increase in price only brought lumber back to where it had been ten years before, however. See Underhill, *From Cowhides to Golden Fleece*, pp. 163–64.

patched to San Francisco in order to take advantage of them.[15]

As early as July 1846, Larkin had expressed a desire "to be entirely free from trade," as real estate schemes seemed to offer a quicker road to wealth.[16] But in a period of rising prosperity, others were ready to take over Larkin's trade as he withdrew from it. Owners of Rousillion and Sainsavain's mill on the San Lorenzo, John Hames's mill on Soquel Creek, and William Blackburn's mill on Branciforte Creek had sold to Larkin; now they began to ship directly to San Francisco, where they found ready buyers in William Heath Davis and Leidesdorff. Meanwhile, lumber production resumed in the redwoods of the East Bay region; much of this was also marketed in San Francisco.[17]

On 29 May 1847, Leidesdorff announced in the San Francisco *Californian* that he had "made arrangements for supplying the Town with Lumber, persons wishing any kind of lumber can have their orders executed by leaving them at his store." The replacement of Monterey as the primary center of the lumber trade of California and of Larkin as the dominant lumber dealer was under way.

Stephen Smith was among those sending lumber to Leidesdorff. The financial relationship between Leidesdorff and Smith is not clear, but it would appear that Leidesdorff furnished the capital with which Smith opened a lumberyard in San Francisco, acquired an interest in vessels to link his mill and the yard, and made necessary purchases of equipment and supplies. Leidesdorff acted as Smith's agent in selling his lumber in the Bay Area, in managing his vessels, and in obtaining supplies and workmen for the mill.[18] In doing all this, Leidesdorff was acting in a manner common among merchants of the day. Merchant-capitalists regularly supplied investment capital to manufacturers,

15. Bancroft, *History of California*, 7:76, 102–3; San Francisco *California Star*, 1 Apr. 1848. Sanford, "Short History of California Lumbering," p. 25, blows these shipments out of all proportion, arguing that most of the lumber used in early Monterey was from Australia.
16. Larkin to Leidesdorff, 11 July 1846, Larkin, *Papers*, 55:128.
17. Burgess, "Lumbering in Hispanic California," p. 244; Burgess, "Forgotten Redwoods of the East Bay," pp. 4–6. See also [J. P. Munro-Fraser], *History of Alameda County, California* . . . (Oakland, Calif., 1883), p. 412; Harry Noyes Pratt, "Oakland, Lumber Town," *Oakland Tribune*, 26 Aug. 1923. Larkin, who had not yet entirely abandoned the lumber trade, also dispatched some cargoes to San Francisco. See Larkin to J. Paty, 6 Sept. 1847, Larkin, *Papers*, 6:316–17; Larkin to R. Sherman, 6 Sept. 1847, ibid., 6:317; J. Paty to Larkin, 12 Sept. 1847, ibid., p. 336.
18. Smith to Leidesdorff, 26 Apr. 1846, LP, box 2; 5 Nov. 1846, 25 Jan., 10, 20 Feb. 1847, LP, box 3; 29 May, 5 June, 15, 27 July 1847, box 4; 11 Feb., 17 Apr. 1848, box 5; Smith to Larkin, 15 Aug. 1847, Larkin, *Papers*, 6:278; Smith to A. Stearns, 15 Nov. 1847, SP, box 60; Brown, *Reminiscences*, p. 67. Leidesdorff may have been a partner of Smith's in the lumberyard and vessels.

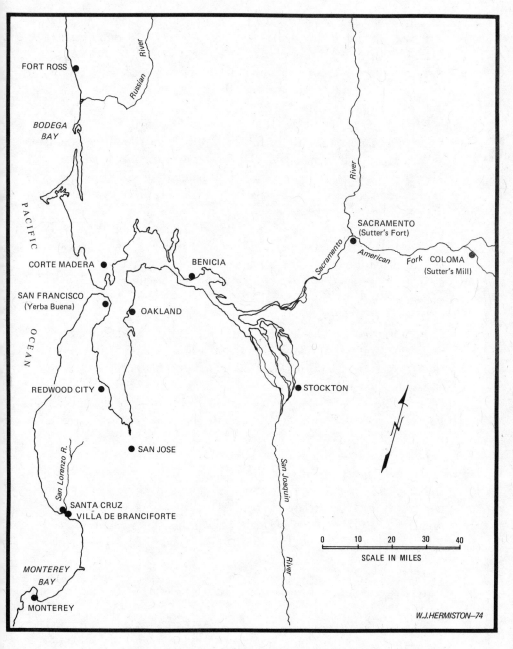

Map 1. San Francisco Bay and Vicinity in the Mid-Nineteenth Century

though others still hesitated to do so, and, in their role as agents, furnished an essential link between producers and their customers. Indeed, acting in these and related capacities, they were more in control of the American economy down to the mid-nineteenth century than were the manufacturers whom they supported and who came to have so much power later.[19]

William A. Leidesdorff was already well known in San Francisco when he began to move to the forefront of the area's lumber trade. In 1845 Larkin had named this Danish-born, naturalized citizen of Mexico the American vice-consul for the port of San Francisco.[20] Leidesdorff's home appears to have been something of a social center in the growing community. When Commodore James Biddle arrived from China in the summer of 1847, he was a guest in the home of the vice-consul. That September Leidesdorff was chosen town treasurer and named to a committee of three to investigate the feasibility of establishing a public school. Like Larkin, Leidesdorff was a major speculator in San Francisco real estate.[21]

Leidesdorff's successes came as no surprise to those who knew him well. Larkin once characterized him as "active, bold, honourable, passionate, and liberal." John H. Everett, a Boston sea captain who had traded along the California coast for years, provided a more revealing assessment when he asked Larkin, "Is Leidesdorff married yet?" and added, "He will be one of the leading men in the country before long—he does not want for bronce [bronze, i.e., he has plenty of brass]." [22]

Unlike Leidesdorff, William Heath Davis moved into the lumber trade with a great deal of reluctance. Davis was by nature a speculator; trade in lumber was too pedestrian to appeal to a person with his ambition. Davis had dreams of entering the risky trade with China and South America, though he apparently never thought of sending lumber there, and of growing wealthy through land schemes and development

19. Porter and Livesay, *Merchants and Manufacturers*, pp. 2, 5–10, 13–36.
20. Larkin to Leidesdorff, 29 Oct. 1845, Larkin, *Papers*, 4:73, 75. Leidesdorff was an interesting choice for the position, for he, together with other merchants in Yerba Buena, had been engaged in smuggling. See Andrew F. Rolle, *An American in California: The Biography of William Heath Davis, 1822–1909* (San Marino, Calif.: Huntington Library, 1956), p. 156.
21. Frank Soulé, John H. Gihon, and James Nisbet, *The Annals of San Francisco* (New York, 1855), pp. 186, 196–97, 307; William Heath Davis, *Seventy-five Years in California*; . . . (San Francisco: J. Howell, 1929), pp. 212–13; Richardson to Stearns, 23 July 1847, SP, box 51.
22. Everett to Larkin, 12 Dec. 1845, Larkin, *Papers*, 4:119; Larkin, notes on principal men, n.d., ibid., 4:330.

projects. His partners, Hiram and Eliab Grimes, were more conservative. They opposed Davis' plans for dispatching their vessel, the
Euphemia, to China and urged Davis to continue to run it, as he had
for some time, in the lumber trade. More from his partners' insistence
than his own choice, Davis soon became one of the leading lumber
dealers in California. The *Euphemia* made regular runs from Bodega
Bay and San Jose to the Hawaiian Islands with cargoes of lumber, a
trade that prospered as commerce between California and the
Hawaiian Islands grew during the last half of the decade. In addition,
Davis' partners chartered the *Mount Vernon* to carry lumber from the
Columbia River to the islands.[23]

Like Leidesdorff, Davis soon found himself in a position of
community leadership. Neither man was above using public office to
advance private business. The two drew criticism when, while city
officials, they sold San Francisco large quantities of lumber for sidewalk
construction.[24]

Leidesdorff and Davis took much of the cut of Stephen Smith's mill
at Bodega Bay. But, though lack of demand was not a problem, other
difficulties plagued the captain's operation. Smith had difficulty
obtaining adequate help and equipment, rough seas sometimes kept
vessels from transporting his cut to market, and during the rainy season
it sometimes became impossible to get lumber from his inland millsite
to the bay because the carts used to transport the lumber became mired
in the mud.[25]

Though Leidesdorff and Davis relieved Smith of the bulk of the
lumber sawed at his mill, the two were not in direct competition in
disposing of it. Leidesdorff acted as Smith's agent in selling lumber in
San Francisco, a market which he and Smith hoped to dominate. They
landed their first shipment at an old hide house on the waterfront at the
foot of present-day Pacific Street; subsequent cargoes went to a wharf
Leidesdorff had constructed near the foot of Pine. There the two
developed what could be called San Francisco's first lumberyard.[26] On

23. S. Smith to Stearns, 10 Feb., 27 June 1847, SP, box 60; Smith to Leidesdorff, 25 Jan., 20 Feb.
1847, LP, box 3; Smith to W. H. Davis, 25 Jan. 1847, Davis Papers, box 8; Harold Whitman
Bradley, "California and the Hawaiian Islands, 1846–1852," *PHR* 16 (1947):19.
24. Rolle, *An American in California*, pp. 61–63, 77; San Francisco *California Star*, 17 Apr. 1847
and other dates.
25. Smith to Stearns, 10 Feb., 27 June, 15 Nov. 1847, SP, box 60; Smith to Leidesdorff, 26 Apr.
1846, LP, box 2; 25 Jan., 10, 20 Feb. 1847, LP, box 3; 29 May, 5 June 1847, LP, box 4; Smith to
Leidesdorff, 25 Dec. 1847, Leidesdorff Papers, California Historical Society Library.
26. Brown, *Reminiscences*, p. 67; Smith to Larkin, 15 Aug. 1847, Larkin, *Papers*, 6:278; Smith to

the other hand, Davis shipped most of the lumber he purchased to the Hawaiian Islands. Indeed, Smith appears to have sold to Davis, Captain Jean Jacques Vioget, and other potential competitors only after they agreed not to ship to San Francisco.[27]

Such efforts were not entirely successful. Others continued to ship lumber to San Francisco. When Robert A. Parker and John Henry Brown set about building the Parker House in San Francisco, they turned not to Smith and Leidesdorff for the lumber they needed, but to pit sawyers operating in the forests near San Francisco Bay. Brown later reported that ten sawpits in the Corte Madera area were kept busy in order to supply lumber needed in building the hotel.[28]

Smith's success may have been limited and his business small, but by 1847 his operations had assumed a pattern that was to become common in the Pacific lumber trade in the years that followed. Like Smith, later leaders in the Pacific Coast lumber industry had mills located in forests to the north of San Francisco, lumberyards in that city, and vessels to link the two. Also, San Francisco was their primary market, while customers in Hawaii and the region immediately tributary to the mills supplied important secondary outlets.[29] This pattern, once perfected, served later millmen even better than it did Captain Smith.

Smith, Leidesdorff, Davis, and others had confidence that California would develop rapidly once it passed under American control; but for all their expectations, it soon began to appear that growth was to be slower than they had anticipated. Some found themselves financially overextended. One cause of their difficulty was the arrival in California of a new kind of merchant. The generation of merchants that had dominated during Mexican times had been easygoing. Generally they cooperated among themselves. Now, many newcomers were going into business and often resorting to price-cutting and similar practices in order to compete with the established firms. Davis lamented,

---

Stearns, 10 Feb. 1847, SP, box 60; Smith to Leidesdorff, 10 Feb. 1847, LP, box 3; 29 May 1847, LP, box 4.

27. Smith to Davis, 25 Jan. 1847, Davis Papers, box 8; Smith to Stearns, 10 Feb., 27 June 1847, SP, box 60; Smith to Leidesdorff, 25 Jan. 1847, LP, box 3; 5 May, 15 July 1847, LP, box 4.

28. Brown, *Reminiscences*, pp. 78–82. Brown and Parker found James Murphy of Corte Madera to be their most dependable whipsawyer, but shallow waters there made it impossible to take on full cargoes at his anchorage.

29. Smith sold not only to Davis and in San Francisco but to buyers in the Sonoma area as well. He also shipped to southern California and considered shipping to Peru, although it is not clear whether he actually did so. See Smith to Stearns, 10 Feb., 27 June, 15 Nov. 1847, SP, box 60; Smith to Leidesdorff, 27 July 1847, LP, box 4.

"Oh what a pity that them chaps has come to San Francisco. . . ." [30]

But these newcomers were probably not the primary cause of the financial difficulties experienced by many of those who had arrived in California earlier. The speculations into which many of the latter had sunk their capital were paying returns more slowly than had been anticipated. In 1848, when a business slump began to appear on top of these other problems, William Leidesdorff found himself on the verge of bankruptcy. On 15 March, Leidesdorff was forced to cancel both his advertisements in the *Californian* and his subscription to it. The editor of the paper, observing conditions, commented, "We hope . . . the heretofore large commercial business of the town will not sustain, by this decline, more than a temporary depression." A small item at the bottom of the page announced that gold had been found "in considerable quantities" at John Sutter's new mill on the American Fork. The end of the depression was nearer than the editor of the *Californian* imagined. But the upturn came too late to be of help to Leidesdorff. On 18 May he died of "brain fever," leaving an estate burdened with some $50,000 in debts. [31]

In the meantime, an end had also come to the mid-decade doldrums in which the economy of the Oregon country had been caught. Poor harvests, increased immigration, and military operations created a shortage of flour, other foodstuffs, and building materials in California. During the winter of 1846 the price of flour reached $30 per barrel. In the spring of 1847 the *Commodore Stockton* brought word of this to the Northwest, after David McLoughlin had chartered the brig from Leidesdorff and Smith and dispatched it with a load of lumber to California. [32] Nathaniel Crosby, Jr., sailing for F. W. Pettygrove, quickly loaded the bark *Toulon* with flour, lumber, and produce and headed for San Francisco. The cargo did not bring great returns—the flour sold for $10 per barrel, the lumber for $50 per thousand board feet—but the new market was most welcome, nonetheless. [33] This first cargo of products of the Oregon country to be dispatched directly to

30. Quoted in Rolle, *An American in California*, p. 64. For information on some of the early firms of San Francisco, see Bancroft, *California Pastoral*, pp. 732–39.
31. Soulé, Gihon, and Nisbet, *Annals of San Francisco*, pp. 201, 307.
32. Smith to Leidesdorff, 20 Feb. 1847, LP, box 3; account, lumber consigned to Leidesdorff by D. McLoughlin, 27 May 1847, LP, box 4. See also Leidesdorff to P. Ogden and J. Douglas, 16 Apr. 1847, LP, box 3.
33. Oregon City *Oregon Spectator*, 10 June 1847; Bancroft, *History of Oregon*, 2:17–18; Throckmorton, *Oregon Argonauts*, pp. 61–62; Rev. Ezra Fisher to [?], n.d., reprinted in *OHQ* 16 (1915):303–4; Crosby, accounts, Oregon Historical Society.

California by Americans was soon followed by others. Some were remarkably diverse; one included "lumber, flour, salmon, beef, potatoes, butter, cheese, cranberries, turnips, cabbage and onions, also a small invoice of almanacs adapted to the meridian of Monterey . . . [and] nine passengers." [34]

Occasional shipments to Hawaii and to the Russians in Alaska also continued. Indeed, the Hawaiian market was also showing signs of improving. Shipments from the Columbia River to Honolulu had brought only $23,101 in 1846; in 1847 the returns rose to $54,784.99. Around 1840 only four or five vessels a year had carried cargoes from the Columbia River to the islands; by 1847 there were twelve vessels engaged in the trade. [35] In April 1847 a total of 1,736 barrels of flour, 171,000 board feet of lumber, and 96,000 shingles was exported from Oregon to Hawaii, California, and Alaska. [36]

Also contributing to the brightening of the economic picture in the Northwest was growth of the local market. Some 4,500 immigrants, the largest influx of the decade, arrived in the fall of 1847. The settlement the preceding year of the boundary dispute with England and the widespread belief that Congress would soon pass a law confirming the titles to sizable tracts of land for those actually settled in the Oregon country did much to instigate this movement. The economic impact of these new arrivals was enhanced by the fact that many were of comfortable means. [37]

In the face of increasing demand, prices soon rose. Indeed, by late 1847 there was an acute shortage of merchandise and supplies in the Northwest. Captain William K. Kilbourne arrived from Newburyport with a cargo of secondhand merchandise and put his brig *Henry* into the San Francisco–Columbia River trade in competition with Crosby's *Toulon*. [38] These and the few other vessels that brought goods to the Oregon country were unable to meet the demand. There was still a shortage of specie, and complaints about monopolies—usually, but not always, meaning the Hudson's Bay Company—were still heard; but

34. Monterey *Californian*, 17 Nov. 1847.
35. Honolulu *Friend*, 1 Jan. 1847, 1 Feb. 1848; Meany, "History of the Lumber Industry," pp. 70–73. Throckmorton places the number of vessels trading out of the Columbia River in 1847–48 slightly lower. See *Oregon Argonauts*, p. 62.
36. Oregon City *Oregon Spectator*, 13 May 1847.
37. Throckmorton, *Oregon Argonauts*, pp. 62–63.
38. Ibid., p. 63; Oregon City *Oregon Spectator*, 27 Mar. 1847; Monterey *Californian*, 17 Nov. 1847; E. W. Wright, ed., *Lewis and Dryden's Marine History of the Pacific Northwest* (Portland, 1895), p. 22.

the worst was already over when news arrived of the discovery of gold in California.[39]

It is appropriate that the discovery of gold, which was to usher in a new age in the history of the Pacific lumber trade, was made at the sawmill of John August Sutter. As C. Raymond Clar has put it, in "a way Sutter was the most prominent lumberman of the Mexican era although the aura of gold tends to hide that simple fact." [40] Sutter was not a dealer in lumber as were Larkin, Leidesdorff, and Davis, but he had many interests. "In a word," his associate John Bidwell explained, "Sutter started every business enterprise possible." [41] In the diversity of his operations he even outstripped Larkin.

Sutter's enterprises created a demand for large quantities of lumber, and his men ranged over much of northern California to obtain it. He purchased much additional lumber on the open market, from Antonio Suñol of Pueblo San José, from sawyers in the East Bay hills and the Santa Cruz area, and from Leidesdorff in San Francisco.[42] "He kept his launch running to and from the bay," Bidwell recalled some years later, "carrying down hides, tallow, furs, wheat, etc., returning with lumber sawed by hand in the redwood groves nearest the bay, and other supplies. On the average it took a month to make such a trip." [43]

Regular though they were, even these purchases failed to satisfy Sutter's needs. One reason for his purchase of the property of the Russians at Fort Ross was his need for lumber. Following the purchase, buildings at the Russian settlement were dismantled and the lumber from them shipped to Sutter's headquarters at New Helvetia, as Sacramento was then known. In addition, Sutter had hopes of taking over Stephen Smith's mill at Bodega Bay; it was, he insisted, built on land that was legally a part of the Russian claim he had purchased.[44]

39. Throckmorton, *Oregon Argonauts*, pp. 64–66; Johansen and Gates, *Empire of the Columbia*, pp. 276–82.

40. Clar, *California Government and Forestry*, p. 42. See also C. Raymond Clar, "John Sutter, Lumberman," *Journal of Forestry* 56 (1958):259–65; Aubrey Neasham, "Sutter's Sawmill," *CHSQ* 26 (1947):109–33.

41. John Bidwell, "Life in California before the Discovery of Gold," *Century Magazine* 41 (1890):170.

42. J. Sutter to Leidesdorff, 31 July 1845, LP, box 1; 11 May, 1 June, 14 Aug. 1846, LP, box 2; Leidesdorff to Sutter, 10 Mar. 1846, LP, box 2; Sutter, account with Leidesdorff, 1 Aug. 1844–27 Jan. 1846, LP, box 1; Clar, *California Government and Forestry*, pp. 42–43; Clar, "John Sutter, Lumberman," p. 261.

43. Bidwell, "Life in California," p. 169.

44. Clar, "John Sutter, Lumberman," pp. 261–62; Duflot de Mofras, *Travels*, 1:248; John Sutter, *New Helvetia Diary* (San Francisco: Grabhorn Press, 1939), p. xii; Sutter to Leidesdorff, 14 Aug. 1846, LP, box 2; Clar, *California Government and Forestry*, p. 39.

Eventually Sutter's quest for lumber led him into production of his own.[45] As early as 1840 he had his men attempt to raft logs down the Sacramento and the American Fork for use at his fort, but the streams proved poorly adapted for the purpose and the project was discontinued. By 1845 he had turned seriously to whipsawing pine in the Sierra and hauling it to the fort. Indeed, Sutter may have been considering selling as well as sawing lumber, for he wrote Larkin: "In this cedar and pine wood [of the Sierra] I intend to build a sawmill. I think this kind of lumber would sell better" than redwood.[46]

Sutter's search for a suitable site for a sawmill was to be a long one, however. John Bidwell spent the spring and summer of 1844 looking for an appropriate location. He was not successful, but Sutter did not abandon the project. "Sutter's many enterprises," Bidwell explained, "continued to create a growing demand for lumber. Every year and sometimes more than once, he sent parties into the mountains to explore for an available site to build a sawmill on the Sacramento River or one of its tributaries, by which lumber could be rafted down to the fort. There was no want of timber or water power in the mountains, but the cañon features of the streams rendered rafting impracticable." [47]

By 1847 Sutter's needs were greater than ever, and, though he was already financially overextended, Sutter once again sent out a party to search for a millsite.[48] James Marshall—described by Bidwell as "certainly eccentric," but "harmless"—was chosen as leader. Marshall soon returned and enthusiastically recommended a site on the American Fork at present-day Coloma. After examining the site, Sutter quickly entered into an agreement with Marshall whereby the latter was to build a mill there and receive half the lumber cut at the mill as compensation for his work. Apparently Marshall planned to raft his share of the lumber down the Sacramento River for sale in San Francisco.[49]

45. Bidwell, "Life in California," p. 179; Clar, "John Sutter, Lumberman," p. 262; Sutter, *New Helvetia Diary*, pp. 2, 3, 5, 55, 57.
46. Sutter to Larkin, 22 July 1845, Larkin, *Papers*, 3:282.
47. Quoted in Clar, "John Sutter, Lumberman," p. 263.
48. Sutter apparently owed money to Leidesdorff, but could only even accounts by hiring his launch to Leidesdorff for his use in hauling lumber to San Francisco. See Leidesdorff to Sutter, 10 Mar., 14 Aug. 1846, LP, box 2; Sutter to Leidesdorff, 13 Oct. 1847, LP, box 4.
49. Bidwell, "Life in California," p. 181. Even Sutter seems to have shared Bidwell's views of Marshall, for he commented: "I was dubious about trusting him out of my sight with all his craziness, but the tools which one finds on a raw frontier are never stable enough to a builders liking. I had no one else so I must needs gamble on the man." *New Helvetia Diary*, p. xxv.

Having made their undercut in a giant cedar, fallers pose for the photographer.
Courtesy of the Forest History Society and the Provincial Archives, Victoria, B.C.

Mendocino Harbor, shown here in 1865, was one of the better "dog holes" on the Mendocino County coast. The two-masted schooners at the foot of the bluff will take on their cargoes from the apron chutes that extend toward them. Cable loaders later replaced chutes at many places along the coast. Courtesy of the San Francisco Maritime Museum, Nannie Escola Collection

Westport Landing on the Mendocino County coast, date uncertain. While the schooner *Ivanhoe* and a second vessel load at the chutes, five other ships wait in the offing for a chance to take on cargoes. Courtesy of the San Francisco Maritime Museum

Bidwell considered the site a poor one. He later wrote that it would be hard to imagine how a sane man could have selected such a place, "and no other man than Sutter would have been so confiding and credulous as to patronize" Marshall by agreeing to build there.[50] Perhaps Bidwell's disappointment at not having found an acceptable site himself lay behind these acid comments. Whatever the case, even Bidwell admitted that the mill Marshall erected on the site was of ingenious design. At first it failed to work properly, however, because the waterwheel was placed too low. "The mill sawed little or no lumber," Bidwell claimed. "As a lumber enterprise the project was a failure, but as a gold discovery it was a grand success." [51]

Bidwell's comment on production at Sutter's mill was not entirely accurate. The mill did turn out lumber, some two thousand board feet every twelve hours. The first planks arrived at New Helvetia in March 1848.[52] Gold fever brought the mill's closure in May, and Sutter soon sold his interest for $6,000. Marshall also sold a part of his interest. The new owners reopened the mill and operated it from December 1848 until a lack of logs forced them to close in May 1849. At the time, lumber was reportedly bringing $300 to $500 per thousand in the area.[53]

The importance of Sutter's mill in the history of the lumber trade of the Pacific Coast does not lie in the boards it sawed, but in its indirect effects. The winds of change that had begun to gather force when the United States extended its dominion to California and the Oregon country in 1846 reached gale proportions following the discovery of gold at Sutter's sawmill. It was to be a gale that swept all before it and brought such basic changes to the Pacific slope of North America that the lumber industry, like everything else there, was never the same again.

50. Bidwell, "Life in California," p. 182. For a revealing description of Marshall, see San Francisco *Call*, reprinted in Seattle *Intelligencer*, 19 Aug. 1867.
51. Bidwell, "Life in California," p. 182. See also Sutter, *New Helvetia Diary*, pp. 61–62, 75, 78.
52. Sutter to Leidesdorff, 10 Sept. 1847, LP, box 4; 25 Mar. 1848, LP, box 5. In his letter of 25 March, Sutter reported that his mill was now cutting lumber, but that he was also planning to form a company to work the gold deposits, "which prove to be very rich." He asked Leidesdorff if he wished to take a share.
53. Clar, "John Sutter, Lumberman," p. 264. For a thorough analysis of the evidence bearing on events at the millsite, see Rodman Paul, *The California Gold Discovery* (Georgetown, Calif.: Talisman Press, 1966). See also Robert P. Heizer, "Archeological Investigation of Sutter's Sawmill," *CHSQ* 26 (1947):134–59.

CHAPTER IV

The
Golden Catalyst

THE discovery of gold in California speeded the evolution of the nascent Pacific Coast lumber trade into a major industry, but the huge demand that was to develop in San Francisco and its hinterland as a result of James Marshall's find did not emerge at once. The initial reaction of San Franciscans, once they became convinced the discovery was large and genuine, was to flock to the mines themselves. By mid-1848, San Francisco was almost a ghost town. The *Californian* found it necessary to shut down for lack of manpower. Thousands arrived in the port in late 1848 and early 1849, but most were just passing through. In August 1849, San Francisco was a tent city of some six thousand people. The demand for lumber in such a city was limited.

With the advent of winter, the picture started to change. Miners began to return from the hills—some disillusioned with mining, some to pass the winter in more comfortable quarters before sallying forth to try their luck once again, and some to spend the gold unearthed during a summer of toil. Meanwhile, new arrivals continued to pour in through the Golden Gate. By late 1849 the population of San Francisco had reached twenty thousand and was climbing rapidly. Accommodations were at a premium. Hotels, boardinghouses, shacks, and tents all seemed to be filled to overflowing. Men with an eye for business decided there was more gold to be won in the burgeoning city by the bay than in the Sierra. The construction of permanent buildings, both residential and business, became increasingly common.

A consequent boom in the lumber industry was soon under way.[1]

The same issue of the *Californian* that had laconically announced the discovery of gold at Sutter's mill gave the price of Oregon pine (as Douglas fir was generally known) as $40 per thousand board feet. California pine and redwood were $40 to $50 per thousand and shingles $6 to $8 per thousand.[2] But in 1849 there were only ten active sawmills in all California, with an estimated annual production of a mere five thousand board feet.[3] As the building boom gained momentum, demand quickly outstripped the limited supply and prices skyrocketed. By 2 August 1849, pine lumber was selling for $300 to $350 per thousand. By the first of October prices had dropped slightly, to $250 to $275 per thousand, but they quickly recovered and two months later had reached $400 to $450 per thousand board feet. Sales as high as $750 per thousand were reportedly made at about midyear.[4] Less than a year later lumber had dropped to $50 per thousand board feet.[5] Contributing to the high and unstable prices was a series of fires that swept San Francisco; the worst, in May 1851, destroyed some fifteen hundred buildings.[6]

Scarcity and high prices were not limited to San Francisco. Stockton, one Argonaut reported in August 1849, consisted of 150

1. F. P. Wierzbicki, *California As It Is, and As It May Be* (San Francisco, 1849), p. 49; William M'Collum, *California As I Saw It* (Los Gatos, Calif.: Talisman Press, 1960), pp. 121–26, 135; Bayard Taylor, *Eldorado, or, Adventures in the Path of Empire* (New York: Alfred A. Knopf, 1949), pp. 152–54; Soulé, Gihon, and Nisbet, *Annals of San Francisco*, pp. 202ff.; San Francisco *Californian*, 29 May 1848. Cf. Judd, *Honolulu*, pp. 137, 141, 149.
2. San Francisco *Californian*, 15 Mar. 1848.
3. U.S. Forest Service, California Forest and Range Experiment Station, *A Century of Lumber Production in California and Nevada*, Forest Survey Release no. 20 (Berkeley, Calif., 1953), p. 12.
4. San Francisco *Alta California*, 2 Aug., 1 Oct., 1 Dec. 1849. See also *Hunt's Merchants' Magazine and Commercial Review* 41 (1859): 45; Menefee, *Historical and Descriptive Sketch Book*, pp. 47–48. Numerous contemporary observers reported sales at $500 per thousand; one leading San Franciscan later recalled that in June 1849 prices reached $750 per thousand; and John Brown had to almost steal lumber to get it at one dollar a foot and "would have been glad to have more at the same low figures." See Caspar T. Hopkins, "The Californian Recollections of Caspar T. Hopkins," *CHSQ* 25 (1946):339; Brown, *Reminiscences*, pp. 106–7; Soulé, Gihon, and Nisbet, *Annals of San Francisco*, p. 254; Mrs. Zachariah Norton, "Voyage of the Sequin, 1849," *OHQ* 34 (1933):256.
5. Benjamin Dore, "The Journal of Benjamin Dore: One of the Argonauts," ed. Charles L. Camp, *CHSQ* 2 (1923–24):114.
6. Edwin T. Coman and Helen M. Gibbs, *Time, Tide and Timber: A Century of Pope & Talbot* (Stanford, Calif.: Stanford University Press, 1949), p. 30; Soulé, Gihon, and Nisbet, *Annals of San Francisco*, pp. 329, 345, 598–613; Josiah Royce, *California from the Conquest of 1846 to the Second Vigilance Committee*, ed. Robert Glass Cleland (New York: Alfred A. Knopf, 1948), pp. 301, 303–7; E. S. Harrison, *History of Santa Cruz County, California* (San Francisco, 1892), p. 60; San Francisco *Alta California*, 4 May 1851. Even as late as 1859, 8,603 of the 10,123 buildings in San Francisco were of wooden construction. See *Hunt's Merchants' Magazine* 43 (1860):647.

tents and one or two wooden houses. "Everybody here lives in tents, as lumber is too expensive and ordinarily no boards are obtainable." [7]

But not every seller of lumber obtained inflated prices. In September 1849, the wife of Captain Zachariah Norton, just arrived in San Francisco from Bath, Maine, lamented because her husband had stopped en route to sell the lumber aboard his vessel and had taken on in its place a group of gold hunters. "Had he come here direct from Bath . . . with his cargo of lumber, he would have made an independent fortune. Lumber is now worth $350 per thousand, and at the time he would have arrived here . . . it was selling for $500 per thousand. . . ." [8]

Even more unfortunate was William Gray, master of the American packet *Sylvia de Grasse*.[9] With lumber prices nearing their all-time peak in San Francisco and his vessel laden with lumber purchased at mills on the Willamette and Columbia rivers for $15 per thousand, Gray was in an enviable position. While still in the Columbia, however, his vessel struck a rock and foundered. By the time the lumber was removed from the stricken ship, loaded aboard three smaller craft, and ferried to San Francisco, it was too late. The bottom had fallen out of the market. Gray not only lost a vessel but, having incurred the added expenses of transferring the lumber to other vessels, he also failed to turn a profit on a cargo which, a short time before, would have brought him a fortune.[10]

The owners of the bark *Suliot* were more fortunate. The vessel carried a wide variety of goods and some forty gold seekers from Maine to San Francisco in early 1849. Though there was no lumber cargo, the berths of the passengers were constructed of hemlock boards

7. Clotilde Grunsky Taylor, ed., "Dear Family: The Story of the Lives of Charles and Clotilde Grunsky, 1823–1891, As Revealed in Their Letters to Their Respective Families," trans. Carl Ewald Grunsky and Clotilde Grunsky Taylor (n.p., n.d.; mimeographed copy, California Historical Society, San Francisco), p. 38.

8. Norton, "Voyage of the Sequin," p. 256. See also Wright, ed., *Lewis and Dryden's Marine History*, p. 23.

9. This was one of the many antiquated vessels pushed into service to meet the unprecedented demands of the Gold Rush. Reportedly, it had been the first vessel to bring news of the French Revolution to the United States.

10. Wright, ed., *Lewis and Dryden's Marine History*, pp. 24–25; James A. Gibbs, Jr., *Pacific Graveyard: A Narrative of Shipwrecks Where the Columbia River Meets the Pacific Ocean*, 3rd ed. (Portland: Binfords & Mort, 1964), p. 117. Similarly, the brig *Wigeon*, carrying passengers and a mixed cargo which included some lumber from Tasmania, ran into difficulties at sea and failed to complete her voyage. See Johnson Dean, *A Trip to California in 1850–3* (Hobart, Tasmania, n.d.), pp. 9–12.

purchased in Maine for $10 per thousand. Upon arrival in San Francisco, the boards were removed and sold for $300 per thousand.[11]

With lumber in demand and bringing good prices, new faces began to appear in the ranks of the lumber dealers of San Francisco. In the *Alta California* of 16 January 1850, the firm of Finch and Johnston announced it had opened a lumberyard at the corner of Battery and Market streets. The owners had for sale lumber received from New York on the ship *Probus*. A good supply would be kept on hand at all times, the advertisers assured potential customers. In the same issue, seventeen other firms and two auctioneers advertised lumber. These included Robinson, Arnold and Sewall, which had a yard at the foot of Pine Street, and Simmons, Hutchinson and Company, which had a yard in the North Beach section of the city.

Nor was the list in the newspaper a complete roster of the city's lumber dealers. Edward C. Williams and Henry Meiggs, among others, had yards in North Beach. Later in the year Williams added a planing mill.[12] By the summer of 1850, the Happy Valley sector on the south edge of town had become a "hive of manufacturing industry" where, in addition to shipyards, foundries, and docks, many lumber-yards were to be found.[13] Growth continued. By 1852 a directory for the city listed thirty-six individuals as lumber dealers and twenty-five lumberyards.[14]

Also present among the lumber dealers in 1850, though just getting started and not yet advertising, were Andrew Jackson Pope and Frederic Talbot. When Pope and Talbot stepped ashore for the first time in San Francisco on 1 December 1849, they had behind them personal experience and a long tradition of family involvement in the

11. William A. Rowe, *The Maritime History of Maine* (New York: W. W. Norton & Co., 1949), p. 168; Brian H. Smalley, "Some Aspects of the Maine to San Francisco Trade, 1849–1852," *Journal of the West* 6 (1967):594.

12. Howard Brett Melendy, "One Hundred Years of the Redwood Lumber Industry, 1850–1950" (Ph.D. diss., Stanford University, 1952), p. 249. *Pacific Coast Wood and Iron* 25 (1896):1; 26 (1896):1; 27 (1897):138. John S. Hittell, *A History of the City of San Francisco and Incidentally of the State of California* (San Francisco, 1878), p. 218. *The Pioneer: or, California Monthly Magazine* 2 (1854):16ff.

13. Soulé, Gihon, and Nisbet, *Annals of San Francisco*, p. 302. Meiggs's was the second planing mill in the city. Hulton's planing mill on Market Street had been erected shortly before. See *The Bay of San Francisco, the Metropolis of the Pacific Coast and Its Suburban Cities*, 2 vols. (Chicago: Lewis Publishing Co., 1892), 1:301.

14. [James M. Parker], *The San Francisco Directory for the Year 1852–53* . . . (San Francisco, 1852). Additional entries for wood yards and lumber and iron dealers raise to fifty-five the total number of firms in San Francisco dealing in forest products.

shipping and lumber industries of Maine.[15] These two young men from East Machias had no intention of going to the mines.[16] Rather they planned to make their fortunes as merchants.

Forming a partnership with two other emigrants from East Machias, Captain J. P. Keller and Lucius Sanborn, they first made their living by lightering cargoes ashore in an old ship's boat purchased for $500. When the boat proved too small to be profitable, they bought an old scow for $2,000. Finding that it brought them good returns, they quickly added a second scow, a yawl, and the little sloop *Bostonian*. They used the latter to carry goods upriver to Stockton, entrepôt to the southern mines. Though the lighterage business was proving profitable, it had little future. The needs of San Francisco could not be adequately served by lighters, however numerous, and the construction of docking facilities was moving forward rapidly. Lighterage consequently began to fall off, but by then Pope and Talbot had gone into a second line of business.[17]

On 4 January 1850, the two rented a beach lot and opened a lumberyard under the name of Pope and Talbot. Since rent was on a month-to-month basis, they dared not build up a large stock; but business was so brisk that "as much could be done in a month as ordinarily in six" in East Machias.[18] In the beginning, the young entrepreneurs received lumber on commission. Rather than bring it ashore, they sometimes sold it directly from rafts in the bay.[19] So, too, did competitors. Indeed, the practice was still common as late as 1854. Workers "taking the rafts of piles . . . to those who have bought them—through shipping—under wharves, against the tide & c," and the fact that these rafts were sometimes broken up by the wind, served to add a special element of confusion to the bustling life along San Francisco's waterfront.[20]

15. Coman and Gibbs, *Time, Tide and Timber*, pp. 36–49.

16. The two were hardly young by Gold Rush standards, however. Pope was 29, Talbot 33.

17. Coman and Gibbs, *Time, Tide and Timber*, pp. 10–11. The first docks in San Francisco were started before the Gold Rush by Thomas O. Larkin, William Heath Davis, and others. Their enterprise proved inadequate after the rush started, thus creating the necessity of lightering. See Rolle, *An American in California*, pp. 59–60.

18. East Machias *Union-Republican*, 23 Sept. 1926.

19. Coman and Gibbs, *Time, Tide and Timber*, pp. 11–13, 368, 428, and *passim*. The name Pope & Talbot was replaced by W. C. Talbot & Co. in 1855, but was restored in 1862 and continued in use into the twentieth century.

20. L. Smith to H. Huntting, 30 and 31 May 1854, Lura Case Smith Papers, Henry E. Huntington Library, San Marino, Calif.; see also letter of 29 June 1854.

Upon their arrival in San Francisco, Pope and Talbot had taken up residence in a cottage being vacated by Captain Lafayette Balch, who was returning to sea in his brig *George Emery*. Balch, a down-Easter related to the Talbots by marriage, had been active in the lumber trade while a resident of Maine, so his return in February with a cargo of lumber probably occasioned little surprise. Balch disposed of a portion of his cargo in San Francisco; the rest he sent up to Sacramento.

Pope and Talbot agreed to sell the lumber dispatched upriver, and on 26 February Talbot embarked for Sacramento to handle the sale. By the time he returned a few days later, his brother, Captain William C. Talbot, had arrived from East Machias with his brig, the *Oriental*. On board were gold hunters and an assortment of lumber, shingles, and house frames from the woods of Maine. This cargo was also sold in Sacramento, where, for the moment, prices were better than in San Francisco. Indeed, sales in Sacramento were quickly becoming a major adjunct of the business. Floods in the area created problems, however. As Frederic Talbot noted, "doing business at Sacramento [was difficult], as the City was almost totally under water, from the overflow and backing up of the water in the rivers; the levee along the banks *only* being out of water; on this we landed our cargoes, which were taken away in boats or rafts." [21]

Meanwhile, in San Francisco Pope was receiving lumber shipped by relatives in Maine and purchasing additional cargoes locally whenever they could be obtained at prices that seemed to offer chance of a profit.[22] In July, when Pope purchased a cargo of lumber from New Zealand for $52.50 per thousand, the firm had already cleared some $2,000. Shortly after, Captain Keller loaded the *Bostonian* with lumber and took her on two trips to Sacramento. Pope and Talbot profited doubly on the venture. As part owners of the vessel, they shared in the revenue the sloop earned from the freight; as lumber dealers, they profited from the sale of the cargoes. Since freights from San Francisco ranged from $60 to $120 per ton, the profit from freight alone must have been considerable.[23] Lumber prices had already passed their peak and were notoriously unstable, but the firm of Pope and Talbot prospered.[24] In time it became one of the giants of the Pacific Coast

21. Journal of Frederic Talbot, quoted in Coman and Gibbs, *Time, Tide and Timber*, pp. 25–26.
22. A. Pope to Wm. Pope & Sons, 29 July 1850, and other dates, PTP, Pope file.
23. Rowe, *Maritime History of Maine*, p. 170.
24. Coman and Gibbs, *Time, Tide and Timber*, pp. 11–13, 24–31.

lumber industry. Unlike most other firms that sprang up during the Gold Rush, Pope and Talbot continues in operation today.

The high price of building materials in California soon gained attention in many quarters. Thousands of bricks were dispatched to San Francisco from the East Coast, New Zealand, and Tasmania in hopes of profiting from the situation.[25] A writer in *The Economist* (London), noting a growing number of shipments of granite from Hong Kong to San Francisco, suggested that the "cities of the new El Dorado may not improbably be built out of the hills of Hong Kong." [26]

However, lumber, not bricks and granite, formed the bulk of the construction materials that entered the Golden Gate. It came from all corners. Lumber, as well as brick, was shipped from New Zealand and Tasmania.[27] Approximately ninety vessels sailed from Maine to California in the years 1849–52; nearly all carried at least some timber products. Typical was the bark *Condor*, which entered San Francisco Bay on 17 February 1850 with "8,500 clapboards; 6 house and store frames, 25x42 feet; 25 grindstones; 100 barrels brandy; 50 barrels gin; 50 barrels whisky; 71,516 feet planed boards; 55,458 feet joists; 25 casks cut nails; 10 kegs saleratus; 30 cooking stoves; 30 kegs tobacco; window frames, doors and sashes." [28] Additional lumber from Eastern forests arrived on vessels dispatched from New York, Boston, and other East Coast ports.[29] Such shipments continued, on a limited scale, at least until 1855.[30] Advertisements in the *Alta California* seem to indicate that lumber also was arriving from Norway, Australia, Panama, and Chile. Some of the lumber was shipped as prefabricated houses.[31]

25. Smalley, "Maine to San Francisco Trade," pp. 595–96, 599.
26. *The Economist*, 8 Mar. 1851, reprinted in Hong Kong *China Mail*, 29 May 1851. While this trade never flourished to the degree anticipated by the writer in *The Economist*, a Hong Kong–West Coast trade did emerge, and granite was one of the items carried on eastward crossings. See below, chap. 7.
27. Smalley, "Maine to San Francisco Trade," p. 599.
28. San Francisco *Alta California*, 18 Feb. 1850. Also, see descriptions of the cargoes of the brigs *Fawn, North Carolina*, and *Denmark* in the San Francisco *Alta California*, 5 Feb., 15 June, 20 Sept. 1850. The consignees stated that the building frames on the *Fawn* were the "heaviest timbered & best framed" ever offered in San Francisco.
29. San Francisco *Alta California*, 16, 21 Jan. 1850; *Hunt's Merchants' Magazine* 41 (1859):50; Smalley, "Maine to San Francisco Trade," pp. 594–96. David Warren Ryder goes too far, however, when he claims that during the Gold Rush most of San Francisco's lumber came around Cape Horn, or from Hawaii. See Ryder, *Memories of the Mendocino Coast* (San Francisco: Union Lumber Co., 1948), p. 51.
30. A. Pope to Wm. Pope & Sons, 23 Jan. 1855, PTP, Pope file.
31. San Francisco *Alta California*, 14 June 1849; 16, 21 Jan. 1850; Charles E. Peterson, "Prefabs in the California Gold Rush, 1849," *Journal of the Society of Architectural Historians* 24

Forest products were also coming from more local sources. To the east of San Francisco Bay, logging and lumbering slowed to a near halt as the mines drew away workers in 1848; but in 1849, as prices rose, many loggers returned to the woods. Attracted by the promise of high prices for their cut, many amateurs joined them and "cleared good wages." [32] In 1849 Henry Meiggs sent a "small army" of loggers into the woods of the East Bay area in order to obtain lumber for his yard in North Beach. Reportedly, he made half a million dollars on the venture. [33] Meiggs may well have done just that, for a hardened woods worker could turn out prodigious quantities of lumber, rails, and shingles. [34] The first steam sawmill in the redwoods of the East Bay area appeared in 1850. [35]

Operations were also springing up elsewhere around the bay. In 1849 Edward A. T. Gallagher established two lumber camps in the redwoods of the San Francisco Peninsula, which together employed 115 men and two full-time deer hunters to feed them. In 1850 Bayard Taylor noted that John C. Frémont had a steam sawmill near San Jose. New mills began to spring up all along the peninsula from there to San Francisco. Fifteen sawmills were cutting on John Coppinger's Cañada Raimundo land grant, near present-day Woodside, before Coppinger was able to clear title to his land in 1853. By that year Redwood City had emerged as a major shipping point from which lumber was ferried up the bay to San Francisco. San Francisco County Supervisor Thomas Swales claimed that during a single day in 1853 he counted fifty wagons bringing lumber to the port for shipment. [36] Lumber was also being dispatched to the city from Marin and Santa Cruz counties. Many vessels—including the *Oriental, David Henshaw,* and *Minerva—*

(1965):318–24; Charles Bateson, *Gold Fleet for California* (East Lansing: Michigan State University Press, 1963), p. 102.

32. Burgess, "The Forgotten Redwoods of the East Bay," p. 5; Ananias Rogers Pond, "Journal, 1852–1862," MS, Henry E. Huntington Library, San Marino, Calif., entry of 18 Dec. 1852.

33. Burgess, "The Forgotten Redwoods of the East Bay," p. 5; Samuel Upham, *Notes on a Voyage to California* (Philadelphia, Pa., 1878), pp. 145–46.

34. Ananias Pond's journal offers a rare glimpse into the life of one such worker. See, especially, the entries of 18 Dec. 1852, 13 Feb., 25, 28 Mar., 10, 17 Apr., 31 May 1853.

35. Burgess, "The Forgotten Redwoods of the East Bay," p. 7. A steam-powered mill appeared even earlier in the Sierra (1849). Its cut sold locally for prices as high as $650 per thousand. See William H. Hutchinson, *California Heritage: A History of Lumbering in Northern California* (n.p., n.d.), p. 6.

36. Taylor, *Eldorado*, p. 70; Clar, *California Government and Forestry*, pp. 43–45, 55–57; Alley, *History of San Mateo County*, p. 145; Stanger, *Sawmills in the Redwoods*, pp. 4, 7–12; Brown, *Sawpits in the Spanish Redwoods*, pp. 13, 15. See also San Francisco *Alta California*, 31 May 1859.

made trips to Santa Cruz to bring back cargoes of lumber.[37]

An even larger fleet—including the *Joven Guypuscoana*, *Anita*, *Undine*, *Starling*, and *Voladora*—sought out lumber at mills along the Columbia.[38] As one observer succinctly reported, "A great many vessels are going to Oregon for lumber." [39] As a result, Andrew Pope saw little future in shipping lumber to San Francisco from the East Coast. He predicted that Oregon would furnish all the city's lumber except pine, adding that there were quite a few mills springing up in California, too.[40] The forest industries of Oregon Territory were soon prospering as never before.

Accounts differ as to how word of the discovery of gold in California first reached the Northwest, but the news was clearly abroad by August 1848. Oregonians were soon flocking to the mines. The Provisional Government was unable to obtain a quorum, businesses closed, and towns were virtually deserted.[41] The *Oregon Spectator* was among those affected. On 12 October 1848, when the paper reappeared after a long absence, the editor wrote: "The Spectator, after a temporary sickness greets its patrons, and hopes to serve them faithfully, and as heretofore, regularly. The 'gold fever' which has swept about 3000 . . . officers, lawyers, physicians, farmers and mechanics . . . from the plains of Oregon into the mines of California, took away our printers also—hence the temporary non-appearance of the Spectator."

Sawmill workers were not immune to the excitement. Many of the mills that supplied lumber to fill the first orders after the discovery in California were soon idled by labor shortages.[42] Labor problems were such, one Hudson's Bay Company official reported, that "responsible

37. Sanford, "Short History of California Lumbering," p. 11; San Francisco *Alta California*, 25 Jan., 22 Feb. 1849; Coman and Gibbs, *Time, Tide and Timber*, p. 25. Sawmills were also at work in the Sierra foothills, but, unable to meet local demand, they did not ship to San Francisco. See San Francisco *Alta California*, 21 Sept. 1852, supplement; 19 Nov. 1854, reprinted from *Sacramento Union*.

38. Oregon City *Oregon Spectator*, 25 Jan. 1849; San Francisco *Alta California*, 22 Feb. 1849.

39. Quoted in Smalley, "Maine to San Francisco Trade," p. 598. For early accounts of voyages to the Columbia River for lumber see Norton, "Voyage of the Sequin," pp. 255–58; Stephen Chapin David, *California Gold Rush Merchant: The Journal of Stephen Chapin David*, ed. Benjamin B. Richards (San Marino, Calif.: Huntington Library, 1956), pp. 25–31.

40. A. J. Pope to E. Pope, 13 Dec. 1850, PTP, Pope file.

41. Portland *Oregonian*, 1 Jan. 1886; Peter Burnett, "Recollections of an Old Pioneer," *OHQ* 55 (1904):371; Charles H. Carey, *History of Oregon* (Portland: Pioneer Historical Publishing Co.), pp. 505–6; Norton, "Voyage of the Sequin," p. 257.

42. Meany, "History of the Lumber Industry," pp. 84–85; J. Minto to G. Hines, 2 Dec. 1910, MP, file 2.

men are unwilling to undertake heavy contracts [for supplying lumber] at any price." [43]

But the shutdowns were only temporary. By 4 October 1849, the *Oregon Spectator* was hailing the rise of trade during the preceding months. "The chief business . . . has been in the lumber trade. Already the trade has assumed the form of a regular commerce." During the summer of 1849, the editor noted, over five million board feet of lumber had been exported. The Oregon Milling Company alone had shipped more than one million. In the past a handful of vessels had entered the Columbia each year. In 1849 the number jumped to fifty. In October twenty vessels were in the river to take on cargoes. When a correspondent wrote to the *Oregon Spectator* the following August noting that the population drain to California had not hurt lumber production in the Northwest, he was stating what had long since become evident to residents of the area. In the face of rising demand, prices climbed too; at least once, they reached $150 per thousand at the mill.[44]

In December, Eden Colvile reported that Peter Skene Ogden, chief factor at Fort Vancouver, had leased the mill there to the local U.S. Army quartermaster for $14,000 per year. "In ordinary times such a mill might be built for about a thousand dollars," he wrote, "so this will give you some idea of the state of things." [45]

As demand and prices rose, new mills were constructed. At least five were built in 1849 and 1850 for the express purpose of exporting to San Francisco: Marland's at Tongue Point; Abernethy and Clark's at Oak Point; Whitcomb, Kellogg, and Torrance's at Milwaukee; and James Welch's and Abrams and Reed's in Portland. All were water-powered except the latter, the first steam-powered mill in the Northwest.[46] The census of 1850 listed thirty-seven sawmills in Oregon Territory with an annual production of 20,932,000 board feet. The cut was valued at over one million dollars.[47]

43. J. Douglas to J. Folsom, 10 Feb. 1849, LP, box 6.
44. Oregon City *Oregon Spectator*, 4 Oct., 1 Nov. 1849, 22 Aug. 1850; Meany, "History of the Lumber Industry," pp. 105–9.
45. E. Colvile to G. Simpson, 7, 8 Dec. 1849, in Eden Colvile, *London Correspondence Inward from Eden Colvile, 1849–1852*, ed. E. E. Rich, Publications of the Hudson's Bay Record Society, 19 (London: Hudson's Bay Record Society, 1956), p. 189.
46. Clark, "Analysis of Forest Utilization," pp. 23–24; Wright, ed., *Lewis and Dryden's Marine History*, p. 23; *Oswego* (Oregon) *Review*, 28 May 1959. Cf. Meany, "History of the Lumber Industry," p. 90.
47. Meany, "History of the Lumber Industry," p. 89.

Lumbering had become the territory's leading industry, though above Willamette Falls flour milling, not the sawing of lumber, remained dominant. Even there, however, lumbering was on the rise. As farmers prospered from the markets for produce that had developed in California, they came to use more building materials. Small mills, catering to local markets, were appearing with increasing frequency. The bulk of Oregon's lumber was sawed below the falls, however; over half of the total cut—73 percent, by value—came from the area near the confluence of the Willamette and Columbia rivers, where the largest mills were located. The Hudson's Bay Company sawmill at Fort Vancouver was by this time valued at $96,000; the Oregon Milling Company's at $90,000. Both reportedly cut 1.2 million feet a year.[48]

As the lumber industry of Oregon grew, new men were attracted to it. Joseph Lane, an Indianian recently appointed governor of the territory, was one. In 1852 he bought George Abernethy's Oregon Milling Company at Oregon City.[49] William Penn Abrams was another. Abrams had come west to make his fortune in the gold fields of California, but within a month of his arrival, he was writing "[I] wish myself anywhere but here." By February 1850 he was in Portland, busily at work erecting the Northwest's first steam sawmill. Portland, he felt, had better prospects than Oregon City. When his mill was completed, he noted with pride that his lumber was as good as any sawn at the older center.[50]

Even with established mills increasing their cut, new mills being built, and new men being attracted into the lumber industry of the Columbia River country, many lumber dealers and sea captains began to look for other sources of supply. In part this search resulted from the nature of the Columbia River. Though Oregonians protested that the Columbia was *not* a graveyard of ships, the fact remained that many a vessel went aground there and others waited days, or even weeks, to cross the bar. The voyage upstream against the current was time-consuming, costly, and also hazardous on account of the river's shifting shoals. When residents of the area argued that the river's reputation was unjustified and pointed out that Nathaniel Crosby had been

48. Ibid., pp. 80–83, 89–93.
49. J. McLoughlin, Memorial to Congress, appendix 13, McLoughlin Private Papers.
50. William Penn Abrams, diary, MS, Bancroft Library, Berkeley, Calif., entries of 10 Sept. 1849, 9 Feb., 13, 14 May 1850.

trading in and out of the river for years and at all seasons without mishap, they were unconvincing. Too many ships had come to grief in the river. In arguing that all that was needed was for the government to mark the channel properly, or for a pilot service to be established, or for tugs to go into operation on the bar, Oregonians only reinforced the belief that the river actually was a hazardous place to navigate. Surely there must be a better place to obtain lumber, many were thinking.[51]

There was. Trees in profusion stretched around the quiet waters of Puget Sound. In John Muir's words, "as if courting their fate . . . , [they came] down from the mountains far and near to offer themselves to the axe." [52] It was only a matter of time before lumbermen began to tap the area. Employees of the Hudson's Bay Company had constructed a sawpit at Nisqually House as early as 1833 and, as has been noted, Michael T. Simmons and his associates had a water-powered mill near present-day Olympia by the mid-forties; but the area around Puget Sound remained largely uninhabited, the timber untapped, until after the Gold Rush was well under way. In 1850 the entire white population around Puget Sound was less than one hundred.[53] But when development of the area's lumber industry did begin, it proceeded apace.

The Hudson's Bay Company apparently was the first concern to export forest products from the sound. In 1848 the company shipped lumber and shingles produced by Michael Simmons' Puget Sound Milling Company at Newmarket (now Tumwater). In an effort to sell the production of the American enterprise, William F. Tolmie later recalled, the Hudson's Bay Company sometimes glutted the Hawaiian market.[54]

51. San Francisco *Alta California*, 7 Sept. 1858; Gibbs, *Pacific Graveyard, passim*; Meany, "History of the Lumber Industry," pp. 94–96. For a graphic description of a lumber vessel coming to grief on the bar, in this case the brig *Potomac*, see U.S. Customs House, Astoria, Ore., "Protests noted, Oct. 21, 1850–Feb. 8, 1881," MS, Oregon Historical Society, Portland, pp. 9–15.
52. John Muir, *Steep Trails*, ed. William Frederic Badé (Boston: Houghton Mifflin Co., 1918), p. 206.
53. Coman and Gibbs, *Time, Tide and Timber*, p. 33; C. B. Bagley, ed., "Journal of Occurrences at Nisqually House," *WHQ* 6 (1915):269; Meany, "History of the Lumber Industry," pp. 97–98; Clark, "Analysis of Forest Utilization," pp. 23, 28–30; Michael T. Simmons, brief in relation to the Puget Sound Agricultural Co.'s claims against the United States, FNP, 1206.
54. Meany, "First American Settlement on Puget Sound," pp. 139–41; Snowden, *History of Washington*, 2:438, 4:349; Buchanan, "Economic History of Kitsap County," pp. 82–83. Cf. *West Shore* 8 (1882):181; Portland *Oregonian*, 5 Jan. 1880.

Late in 1849 the situation began to change. Simmons sold the Hudson's Bay Company one hundred thousand feet of lumber at $16 per thousand. Company officials had intended to ship the lumber to California themselves, but prices rose so rapidly they were able to sell it to the supercargo of a visiting merchantman for $60 a thousand. It was, as Eden Colvile called it, "a tolerably good speculation." [55]

The new developments posed problems, however. Lumber was now worth more at the point of production than in Hawaii. Company vessels present in the Northwest could be kept profitably employed only by dispatching them to San Francisco, but crews then had to be paid high San Francisco wages. The decision to undertake the risky business of shipping to California was not made lightly; nevertheless, by November 1849 the company was shipping lumber to San Francisco from a water-powered mill on Vancouver Island, and Eden Colvile was reporting: "Lumber . . . sells wel, I wish we [had] half a dozen sawmills here instead of one." [56]

Two years later, Hudson's Bay Company employees were operating a steam-powered mill, the Vancouver Island Steam Mill Company. In 1853 eighteen of the nineteen vessels that departed from Sooke Sound and Victoria with lumber were bound for San Francisco.[57] But, in spite of the head start of the British, leadership in the exportation of forest products from the Northwest soon passed to Americans. Both labor and logs were cheaper and more readily available at the excellent anchorages on the American side of the border.[58]

When the Hudson's Bay Company began exporting lumber from its mill on Vancouver Island, Michael Simmons chartered the brig *Orbit* to haul the cut of his own plant to market. The vessel arrived in Puget

---

55. Colvile to G. Simpson, 15 Oct. 1849, Colvile, *Correspondence Inward*, p. 181. See also Colvile to J. Pelly, 14 Jan. 1850, ibid., p. 15.

56. Colvile to G. Simpson, 7, 8 Dec. 1849, 14 Jan. 1850, ibid., pp. 188–90; Colvile to J. Pelly, 26 Oct., 8 Dec. 1849, 14 Jan., 6 Feb. 1850, ibid., pp. 6–8, 12–16, 19; W. Kaye Lamb, "Early Lumbering on Vancouver Island," *British Columbia Historical Quarterly* 2 (1938):39; S. W. Jackman, *Vancouver Island* (Newton Abbot, England: David & Charles, 1972), p. 63.

57. Records in the Fort Nisqually Papers show that the Puget Sound Agricultural Co., a subsidiary of the Hudson's Bay Co., continued to buy lumber from producers on the American side of the border, especially the Sequalachew (*sic*) Mill Co. The quantities purchased would indicate these purchases were for local use rather than export, however. See FNP, 51, 190, 861, 1068, and 1206.

58. Lawrence, "Markets and Capital," pp. 5, 7; Carrothers, "Forest Industries of British Columbia," pp. 255–56. Apparently the managers of the mill on Vancouver Island found it necessary to purchase logs on the American side of the border almost from the beginning. See James Cooper, receipts of payment for saw logs, 29 Mar. 1854, FNP, 183.

Sound in January 1850. At this point Simmons apparently changed his plans. He sold his mill to Clanrick Crosby, brother of Nathaniel, and invested the proceeds in a controlling interest in the little brig. *Orbit* thus became the first vessel in what was to become the sizable fleet owned by residents of the Puget Sound area. Simmons quickly dispatched it to San Francisco with a cargo of piles.[59]

Other vessels also appeared in the sound to pick up piles during 1850. The *George Emery*, Captain Balch, made several trips.[60] The ship *G. W. Kendall*, dispatched to the area by owners who thought a cargo of ice could be obtained at that latitude, carried piles and squared timbers back to San Francisco instead. Piles that sold for one dollar a lineal foot in California, where they were much in demand for the construction of docks and waterfront structures, could be obtained for eight cents a foot at shipside in the north. The promise of such a markup had attracted some ten or twelve vessels into the trade on a regular basis by 1851.[61]

The trade soon led to the establishment of embryonic settlements. In 1850 Captain Balch staked out a donation land claim, built a store, and thus founded Steilacoom. Henceforth, he carried merchandise for the store on the northbound voyages of the *George Emery*; reportedly it was the best-stocked store on the sound.[62]

Balch was also indirectly responsible for the establishment of Port

59. Snowden, *History of Washington*, 2:451–52; Herbert Hunt and Floyd C. Kaylor, *Washington West of the Cascades*, 3 vols. (Chicago: S. J. Clarke Publishing Co., 1917), 1:114; Joseph Schafer, *History of the Pacific Northwest* (New York: Macmillan, 1905), pp. 209–10; Buchanan, "Economic History of Kitsap County," pp. 75, 81; Hubert Howe Bancroft, *History of Washington, Idaho and Montana* (San Francisco, 1890), pp. 15, 17; T. Chambers to W. Tolmie, 7 Aug. 1849, FNP, 147.

60. Coman and Gibbs, *Time, Tide and Timber*, pp. 33–34. The *George Emery* apparently was the first American vessel to carry forest products to San Francisco from Puget Sound. It departed on 19 Nov. 1851, ten days before the *Orbit*. The latter was listed as bound for Hawaii, but seems to have gone to San Francisco instead. See U.S. Collector of Customs, Puget Sound District, "Registers of Entrances and Clearances of Vessels, 1851–1913," MSS, Record Group 36, Series 17, Federal Record Center, Seattle, vol. 1.

61. Hunt and Kaylor, *Washington West of the Cascades*, 1:183–84; Meany, "History of the Lumber Industry," pp. 57, 98–100; Coman and Gibbs, *Time, Tide and Timber*, pp. 33–34; Snowden, *History of Washington*, 4:350–51; Olympia *Pioneer and Democrat*, 3 July 1857; Olympia *Columbian*, 9 July 1853; Portland *Oregonian*, 4 June 1879, 5 Jan. 1880; W. G. Ballard, "A. D. Ballard," Oregon Pioneer Association *Transactions* (1897), pp. 105–6; Victor J. Farrar, ed., "Diary of Colonel and Mrs. I. N. Ebey," *WHQ* 8 (1917):132, 146–47; U.S. Collector of Customs, Puget Sound, "Registers of Entrances and Clearances," vol. 1, 9 Jan. 1852. By 1852 eight schooners and seven brigs were making regular runs between the sound and San Francisco. See Wright, ed., *Lewis and Dryden's Marine History*, p. 39.

62. Coman and Gibbs, *Time, Tide and Timber*, p. 34; Snowden, *History of Washington*, 2:454–55.

Townsend. During one of his trips to San Francisco, Balch met Alfred Plummer and Charles Bachelder, who took passage on the *George Emery* to Olympia. As the vessel passed the future site of Port Townsend, Balch is said to have pointed out its excellent location for hewing timbers and its good anchorage. Plummer and Bachelder promptly made a contract with the captain to supply him with piles and timbers and then filed claims and built cabins on the site. Port Townsend, destined to be a key port on the sound throughout the days of sailing ships, was founded on 24 April 1851. Asa Lovejoy, sensing greater opportunity around Puget Sound than beside the Willamette, left the Portland he had helped found in order to join in building Port Townsend.[63]

The new town prospered. By 1853 a resident was writing: "No branch of business vigorously pushed along can fail to pay [here], whether it be Agriculture, Fishing, Lumbering, or Mining." The main reason for the prosperity, the commentator admitted, was the mushrooming lumber industry around the sound. Port Townsend, as entrepôt to the entire area, inevitably benefited.[64]

Other settlements also put in their appearance in 1851. John L. Tukey came to Discovery Bay and put "all the Indians and white men he could hire" to work hewing ship's knees. A sizable settlement developed at Port Discovery, though certainly never the largest in the Northwest, as one historian has claimed.[65]

The first settlers of Seattle arrived the same year that Tukey was beginning operations at Discovery Bay. Within a month after their arrival the brig *Leonesa* dropped anchor beside the embryonic settlement to take on a cargo of piles for San Francisco.[66] Asked how the little community was able to make so much progress during its first year, Arthur A. Denny, leader of the settlers, replied, "By muscle and

63. Clark, "Analysis of Forest Utilization," p. 35; W. D. Walsh, *A Brief Historical Sketch of Port Townsend* (Port Townsend, Wash., 1941), pp. 5–7; Harvey W. Scott, *History of the Oregon Country*, 6 vols. (Cambridge, Mass.: Riverside Press, 1924), 2:278. Cf. Edmond S. Meany, *History of the State of Washington* (New York: Macmillan, 1909), p. 227.

64. H. Wilson to J. Wilson, 24 Apr. 1853, Bushrod W. Wilson Papers, University of Oregon library, Eugene. Twenty cargoes of piles were dispatched to San Francisco from the sound in 1853 and, with the sawmills still largely unfinished, only two cargoes of lumber. There were also two cargoes of lumber shipped from the sound the year before. That on the *Jane*, which departed for San Francisco on 26 July 1852, was the first shipment of lumber from the sound, aside from those of the Hudson's Bay Co. See U.S. Collector of Customs, Puget Sound, "Registers of Entrances and Clearances," vol. 1 *passim*.

65. Clark, "Analysis of Forest Utilization," pp. 35–36.

66. Archie Binns, *Northwest Gateway* (Garden City, N.Y.: Doubleday, 1941), p. 49.

timber." [67] When Henry Yesler built the first steam sawmill north of the Columbia at Seattle in 1852, the continued growth of the settlement seemed assured.[68]

The number of villages around the sound grew in 1852 and 1853 as demand for forest products continued in San Francisco and more and more entrepreneurs, mostly from the Bay Area, detected the potential for lumber production around Puget Sound. The settlement of Tacoma was initiated when Nicholas Dellin built a small mill on the site in 1852. Farther to the north, at the good millsite offered by Whatcom Falls, Bellingham appeared.[69]

By late 1851, Pope and W. C. Talbot had decided that, if their lumber business in San Francisco were to remain profitable, they would have to obtain a source of supply nearer than the mills of relatives in Maine and more dependable than the occasional cargoes offered for sale by enterprising sea captains. Agreeing that they needed a mill of their own, they decided, perhaps on the advice of their friend, Captain Balch, to construct one on Puget Sound. In 1852, to pave the way for building their own manufactory, they created the Puget Mill Company.[70]

In building their own mill, Pope and Talbot, the lumber dealers, commenced backward vertical integration of their holdings. This course was to prove common in the Pacific Coast lumber industry, in contrast to most other industries in the United States where vertical integration was more often instigated by producers seeking reliable shipping, marketing facilities, or sources of raw materials, than by merchants.[71]

Their first step was a trip to Maine to make preliminary arrangements. By the summer of 1853, when Talbot finally arrived at the

67. Quoted in Roberta Frye Watt, *Four Wagons West: The Story of Seattle* (Seattle: Lowman & Hanford, 1931), p. 84.
68. John R. Finger, "Seattle's First Sawmill, 1853–1869: A Study in Frontier Enterprise," *Forest History* 15 (1972):24–31; Henry Yesler, "The Settlement of Washington Territory," MS, Bancroft Library, Berkeley, Calif. As mentioned above, a steam sawmill was apparently in operation earlier on Vancouver Island, but Yesler's was the first on American soil north of the Columbia and the first on the mainland north of the Columbia.
69. Meany, *History of Washington*, p. 231; Thomas W. Prosch, *McCarver and Tacoma* (Seattle: Lowman & Hanford, 1906), p. 156; Clark, "Analysis of Forest Utilization," pp. 36–37; Donald H. Clark, *Eighteen Men and a Horse* (Seattle: Metropolitan Press, 1949), p. x.
70. A third partner in the milling venture was Capt. J. P. Keller. Frederic Talbot had returned to East Machias in December 1850 and played no further role in the family's West Coast operations.
71. For a useful summary, see Glenn Porter, *The Rise of Big Business, 1860–1910* (New York: Thomas Y. Crowell Co., 1973), pp. 43–54.

sound aboard the *Julius Pringle* to search out a site for the mill, J. J. Felt already had a steam sawmill in operation at Appletree Cove, while William P. Sayward was busily constructing another at Port Ludlow. After seriously considering Port Discovery, Talbot selected Teekalet (renamed Port Gamble in 1869) as the site for the mill. It was erected and ready to do business by September. At almost the same time Captain William Renton opened a steam sawmill on Alki Point, across Elliott Bay from Yesler's mill in Seattle.[72]

The area around Puget Sound was developing rapidly. "Each of these [new] mills brought hundreds of people to the Territory," Donald H. Clark has observed, "and the towns were trading centers for large areas." [73] Seattle, especially, was prospering. The Olympia *Columbian*, as yet the only newspaper in the territory, reported:

Seattle is thriving. All accounts that we receive from thence tell us of new buildings and other improvements. Yesler's steam sawmill is working finely. Alki is full of vigor and goaheaditiveness; her commerce is increasing and her men of business are doing well. Renton's steam sawmill will be in operation in a few days. The enterprising inhabitants of these two places, near together as they are, seem determined that their full, high and important destiny shall be achieved as soon as possible. Success attend them.[74]

The growth of Seattle was hardly surprising. Three of the five steam sawmills in operation around Puget Sound by the end of 1853 were in the area immediately tributary to the settlement. Only the mills at Port Ludlow and Teekalet were some distance removed and, being near the northern end of the sound, even these were tributary to Seattle rather than to older Olympia to the south. For the time being, however, Olympia was still the larger settlement; indeed, it was over twice the size of Seattle and nearby Alki combined, while the other main center on the southern part of the sound, Steilacoom, was about as large as the combined settlement on Elliott Bay.

72. Coman and Gibbs, *Time, Tide and Timber*, pp. 31–35, 53–60; Clark, "Analysis of Forest Utilization," p. 37; Binns, *Northwest Gateway*, p. 80; partnership agreements, PTP, A. J. Pope business file; Olympia *Columbian*, 20 Aug., 24 Sept. 1853. Even before September, Talbot was shipping forest products from Puget Sound to San Francisco. After landing men and supplies at Teekalet, the *Julius Pringle* crossed to Yesler's mill and took on a cargo of lumber for the return voyage. The *L. P. Foster*, which brought additional men as well as mill machinery to Teekalet, returned with a cargo of piles. See E. G. Ames, "Story of Pope & Talbot as Told by Mr. E. G. Ames" (mimeographed; n.p., n.d.), copy in PTP, historical file.
73. Clark, "Analysis of Forest Utilization," p. 37.
74. Olympia *Columbian*, 20 Aug. 1853. A year before, on 16 Oct., the paper had said that Washington needed territorial status, more people, better communications, and steam sawmills in order to develop.

The impact of the discovery of gold in California had, in short, been felt all around Puget Sound. The entire region was growing. When federal investigator J. Ross Browne visited the territory of Washington in 1854 and recommended that the port of entry be moved from Nisqually in the south to Port Townsend beside the Strait of Juan de Fuca, he was recognizing that henceforth all the area around the sound would be contributing to trade. The recommendation was just one more bit of evidence of the profound changes that events in California had worked in the far Northwest.[75]

Similar changes were taking place elsewhere. The Spanish frigate *Tres Reyes*, one of three vessels in a squadron of exploration under Sebastián Vizcaíno, had entered Humboldt Bay as early as 1602, but the area remained unsettled, its resources undeveloped, until 1849, when the bay was rediscovered both by land and by sea. The search for a route to gold mines on the Trinity River, not a search for new sources of lumber, lay behind the rediscovery.[76] Nevertheless, forest products were soon being exported. The brig *Cameo*, one of the fleet that rediscovered the bay, apparently was the first to engage in the trade. Returning from her voyage of exploration, the *Cameo* carried piles from Trinidad Bay (just north of Humboldt Bay) to San Francisco. Shipments of piles and timbers continued throughout the summer. Most were by-products of the traffic in passengers headed for the Trinity mines. Sawmills soon began to appear in this sector of northern California, however, and with their establishment the traffic quickly became self-generating.[77]

The first sawmill to be erected in the area went into production in 1850.[78] By early 1851 lumber from what was to become Humboldt

75. David M. Goodman, *A Western Panorama, 1849–1875: The Travels, Writings, and Influence of J. Ross Browne* . . . (Glendale, Calif.: Arthur H. Clark Co., 1966), pp. 58–59.
76. Herbert Eugene Bolton, ed., *Spanish Explorations in the Southwest, 1542–1706* (New York: C. Scribner's Sons, 1916), pp. 101–2; Chad L. Hoopes, *Lure of Humboldt Bay Region* (Dubuque, Iowa: W. C. Brown Book Co., 1966), pp. 2–31; Owen C. Coy, *The Humboldt Bay Region, 1850–1875: A Study in the American Colonization of California* (Los Angeles: California State Historical Association, 1929), pp. 17–48.
77. San Francisco *Alta California*, 28 Apr., 8, 10, 15, 21 Aug., 25 Sept., 21 Oct. 1850; San Francisco *Herald*, 19 Feb. 1851; Sanford, "Short History of California Lumbering," pp. 44–45; Coy, *Humboldt Bay Region*, pp. 117–18, 120–21, 124–25. When the quest for gold ceased to be the main stimulus of the traffic, Humboldt Bay quickly replaced Trinidad Bay as the leading center of the trade.
78. Accounts differ. According to one version, J. M. Eddy and Martin White established the Pioneer Mill in Eureka in August and, in addition, White erected a second mill at nearby Bucksport. The second, or Papoose, mill was in operation and selling lumber by November. Others indicate that the two mills were one and the same, that the Pioneer Mill was also known as

County was appearing in San Francisco. During 1851 two other mills were erected in the area. One, built on Little River by the Baron von Loeffelholz, was equipped with machinery designed for use in the forests of Germany. The second was powered by machinery of the steamer *Chesapeake*. Found leaking on a trip down the coast, the vessel was run ashore at Bucksport and the machinery converted to sawmill use.[79]

Though they showed the way to others, these first mills were not particularly successful. Too small to handle redwood, they were dependent upon pine, fir, and spruce for their sawlogs. Mills incapable of handling the main species of timber in the area had but little potential. Their small size created other problems as well: loss of economies of scale, for example.[80] The treacherous bar at the mouth of Humboldt Bay was also a handicap. A high percentage of vessels making the crossing was lost before a steam tug went into operation on the bar in 1852. Even after that date the bar was a source of trouble for shippers.[81]

the Papoose Mill. Extant evidence yields no clear indication of which view is correct, though one cannot help but wonder how White could have found time to oversee the simultaneous erection of two sawmills in so newly opened a region. See Hyman Palais and Earl Roberts, "The History of the Lumber Industry in Humboldt County," *PHR* 19 (1950):1; Bancroft, *History of California*, 6:503; Coy, *Humboldt Bay Region*, p. 118; W. H. Wilde, "Chronology of the Pacific Lumber Company, 1869 to 1945" (San Francisco, n.d.; typescript, Bancroft Library, Berkeley, Calif.), p. 6.

79. Coy, *Humboldt Bay Region*, p. 118; Melendy, "One Hundred Years of the Redwood Industry," p. 146; Wilde, "Chronology of the Pacific Lumber Company," p. 7; Palais and Roberts, "Lumber Industry in Humboldt County," pp. 1–2; W. W. Elliott & Co., *History of Humboldt County, California* (San Francisco, 1882), p. 141. There is no agreement on what vessel carried the first cargo. Palais and Roberts and W. W. Elliott & Co. indicate it was the *James R. Whiting*; Coy credits the schooner *Odd Fellow*.

80. The potential for economies of scale—the lower per-unit costs of production that come with larger, more efficient plants—may have been quite limited. Lance Davis and Douglass North have argued that before 1870 most industries enjoyed only constant returns to scale, that is, their per-unit costs remained roughly the same regardless of plant size. Moreover, Walter J. Mead has demonstrated that in the mid-twentieth century sawmills benefited from few economies of scale once they reached a rather modest size and beyond a certain point became less, not more, efficient. Still, it seems clear from the nature of these pioneer mills on the redwood coast that the larger mills then possible would have produced lumber more cheaply on a per-unit basis. See Davis and North, *Institutional Change and American Economic Growth* (Cambridge: Cambridge University Press, 1971), p. 170; Walter J. Mead, *Competition and Oligopsony in the Douglas Fir Lumber Industry* (Berkeley: University of California Press, 1966), pp. 11–31.

81. Palais and Roberts, "Lumber Industry in Humboldt County," pp. 1–2; Coy, *Humboldt Bay Region*, pp. 118, 120–22; Melendy, "One Hundred Years of the Redwood Industry," pp. 146, 151–52, 279; Wilde, "Chronology of the Pacific Lumber Company," p. 7; W. W. Elliott & Co., *History of Humboldt County*, pp. 130, 176; D. Jones to J. Kentfield, 2 Feb. 1872, JKCP, 1:1; C. Nelson to Kentfield, 25 Mar., 5 Oct. 1875, JKCP, 1:4 and 1:7; C. Reel to Kentfield & Co., 6 Nov. 1875, JKCP, 1:7; W. Woodley to Kentfield, 27 Nov. 1878, JKCP, 3:6; Arcata (Calif.)

But larger, better designed mills were soon under construction. The first truly successful mill in the area was built by Ryan, Duff and Company. On 24 February 1852, Captain James T. Ryan, the managing partner, arrived at Humboldt Bay with the steamer *Santa Clara*. Ryan ran the vessel ashore at a site he had selected some months before. The hull of the *Santa Clara* became the bunkhouse, its engine the power plant for the sawmill which Ryan and his crew proceeded to erect. Within six weeks the mill was in operation. Its first cargo was loaded out a month later, but the vessel carrying it was lost on the bar, as were the next two dispatched from the mill. Chastened by the experience, the company purchased the tug *Mary Ann* and put it into operation on the bar in November.

Quickly the mill began to prosper. By 1854 it had four gang saws in operation and a capacity of sixty thousand board feet a day. The mill employed some thirty-five men. To market the cut, a lumberyard was established in San Francisco in May 1853.[82] As early as January 1853 one resident of the area wrote that Humboldt's ratio of exports to population compared favorably with that anywhere. Two years before there had been little prosperity, but now four mills were sawing and "Humboldt Lumber now has a Controlling Influence on the Market in San Francisco." [83]

Ryan, Duff and Company was not to ship the first cargo of redwood lumber from the bay, however. As had their predecessors, Ryan and Duff used sawlogs of pine, spruce, and fir. The distinction of shipping the first cargo of redwood went to a tiny muley mill owned by William Carson.

Carson had joined the rush to the Trinity gold fields. Short of food, he and some friends withdrew from the mines and spent the winter of 1850/51 getting out logs for the Pioneer Mill, then returned to the mines the next spring. While there, he heard of the mill Ryan and Duff were building on Humboldt Bay. This time he left the mining area for good. He journeyed to Sacramento, purchased a team of oxen, and

Humboldt Times, 7, 14 Sept. 1854; Eureka *Humboldt Times*, 25 Sept. 1858, 25 Feb. 1895, 31 Dec. 1899; San Francisco *Alta California*, 9 Oct. 1853. The *Humboldt Times* was published in Arcata from 1854 to 1858, in Eureka after 1858.
82. Palais and Roberts, "Lumber Industry in Humboldt County," pp. 2–3; Wilde, "Chronology of the Pacific Lumber Company," p. 7; Eureka *Humboldt Times*, 18 Oct. 1862, 13 June 1863, 7 Dec. 1872; San Francisco *Alta California*, 6 May 1853.
83. J. Martin to T. and A. Bell, 26 Jan. 1853, James O. Martin Papers, California Historical Society Library, San Francisco.

returned to Humboldt Bay, where he went to work as a logging contractor for the new mill. His team of oxen was the first put to use getting out logs from the area's forests. In 1854 Carson purchased a small muley mill. Though it had a capacity of but five thousand board feet a day, Carson threw himself into the operation with a vengeance and was able to turn out some two hundred thousand feet of lumber his first year. He shipped the first all-redwood cargoes on the brigs *Quoddy Belle* and *Tigris*.[84] Carson's operations grew until they were among the largest on the coast.[85]

Other mills were also being built. In 1853 Ridgeway and Flanders, the second largest plant in the area (behind Ryan, Duff and Company), employed thirty-one men. When sold to John Vance, it brought $65,000. May and Brother's plant employed eighteen. In 1852, 30 vessels sufficed to carry the area's cut to San Francisco; by 1853, 143 were required. Twenty-seven lumber-laden vessels departed from Humboldt Bay in September alone.[86] A year later, the *Humboldt Times* reported that there were nine mills on the bay, seven of them in Eureka. Together they produced two hundred twenty thousand board feet of lumber a day, represented an investment of $400,000, and provided employment for 200 loggers and 130 sawmill workers. More than twenty miles of track had been laid, and horse-drawn cars were being used extensively in mill operations in the area.[87]

An economy based on the lumber industry was emerging. As Ulysses S. Grant later recalled, the only way to get to the area was "to take passage on a San Francisco sailing vessel going after lumber." [88] In the *Alta California*, Eureka was described as the lumber center of the

84. Arcata *Humboldt Times*, 16 Sept. 1854; Eureka *Humboldt Times*, 31 Dec. 1899; Palais and Roberts, "Lumber Industry in Humboldt County," pp. 2–3, 5–6; Melendy, "One Hundred Years of the Redwood Industry," p. 156; Leigh H. Irvine, *History of Humboldt County, California* (Los Angeles: Historic Record Co., 1915), p. 608; Wilde, "Chronology of the Pacific Lumber Company," p. 9. Though others soon followed Carson's lead and began to use redwood sawlogs, it was not until the 1860s that equipment was developed that could handle whole logs. Until then they were split, then cut into lumber. See below, chap. 10.

85. The history of Carson's operations deserves a full-length study. The voluminous records of the Dolbeer and Carson Lumber Co. for the years 1884–1940 are available at the Bancroft Library. Unfortunately there is no early correspondence in the collection.

86. Coy, *Humboldt Bay Region*, pp. 118–19, 122; Palais and Roberts, "Lumber Industry in Humboldt County," p. 7.

87. Arcata *Humboldt Times*, 9, 16 Sept. 1854; Wilde, "Chronology of the Pacific Lumber Company," p. 9; Lynwood Carranco and Mrs. Eugene Fountain, "California's First Railroad: The Union Plank Walk, Rail Track, and Wharf Company Railroad," *Journal of the West* 3 (1964):243. Cf. Ryder, *Memories of the Mendocino Coast*, p. 57.

88. Ulysses S. Grant, *Personal Memoirs of U. S. Grant* (New York, 1885–86), p. 207.

West. "The steam that belches forth from our numerous mills," the paper boasted, "together with the sound of the hammer and woodsman's axe, give unmistakable evidence that the spirit of California enterprise is doing its work." [89]

Finally, even to the south of Humboldt Bay, along the rugged Mendocino coast, the effects of the inflation of the lumber market that came with the Gold Rush were being felt. Here there were no good harbors. Mountains rose abruptly from the sea, leaving no coastal plain or spacious harbors. Anchorages existed at the mouths of some of the region's larger streams and in the lee of eroded headlands, but these left much to be desired. Sailors referred to them as "dogholes," presumably because they offered barely enough room for a dog to turn around in.

Yet, the Mendocino coast had advantages. It was closer to San Francisco than Puget Sound or Humboldt Bay and had large stands of redwoods close to the coast. In 1851 a group of businessmen from San Francisco decided to brave this inhospitable coastline in order to tap its forest wealth. Henry Meiggs, Edward C. Williams, Jerome Ford, and Captain David Lansing formed the California Lumber Company to further their scheme.[90]

Accounts vary as to how their attention was first drawn to the Mendocino coast. According to one version, a Chilean vessel laden with silk was wrecked off the Noyo River in 1851. During salvage operations, Meiggs, already a leading figure in the lumber business of San Francisco, was attracted by the fine stand of redwood lining the shore. He proceeded to organize the California Lumber Company as a means of capitalizing on the resource. According to a second version, Williams and Meiggs had received at their lumberyards in San Francisco some redwood from Captain Stephen Smith's mill at Bodega Bay. They liked it but needed more than Smith's small mill could supply. To solve the problem, they decided to build a mill of their own in the untapped forests to the north of Smith's operation.[91] There may be truth in both accounts. Meiggs's action in sending loggers into the redwoods to the east of San Francisco Bay in 1849 indicates he was interested in obtaining more lumber. The wreck off the Noyo may simply have drawn his attention to a new possibility for a source of supply.

89. San Francisco *Alta California*, 21 Jan. 1854. See also 9 Oct. 1853.
90. Ryder, *Memories of the Mendocino Coast*, pp. 5, 19–20, 51.
91. Ibid., pp. 51–52.

Many problems attended the building of the California Lumber Company's mill. At first the group intended to equip their mill with used sawmill equipment purchased locally; but when they viewed the timber on the site they had selected—near the mouth of Big River—they realized that they would need larger machinery. An order was quickly dispatched to the East. When the machinery arrived, it was sent up from San Francisco on the brig *Ontario*. Buffeted by headwinds and leaking badly, the vessel nearly foundered at sea. As soon as it arrived at the Big River site, construction began; but before the roof was on the mill, torrential winter rains set in, stopping nearly all work. When the mill was finally put into operation the next spring, it proved to be inadequate. The capacity of the mill was fifty thousand feet per day, but the company had commitments for one hundred thousand. In 1853 a second mill was added to meet the demand.[92] Others, including William A. Richardson, soon followed the lead of Meiggs and his associates and erected sawmills on the Mendocino coast. The California Lumber Company long remained the dominant mill in the county, however.[93]

In sum, the Gold Rush years were a time of great change for the lumber industry of the West Coast, years during which an industry of regional importance began to emerge to replace the scattered handful of tiny plants that had been present before. Mills had been built all along the coast—from Vancouver Island and Puget Sound in the north to the redwoods of the San Mateo and Santa Cruz areas in the south—for the explicit purpose of shipping their cut to markets far removed from the source of supply, markets that could be reached only by ocean transport. New areas of production had been opened, fleets of

92. Ibid., pp. 20, 52–56; Melendy, "One Hundred Years of the Redwood Industry," pp. 223–26, 248–49; Wilde, "Chronology of the Pacific Lumber Company," p. 8; Palmer, *History of Mendocino County*, pp. 398–99, 409, 429–36; Edward C. Williams, "The First Redwood Operations in California," *Pioneer Western Lumberman* 58 (1912):9–13; Mary Floyd Williams Papers, Bancroft Library, Berkeley, Calif., *passim*; Henry Huntley Haight, "Journal of a Hunting Trip in Sonoma and Mendocino Counties, June 11 to July 1, 1857," MS, Henry E. Huntington Library, San Marino, Calif., 1:24, 28–34; "Reminiscences of Mendocino," *Hutchings' Illustrated California Magazine* 3 (1858–59):179–81. A thorough study of the California Lumber Co. and its successor corporations is needed. The Bancroft Library acquired two truckloads of records and manuscript materials from the Union Lumber Co. in 1967 on which such a study could be based. Unfortunately, they contain no correspondence prior to 1900.
93. Palmer, *History of Mendocino County*, pp. 141, 398, 418, 431. Richardson, the first Anglo resident of San Francisco, had produced lumber in Marin County as early as 1847. See Richardson to Larkin, 30 June 1847, Larkin, *Papers*, 6:229–30; Smith to Leidesdorff, 25 Dec. 1847, Leidesdorff Papers, California Historical Society Library, San Francisco.

lumber droghers had been marshaled, dealers specializing in the lumber trade had appeared in the cities and towns, and San Francisco had emerged as the center from which the budding industry was dominated.

There were both similarities and differences between the rising lumber industry of the Pacific Coast and that of Maine, from whence many of the former's leaders came. Both depended on markets reached by sea and thus demanded knowledge of the sea and ships, as well as of woods and sawmills. Both depended on investment capital from beyond the area of production. Lumber manufacturers thus shared with most industrial producers of the period a dependence on merchant-capitalists in the nation's commercial centers.[94] But, while established lumbermen in Maine obtained backing from sources in Boston, on the Pacific Coast entrepreneurs in San Francisco more often set up plants of their own rather than investing in already existing mills.

The reasons for this difference are readily apparent. The manufacturing of lumber was technologically uncomplicated. Mills in existence in the Far West at the time of the Gold Rush were simple in the extreme; most were so primitive and many so poorly located for large-scale production that the erection of new plants was generally preferable to expansion of the old ones. Moreover, there were so few mills in existence on the eve of the Gold Rush that even if working through established firms had seemed advisable, there were simply too few to meet the demand. Puget Sound and Humboldt Bay, both areas of great potential, were almost completely untapped. San Francisco's merchants quickly moved to utilize the forests of these areas instead of investing solely in existing mills on the Columbia and near Monterey and Bodega bays. The following decades demonstrated the wisdom of their decision. Much remained to be done before the Pacific Coast lumber industry was to reach maturity, but between 1848 and 1853 unprecedented progress had been made.

What had emerged on the West Coast was more than a nascent lumber industry. The foundations of a modern society had been laid. The discovery at Sutter's mill was instrumental in speeding the process. As Sutter himself wrote:

I think now from all this you can mention how many thousands made their fortunes from this Gold Discovery produced through my industry and energy (some wise

94. See Porter and Livesay, *Merchants and Manufacturers*, pp. 62–78; Elisha P. Douglass, *The Coming of Age of American Business: Three Centuries of Enterprise, 1600–1900* (Chapel Hill: University of North Carolina Press, 1971), pp. 172–73, 176.

merchants and others in San francisco called the building of the Sawmill another of Sutter's folly) and this folly saved not only the Mercantile world from Bankruptcy, but even our General Govt. but for me it has turned out a folly, then without having discovered the Gold, I would have become the wealthiest man on the Pacific shore.[95]

Folly it may have been, but stimulus it was, too. Thanks to the catalytic action of Sutter's mill, the embryonic lumber industry of pioneer times quickly developed into a robust youth.

95. John A. Sutter, *The Diary of Johann August Sutter* (San Francisco: Grabhorn Press, 1932), p. 56.

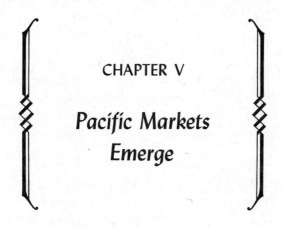

CHAPTER V

Pacific Markets Emerge

THE Gold Rush brought sweeping changes to the manufacturing and marketing of lumber on the Pacific coast of North America, but the results of the discovery at Sutter's mill were felt far beyond the western littoral of the continent. The Gold Rush set in motion forces that helped to tie the coast's lumber industry firmly to the scattered markets of the Pacific Basin. Within a few short years, San Francisco was to emerge not only as the main West Coast market for lumber and the nerve center from which its manufacture was increasingly controlled, but also as the locus from which the maritime commerce that linked the sawmills and their Pacific markets was directed.

However, this eventuality was not at first apparent. Burgeoning demand in San Francisco and its hinterland led initially to a reduction in shipments to such markets as Hawaii. To some it may have seemed that there would never again be a need to seek out buyers in Hawaii, China, or Latin America, that henceforth California would offer enough consumption to keep the region's lumber manufactories profitable and growing. But even while the exportation of lumber was in decline, the foundation of a new order was being laid.

Among the many results of James Marshall's discovery at Sutter's millsite was a revitalization during the early fifties of trade between the West Coast and the Orient. Both British and American observers were heartened by the development. Noting the unwonted surge of activity the revived trade had caused in Hong Kong, an Englishman commented: "The fairest hopes of the colony are founded on the new trade which is springing up between it and California. . . . In these circumstances there is some prospect of Hongkong becoming a useful settlement." [1] In San Francisco the editor of the *Alta California* noted: "Each day adds to the value and amount of our China trade, and each month shows new sources of profit to those who embark in this adventure." [2]

Traders were stirring in the Pacific Northwest as well as in California. Early in 1851 the brig (or schooner) *Emma Preston* stood out from the bar of the Columbia River bound for Hong Kong with a cargo of spars. Six months later the little vessel was back laden with sugar, tea, rice, and assorted Chinese goods. The brig *Amazon*, with a similar cargo, crossed from Whampoa to Portland a few weeks ahead of her. Other vessels soon followed in their wakes.[3] Trade between the Pacific Northwest and China, quiescent since the death of the maritime fur trade, was once again under way.

The *Oregon Spectator* rejoiced in the development. "Already the trade of China has been directed, to a small extent, to this new Territory. It has been the result of enterprise. Just let that spirit become a little more generally diffused among our commercial men, and it will be easy to divine the advantageous consequences that will follow." The editor looked forward to the day when "the whole eastern world" would serve as a market for the forest products of the Northwest.[4]

1. *The Economist*, 8 Mar. 1851, reprinted in Hong Kong *China Mail*, 29 May 1851. See also *Friend of India*, 17 June 1852, reprinted in Hong Kong *China Mail*, 22 July 1852. Thomas O. Larkin was one of the first to enter the trade. See Larkin to F. Atherton, 19 Jan. 1849; Larkin and J. Leese, contract, 12 Feb. 1849; Larkin, Leese, and R. Mason, agreement, 14 Feb. 1849; T. Green to Larkin, 29 June 1849, all in Larkin, *Papers*, 8:xi, 102–5, 138–39, 144–45, 247; "catalog of the brig *Eveline*" (San Francisco, 1849; broadside, Bancroft Library, Berkeley, Calif.).
2. San Francisco *Alta California*, 16 May 1851. In his annual message to the legislature, Gov. Peter Burnett called the attention of the lawmakers of the state to California's growing commerce—especially that with China, the Hawaiian Islands, and the Northwest. See California, *Journals of the Legislature of the State of California . . . 1851* (n.p., 1851), pp. 806–7.
3. Hong Kong *China Mail*, 3 Apr., 22 May, 25 Sept., 2 Oct. 1851, 18 May, 1 Apr., 16 Dec. 1852; Hong Kong *China Overland Mail*, 23 May 1851, 27 Dec. 1852; Oregon City *Oregon Spectator*, 16 Jan., 3 July, 30 Sept. 1851; Portland *Star Marine Journal*, 23 Jan. 1851; Portland *Oregon Weekly Times*, 26 June 1851; U.S. Dept. of State, Consular despatches, Hong Kong, National Archives, Record Group 59, microcopy 108, 1 July 1851; Eldon Griffin, *Clippers and Consuls: American Consular and Commercial Relations with Eastern Asia, 1845–1860* (Ann Arbor, Mich.: Edwards Brothers, 1938), p. 451; Wright, ed., *Lewis and Dryden's Marine History*, pp. 36–39, 53; U.S. Customs House, Astoria, Ore., "Marine List and Memoranda" (MSS, Oregon Historical Society, Portland), vol. 1 *passim*. The first such shipment from Puget Sound was a cargo of spars carried to Singapore in 1853 by the bark *Mary Adams*. U.S. Collector of Customs, Puget Sound, "Registers of Entrances and Clearances," vol. 1 *passim*.
4. Oregon City *Oregon Spectator*, 30 Sept. 1851, 14 Oct. 1849.

Many of the first crossings were but an incidental result of the gold-created demand for tonnage to carry passengers and freight to California. Among the vessels that flocked in through the Golden Gate were a number too valuable to be left riding idly at anchor or rotting on the mud flats of San Francisco Bay. Some, such as W. C. Talbot's brig *Oriental*, joined the fleet of lumber droghers operating along the West Coast in order to supply building materials to San Francisco and its hinterland. When prices suffered temporary declines in San Francisco, the *Oriental*'s owners sometimes dispatched her to Hawaii rather than back to the Northwest. Lumber apparently formed the bulk of the cargo on these voyages to the islands. Sometimes the owners dispatched her to Santa Cruz for forest products, other times they loaded her with excess lumber from the yards in San Francisco.[5]

But such employment was beneath the proud clippers. The owners of these sought more lucrative freights. Few cargoes for the East Coast or Europe were available in California; therefore, many owners of clippers, rather than have their vessels brave the dangerous passage around Cape Horn in ballast, had them dispatched to China to pick up cargo to carry from there around the Cape of Good Hope to Boston, New York, or London.[6] To cut losses on the crossing of the Pacific, they had whatever cargo could be obtained in California put aboard for the transpacific leg of the journey. The crossings generated the trade, not the reverse.[7]

5. Coman and Gibbs, *Time, Tide and Timber*, pp. 26–27; Peter Talbot & Co. to Edward D. Peters & Co., 15 Feb., 3 Mar., 30 Sept. 1851, PTP, letter copybook, business correspondence, 1846–66.
6. The British Navigation Acts were repealed in 1849, so London was by this time open to vessels sailing under the American flag. For a discussion of the impact of the repeal of the Navigation Acts on U.S. maritime interests, see William Armstrong Fairburn, *Merchant Sail*, 6 vols. (Center Lovell, Me.: Fairburn Marine Educational Foundation, 1945–55), 2:1534–41.
7. John G. B. Hutchins, *The American Maritime Industries and Public Policy, 1789–1914: An Economic History* (Cambridge, Mass.: Harvard University Press, 1941), pp. 265–69; Fairburn, *Merchant Sail*, 2:1534–37; 4:2159–2296, 2454–61. See also consular despatches and returns of trade from Hong Kong and Shanghai and commercial intelligence from Hong Kong *China Mail* and Shanghai *North China Herald*, *passim*. These make it clear that the majority of vessels continued to arrive in ballast for some time. As was the case in the Northwest, the California traffic was not all westbound. Attracted by the news of gold in California, many in southern China, both native and foreign, were eager to take passage to the new El Dorado. Cargoes, as well as passengers, were usually carried on these crossings. See "catalog of cargo of the brig *Eveline*"; J. L. Riddle & Co., "Part of the cargo of the 'Lord Cornwall' direct from China" (San Francisco, 1850; broadside, Bancroft Library, Berkeley, Calif.); Hong Kong *China Mail*, 29 May 1851; San Francisco *Alta California*, 12, 16 May 1851; Consular despatches, Hong Kong, 1 July, 11 Apr. 1851; Gunther Barth, *Bitter Strength: A History of the Chinese in the United States, 1850–1870* (Cambridge, Mass.: Harvard University Press, 1964), pp. 59–60 and *passim*.

The situation soon began to change, however. Agriculturists and lumbermen had been able to sell all they could produce for almost unbelievable prices, but by 1853 they faced falling prices and glutted markets. Prices showed no signs of improvement during the early months of 1854. In May one observer reported that "lumber as well as work is very cheap now on account of the dull times." [8] As the depressed prices continued, a search for new outlets developed; agents for vessels found freight being offered; and some producers—eager to move surpluses—even dispatched cargoes as ventures carried on their own accounts.[9]

In July 1854 Nathaniel Crosby, Jr., wrote to Joseph Lane, then territorial delegate for Oregon. Crosby's letter, and a second written three weeks later, clearly revealed the new state of economic affairs. Crosby had been engaged in trade between Oregon and California since 1847, but now, he calculated, trade with San Francisco was "about don[e]." The captain announced that he was loading the bark *Louisiana* with lumber from the Columbia River country and would soon be off to China in quest of new markets for the products of the Pacific Northwest. He would visit all the ports of China that had been opened to trade, and then would stop off in Japan on his return voyage to canvass its potential as well.[10]

Once in the Far East, Crosby was so impressed with the potential of the markets there that he decided to take up residence in Hong Kong where he would act as agent for additional cargoes of lumber that he planned to have dispatched from the Northwest. He returned to Portland to wind up his affairs there and to make arrangements for continuing the trade.[11] Crosby's example was seized upon by others: lumber-laden vessels were soon being dispatched to China from Humboldt Bay and Puget Sound, as well as the Columbia.[12]

8. L. Smith to H. Huntting, 30 and 31 May 1854, Smith Papers. See also letter of 29 June 1854.
9. Coman and Gibbs, *Time, Tide and Timber*, pp. 75–77; Coy, *Humboldt Bay Region*, pp. 119–21; Meany, "History of the Lumber Industry," pp. 107–18; Fairburn, *Merchant Sail*, 2:1535; Carl Cutler, *Greyhounds of the Sea: The Story of the American Clipper Ship* (New York: G. P. Putnam's Sons, 1930), p. 276; Hittell, *History of California*, 3:431–33; Soulé, Gihon, and Nisbet, *Annals of San Francisco*, pp. 519–21; *Hunt's Merchants' Magazine* 41 (1859):45.
10. N. Crosby to J. Lane, 22 July, 13 Aug. 1854, Joseph Lane Papers, Oregon Historical Society, Portland.
11. Mrs. George E. Blankenship, comp. and ed., *Early History of Thurston County, Washington* (Olympia, Wash.: The Thurston Co., 1914), pp. 267–71; Scott, *History of the Oregon Country*, 2:29, 247; Portland *Oregonian*, 5, 18 Aug. 1854; Hong Kong *China Mail*, 17 May 1855, 31 Jan., 11 Nov. 1856; San Francisco *Daily Herald*, 28 July 1855.
12. U.S. Collector of Customs, Puget Sound, "Registers of Entrances and Clearances," vol. 1 *passim*; Hong Kong *Friend of China and Hongkong Gazette*, 26 Jan. 1856; Hong Kong *China*

Lumbermen in the area around Humboldt Bay were also becoming concerned over the condition of the market in San Francisco. It was increasingly clear that California was not going to be able to absorb the cut of the many mills that had been erected during the excitement of the Gold Rush. As a result, the owners of several mills in Humboldt County decided to pool their resources to form a single firm that would actively seek out new markets for their cut. In September 1854 the Humboldt Lumber Manufacturing Company was established with James T. Ryan as president. The firm soon failed, but its formation indicates that a new outlook was developing among the entrepreneurs engaged in the lumber industry on the Pacific Coast.[13]

Others were also seeking new markets. Oregonians had lost the Hawaiian lumber market to traders from California during the Gold Rush, but by 1853 shipments from Puget Sound were regaining a share of that market for the Northwest. The Port Gamble sawmill of Pope and Talbot was built for the express purpose of supplying lumber to San Francisco and its hinterland. In 1854, its first full year of operation, some 3.6 million board feet of lumber were shipped from the mill, but, contrary to expectations, it was necessary to ship well over a third of the cut to foreign ports.[14] The Olympia *Pioneer and Democrat* urged a continued search for outlets in the Pacific Basin: "That we shall have the boundless Pacific for a market is manifest destiny. We were born to command it." [15]

Mail, 7 Dec. 1854, 22 Mar., 21 June, 23 Aug., 13 Sept., 11, 25 Oct. 1855, 31 Jan., 22 May, 9 Oct. 1856, 1 Jan. 1857, 8 July, 26 Aug. 1858, 27 Jan., 21 Apr., 21 July 1859; Arcata *Humboldt Times*, 28 Dec. 1856; Eureka *Humboldt Times*, 1 May 1858; San Francisco *Herald*, 4 May 1857; Coy, *Humboldt Bay Region*, pp. 123–24. Cargoes were also being transshipped from San Francisco. See Hong Kong *China Mail*, 4 Oct., 9, 29 Nov. 1855, 27 May 1858.

13. Melendy, "One Hundred Years of the Redwood Industry," pp. 149–51, 154–55, 277–78; Palais and Roberts, "Lumber Industry in Humboldt County," pp. 3–5; Coy, *Humboldt Bay Region*, pp. 119–20; Sanford, "Short History of California Lumbering," pp. 46–47; Arcata *Humboldt Times*, 9, 23 Sept., 28 Oct. 1854, 10, 17, 24 Feb., 10, 17 Mar., 7, 14, 28 Apr., 9 June 1855, 29 Nov. 1856; W. W. Elliott & Co., *History of Humboldt County*, p. 141. A similar organization was soon being proposed in the Northwest. See Olympia *Pioneer and Democrat*, 3 Mar. 1855.

14. Coman and Gibbs, *Time, Tide and Timber*, pp. 51–60; Olympia *Columbian*, 15 Jan. 1855; Meany, "History of the Lumber Industry," pp. 114–15. Records in the Astoria Customs House Papers, cited elsewhere in this chapter, show that Hawaii all but disappeared as a direct market for northwestern lumber during the Gold Rush. Apparently lumbermen preferred to ship their cuts to San Francisco and then have them transshipped to Hawaii if prices in California were low. During the Gold Rush the residents of Hawaii were much more interested in trade with the new El Dorado than with the Northwest, of course, and thus did nothing to resist the trend. See also Bradley, "California and the Hawaiian Islands," p. 19; Judd, *Honolulu*, pp. 137, 141, 149.

15. Olympia *Pioneer and Democrat*, 18 Nov. 1854. See also 10 Mar. 1855.

Though prices were dropping in San Francisco and lumbermen (and others) were busily searching out new markets for their products, many failed to recognize that a major economic readjustment was due. They were soon enlightened. On 6 October 1854 the jerry-built lumber and promotional empire of "Honest Harry" Meiggs came tumbling down. Meiggs fled from the country, money from forged city warrants in his pockets. Meiggs's flight was a shock to the business community of San Francisco and, as a result of the general reassessment that followed, many of the city's business leaders began to retrench. A wholesale collapse of the price and credit structures soon followed. Among the businesses that failed were the prestigious banking firms of Page, Bacon and Company and Adams and Company. The California Lumber Company that Meiggs and his partners had established was shut down for three years while the legal tangle resulting from the failure and flight of the principal owner was unsnarled.[16]

Of course, Meiggs did not really cause the economic depression that gripped San Francisco in 1854 and 1855. The city's economy was unsound long before the events of 6 October. Indeed, this condition had helped to bring on Meiggs's failure, for he had overextended during the flush days and soon found himself insolvent when recession set in. But, if Meiggs's failure was not the cause of the depression, it did make it abundantly clear, even to the most economically myopic, that the boom days of the Gold Rush were over. The false optimism that had helped to carry the economy through 1853 and the early months of 1854 vanished. The failure of Meiggs, who was a city councilman, builder of a magnificent two-thousand-foot wharf at North Beach, owner of sawmills and lumberyards, and moving spirit behind a number of other projects—a man, in short, who was the epitome of the

---

16. A. Grogan to F. Atherton, 15 Oct. 1854, Grogan & Lent to Atherton, 15 Oct. 1854, Faxon Dean Atherton Papers, California Historical Society Library, San Francisco. Watt Stewart, *Henry Meiggs, Yankee Pizarro* (Durham, N.C.: Duke University Press, 1946), pp. 6–17. Melendy, "One Hundred Years of the Redwood Industry," p. 227. Ryder, *Memories of the Mendocino Coast*, pp. 56–57. Royce, *California*, pp. 335–39. Hittell, *History of California*, 3:434–41. Hittell, *History of San Francisco*, pp. 209, 218–26. Soulé, Gihon, and Nisbet, *Annals of San Francisco*, pp. 432, 461, 519–21, 543–44. *Pioneer: or, California Monthly Magazine* 2 (1854):297; 3 (1855):16–22, 172, 238–39, 368–73. Williams, "First Redwood Operations," pp. 11–13. Adams & Co., "Comments Which Have Been Made on the Closing of Our House" (open letter to creditors) (San Francisco, 1855). For insights into some of the legal complications following Meiggs's flight, see complaint, William Neely Thompson & Co. *v.* California Lumber Manufacturing Company, Adams & Co. Papers, Huntington Library, San Marino, Calif., 23; Alfred A. Cohen, memorandum, [June or July] 1855, Adams & Co. Papers, 39.

A sawpit built by employees of John Sutter in the Sierra foothills in 1845. From here pine planks were transported to Sutter's Fort in New Helvetia (Sacramento). Courtesy of the California State Library

The Puget Mill Company's plant at Port Gamble, around 1890. Originally built in 1854, the sawmill, though much changed, continues in operation today—the longest continually active sawmill in the United States. Courtesy of the Photography Collection, Suzzallo Library, University of Washington

Ten million feet of fine Douglas fir logs in storage near the Puget Mill Company's sawmill at Port Gamble, Washington. Courtesy of the Forest History Society, Weyerhaeuser Timber Company Collection

The Puget Mill Company's plant at Port Ludlow as it appeared some time prior to 1 July 1889. Courtesy of the Photography Collection, Suzzallo Library, University of Washington

speculative spirit that had long dominated in San Francisco—and the attendant closing of two of the city's most respected banks served as a psychological stimulus that speeded the process of adjustment.[17]

As the effects of the economic crisis spread, various scapegoats were found. In San Francisco Meiggs himself was blamed; but when the Humboldt Lumber Manufacturing Company failed in the spring of 1855, residents of the redwood coast tended to blame the resulting closure of most of the area's sawmills on supposed mismanagement of the combine.[18] In the north, the Olympia *Columbian* protested that the area around Puget Sound was suffering because the lumber industry was controlled by speculators in San Francisco who drained profits from the area while doing almost nothing to advance it. The rival *Pioneer and Democrat* agreed: the credit system by which sales in the area were financed was "simply an opening for a batch of thieving banks on the Pacific." [19]

In the face of these developments, confidence in the lumber-producing areas dwindled. In 1854 J. Ross Browne described the area around Puget Sound as of limited potential: it is "a good country for coarse lumber, and nothing more." In 1853 Bushrod Wilson, who earlier had spoken in glowing terms of the prospects of Port Townsend, estimated that "Oregon is a money making country I think [even] better than

17. An analysis of the causes of the depression is beyond the scope of this work. Suffice it to say, it has been argued that it resulted from maladjustments peculiar to the West Coast that stemmed from the Gold Rush boom and, conversely, that it was merely a West Coast manifestation of the national recession of 1857. One thing is clear, however: West Coast lumber markets had not been glutted as a result of the Treaty of Reciprocity of 1854 with Canada as has sometimes been claimed. The total capacity of the Canadian mills that had access to the markets of the West Coast was too tiny to be of significance in the San Francisco lumber market. Among those who have blamed the glut on the treaty have been R. T. Wattenberger and Howard B. Melendy. See R. T. Wattenberger, "The Redwood Lumber Industry of the Northern California Coast, 1850–1900" (master's thesis, University of California, Berkeley, 1931), p. 24; Melendy, "One Hundred Years of the Redwood Industry," pp. 276–77.

18. Arcata *Humboldt Times*, 3 Mar., 4 Aug. 1855. There may have been something to the charge. Andrew Pope noted that the lumber produced by the mills of the combine was of poor quality. See Pope to Wm. Pope & Sons, 23 Jan. 1855, PTP, Pope file.

19. Olympia *Columbian*, 10 Dec. 1853; Olympia *Pioneer and Democrat*, 10 June 1854. That the area around Puget Sound was being exploited by the capitalists of San Francisco was an oft-repeated complaint in the years that followed; similar grumbling was heard in Oregon. Economic exploitation by outsiders is, of course, a common theme in developing areas, for with "foreign" capital comes "foreign" control and the drainage of profits earned in the underdeveloped areas to stockholders living elsewhere. Among the notable exceptions were the Oregon Steam Navigation Co. and the Willamette Steam Mills Lumbering and Manufacturing Co. See Portland *Oregonian*, 23 Apr. 1865. Seattle *Intelligencer*, 11, 17, 21 Oct. 1876, 16, 30 May 1877. *Canada Lumberman* 1 (15 Mar. 1881):9; 1 (1 Apr. 1881):3. Clark, "Analysis of Forest Utilization," pp. 54–55. Buchanan, "Economic History of Kitsap County," p. 62.

California." However, by 1854 he was writing to relatives: "[If] you are doing well in Minasota Dont come here. All the choice land is taken up and times look bad." [20]

Though developments along the Pacific Coast dampened enthusiasm and slowed migration to the region, they also speeded the search for new markets for the products of the area. Sawmills represented fixed investments that entrepreneurs were loath to abandon. Moreover, the rich resource base upon which the mills drew was still present and largely untapped. Locating new markets, not abandoning the industry, seemed the logical solution to the economic problems confronting lumbermen along the West Coast. Outlets in China, Australia, the Hawaiian Islands, and along the west coast of South America were soon being exploited.

Among the first of the markets to which lumbermen turned during the mid-fifties was that in the kingdom of Hawaii. During the first half of 1855, while economic conditions in San Francisco were at their nadir, shipments of lumber from Puget Sound to the islands burgeoned. The total for 1855 was four times greater than the year before.[21]

This sharp increase is hardly surprising. Hawaii was a relatively close market to which forest products from the West Coast had been shipped for years. It had numerous ties with the United States, especially the West Coast. Since Anglo-Americans dominated the economy, the islands provided a relatively easy and congenial place for Westerners to do business. Finally, the Hawaiian Islands offered the prospect of profitable return cargoes. Lumber could be taken to Honolulu from the redwood coast or the Northwest, Hawaiian products—especially sugar, molasses, and rice—carried to California on the return, and general cargoes of goods needed in the lumber-producing areas then carried on the northbound leg to complete the triangle.[22] There was also opportunity for a direct, but limited, two-way trade between the lumber ports and the island kingdom.[23] With so many

20. Browne is quoted in Goodman, *Western Panorama*, p. 58; B. Wilson to J. Wilson, 18 Dec. 1853, 3 Oct. 1854, Wilson Papers.
21. U.S. Collector of Customs, Puget Sound, "Registers of Entrances and Clearances," vol. 1 *passim*.
22. For typical cargoes on the northbound leg, see records of the bark *Nahumkeag*, U.S. Customs House, Astoria, Ore., Papers of the Astoria Customs House, 1848–68 (microfilm copies, Oregon Historical Society, Portland), reel 2:F125. Figures in the Astoria Customs House Papers make it clear that a high percentage of the vessels came north in ballast, however. Far more tonnage was needed to haul bulky cargoes of lumber from the producing areas than to bring in needed manufactured items.
23. The editor of the *Pioneer and Democrat* felt that a direct trade with the islands was preferable

factors working to recommend them, it is no wonder the Hawaiian Islands came quickly to mind as the market in San Francisco contracted.

But in the early 1850s the demand for lumber in Hawaii was severely limited. The boom in the islands' economy that had been ushered in by gold discoveries in California collapsed in 1851. Drought also struck.[24] When Albert A. Durham sent lumber to Hawaii from his mill in Oregon, his agent sold most of it to the Hawaiian government. The monarchy, however, was reluctant to go into debt and thus possibly provide an excuse for foreign intervention. Short on funds, the government could purchase only small quantities despite a desire to push forward various public works projects. When Durham's agent attempted to sell his posts and timbers on the open market, he found they moved slowly. The many abandoned vessels in the islands provided free timbers for the taking. Moreover, lumbermen had already overstocked the market with planking. Discouraged, Durham abandoned efforts to penetrate this market.[25]

The economy of the islands was on the threshold of change, however. Prior to the mid-fifties, money brought by whaling vessels and earned in the re-export trade supported the economy. In the forties, items produced in the Hawaiian Islands constituted only some 10 or 12 percent of the gross exports of the kingdom; the rest was produced elsewhere. But domestic production was rising by the 1850s. Since whalers were staying longer in the waters of the Pacific, Hawaii was becoming more important as a source of provisions. Agricultural production increased in response to the rising demand. In addition, a pattern of trade was developing in which whaling vessels, rather than making the long voyage back to the East Coast, deposited their oil in Hawaii and then returned to the whaling grounds. Speedy transports that called at the islands to pick up oil and deliver it to the East Coast provided a transportation link that Hawaiian producers were soon using to dispatch goods to distant ports. This, too, helped encourage

to a triangular trade that included San Francisco. The latter, he believed, would drain more specie from the Northwest. See Olympia *Pioneer and Democrat*, 25 Mar. 1854.
24. Honolulu *Polynesian*, 18 Jan., 2 Aug. 1851; Merze Tate, *Hawaii: Reciprocity or Annexation* (East Lansing: Michigan State University Press, 1968), pp. 24, 28; Judd, *Honolulu*, p. 170; Bradley, "California and the Hawaiian Islands," p. 25.
25. T. Metcalf to A. Durham, 14 Aug. 1852, 29 Mar. 1854, Durham Papers, Oregon Historical Society, Portland; *Oswego* (Oregon) *Review*, 28 May 1959; Astoria Customs House Papers, 1:F39. It is not clear where these other shipments of lumber were from, but in all probability they, too, were from the West Coast.

Hawaiian agriculture, as did the abandonment of the old feudal system of land tenure that had hindered the rise of commercial agriculture earlier. By the 1870s nine-tenths of Hawaii's exports were domestically produced.[26]

These factors were important in changing the economy of the Hawaiian Islands, but the days of the whaling industry were numbered. In the long run, the greatest change was to be wrought not by whalers but by a new source of wealth: sugar. From 1855 on, sugar exports increased rapidly. The industry's foundations had been firmly laid before the whaling industry began its final decline in 1859. As Theodore Morgan has observed, it "is not possible to take seriously the legend that Hawaiian sugar was in the doldrums, or declining, until rescued by the Reciprocity Treaty" of 1876.[27] The construction that accompanied the establishment and expansion of sugar plantations, and of irrigation systems to water them, created a demand for lumber that producers on the West Coast were eager to fill. As the economy expanded, Honolulu grew. Lahaina and Hilo were also becoming towns of modern frame structures. The result was an even further increase in the demand for building materials.[28]

Pope and Talbot was in the forefront of those who moved to fill the demand. From 1854 until 1860 the firm was dominant in the market and profited immensely as a result. Pope reported the firm received more per thousand board feet in Hawaii than anywhere else. Even after the Port Madison mill entered into serious competition for Hawaiian customers, Pope and Talbot continued to earn significant profits through sales in the islands.

But Pope and Talbot was never alone in dispatching lumber to the

26. Theodore Morgan, *Hawaii, a Century of Economic Change, 1778–1876* (Cambridge, Mass.: Harvard University Press, 1948), pp. 123–53, 159–72, 195–98; *Hunt's Merchants' Magazine* 40 (1859):603–4. See also Bradley, "California and the Hawaiian Islands," pp. 24–25; Tate, *Hawaii*, pp. 21–24.

27. Morgan, *Hawaii*, p. 181. Cf. Judd, *Honolulu*, pp. 202–3; John Kentfield & Co. et al., *Petition to the U.S. Senate and House of Representatives by Ship Builders, Ship Owners, and Lumber Merchants of Pacific Coast States, Relating to the Treaty of Reciprocity between the United States and the Hawaiian Islands* (San Francisco, 1886).

28. Morgan, *Hawaii*, pp. 173–98; Judd, *Honolulu*, pp. 186–87; Ralph S. Kuykendall, *The Hawaiian Kingdom*, 3 vols. (Honolulu: University of Hawaii Press, 1938–67), 2:135–76; Lawrence, "Markets and Capital," pp. 12–13; Kentfield & Co. et al., *Petition*, p. 5; Sylvester K. Stevens, *American Expansion in Hawaii, 1842–1898* (Harrisburg: Archives Publishing Co. of Pennsylvania, 1945), pp. 32–36. Not just foreigners, but natives as well were beginning to erect frame buildings. See Honolulu *Pacific Commercial Advertiser*, 4 Sept. 1856, 28 May 1857, 8 July 1858; A. Pope to Wm. Pope & Sons, 31 Mar., 7, 23 Apr., 29 June, 15, 31 July 1855, 16 May 1857, 20 Dec. 1860, PTP, Pope file.

Hawaiian Islands. During the year ending 30 June 1855, the *Metropolis* (twice), *Eudorus, Mary Reed, S. R. Jackson* (twice), *Agate, Detroit,* and *Kaluna* carried lumber to the islands from Oregon. None was apparently dispatched by Pope or his associates. Other vessels carried lumber from Puget Sound and San Francisco, but not all of that shipped from either place came from the mill of Pope and Talbot.[29] Shipments of other sorts of forest products were also being made. During 1854, piles were shipped to the island kingdom from Sooke Inlet on Vancouver Island. Before the end of 1855, lumber was being dispatched to Hawaii from the redwood coast.[30] During the next nineteen years, at least fifty-seven vessels carried cargoes of lumber to the islands from the Columbia River alone.[31] The merchantmen involved in this trade ranged all the way from such tiny schooners as *San Diego*, a mere fifty-one tons registry, to the clipper bark *Jane A. Falkenburg*, one of the finest vessels on the Pacific Coast. The *Falkenburg*, and the barks *Metropolis, Cambridge,* and *A. A. Eldridge,* made regular runs to the islands.[32]

Materials in the Astoria Customs House Records make it clear that vessels running from Oregon to the Hawaiian Islands carried a wide variety of other products besides lumber: preserved salmon, oats, flour, beans, barley, bran shorts, onions, potatoes, hams, butter, and dry goods. Lumber was the largest single item, both in bulk and in value, however.[33] Cargoes on the return voyage were equally varied: sugar,

29. U.S. Dept. of State, Consular despatches, Honolulu, National Archives, Record Group 59, microcopy 144, reel 6, consular returns, 1 July to 31 Dec. 1854, 1 Jan. to 30 June 1855; Astoria Customs House Papers, 1:F54. By the fall of 1856, the trade with Puget Sound was "becoming quite important, no less than seven vessels of about two thousand tons capacity being regularly employed. . . ." Honolulu *Pacific Commercial Advertiser*, 4 Sept. 1856.

30. Wright, ed., *Lewis and Dryden's Marine History*, p. 54; Arcata *Humboldt Times*, 1 Sept. 1855. However, the first cargo made up entirely of redwood did not arrive until 1858. On 25 Nov. 1858 the *Pacific Commercial Advertiser* commented that redwood is "extensively used at the coast, and by many preferred to the northern fir, on account of its durability and nonliability to shrink"; redwood "is also susceptible to finer dressing for clapboards and flooring than fir and on this account has gradually been growing in favor in these islands the past two years."

31. Astoria Customs House, "Marine List and Memoranda," vols. 1 and 2 *passim*. See also Kentfield & Co. et al., *Petition*, pp. 1, 3–4.

32. Astoria Customs House Papers, 1:F1; 2:F122, F161a; 4:F213; 5:F211, F222, F222a, F223, F223a, F226, and *passim*; Astoria Customs House, "Marine List and Memoranda," vols. 1 and 2, *passim*; Wright, ed., *Lewis and Dryden's Marine History*, pp. 66, 188; Portland *Oregonian*, 28 Mar. 1855. For the details of one of the *Falkenburg*'s voyages, see Captain D. B. Foster, "Journal of a Voyage from Astoria, Oregon to Honolulu, . . ." MS, University of Hawaii library, Honolulu.

33. Astoria Customs House Papers, 1:F6, F54; 2:F122, F130, and *passim*. Cargoes from other lumber-producing areas probably were not as diversified, for these other areas did not have the abundance of good farm land or the breadth of production that Oregon did.

molasses, salt, coffee, pulu, whale oil, rice, sweet potatoes, koa wood, arrowroot, tapioca, bananas, watermelons, pumpkins, oranges, and limes. Sugar, molasses, rice, and salt—the latter needed in large quantities in the Northwest for preserving salmon—were the most important items in the return traffic.[34] In addition to cargoes, the vessels sometimes carried a few passengers either to or from Hawaii.[35]

Many Westerners, eager to encourage trade with Hawaii, pushed for a treaty of reciprocity between the United States and the island kingdom. Tariffs presented no particular problems to those shipping goods to the islands; but many residents of the West Coast were convinced that, if Hawaiian sugar were allowed free entry into the United States, the economy of the islands would boom. Economic expansion in the islands would inevitably mean larger markets for products of the Pacific Coast.[36] Occasionally Westerners went so far as to advocate annexation of the island kingdom as a means of furthering the trading interests of those resident on the Pacific Coast of the United States.[37]

While Hawaii was being cultivated as a market for lumber and other products from the West Coast, Nathaniel Crosby and others were at work trying to find similar outlets along the coast of China. This interest in China was hardly surprising. Since 1783 Northeasterners had been enriching themselves in the old China trade. Many of the sea captains, lumbermen, and merchants operating along the Pacific Coast were transplanted New Englanders who had grown up in a region where China was a frequent topic of discussion, where things Chinese

34. Ibid., 1:F6, F51; 2:F122, F130, F166; 4:F213; 5:F222, F223, F223a, and *passim*. The return cargoes to San Francisco appear to have been largely made up of sugar, especially in later years, while those to Puget Sound appear to have been somewhat smaller and less diversified, doubtlessly because of the smaller population in the area north of the Columbia. The return cargo carried by the *L. P. Foster* on a voyage in 1858 was typical: coffee, molasses, sugar, and several barrels of poi—the latter for the Hawaiian laborers working at the mills on the sound. See Honolulu *Pacific Commercial Advertiser*, 4 Feb. 1858; P. Shepherd to Kentfield, 13, 26 Apr. 1877, JKCP, 2:4; JKCP, boxes 16–17 *passim*; Dorothy Weidberg, "The History of John Kentfield & Co., 1854–1925" (master's thesis, University of California, Berkeley, 1940), pp. 42–44.

35. Astoria Customs House Papers, 1:F5, F51; 2:F161a, and *passim*; Foster, "Journal," *passim*.

36. Tate, *Hawaii*, pp. 47, 48, 66, 116; Kuykendall, *Hawaiian Kingdom*, 2:40, 45, 199, 218–20, and *passim*; Consular despatches, Honolulu, 17 Mar. 1855. Residents of the islands shared these views. See Honolulu *Pacific Commercial Advertiser*, 25 July, 21 Nov. 1874, 16 Jan. 1875; Tate, *Hawaii*, pp. 28–29, 31; Frederick H. Allen, *Commercial Aspect of the Hawaiian Reciprocity Treaty* (Washington, D.C., 1882).

37. Joseph Lane, quoted in Oregon City *Oregon Statesman*, 30 Mar. 1852; Samuel R. Thurston, quoted in Lomax, "Hawaii–Columbia River Trade," p. 329; Judd, *Honolulu*, p. 172. See also Duflot de Mofras, *Travels*, 2:35.

were to be found in many a home, and where more of the young men
had seen Canton than had visited Chicago or St. Louis.[38] In addition,
the long-standing American idea that expansion to the Pacific Coast
would provide the nation with gateways to the Orient made it natural
for Americans, now that they had arrived in large numbers on the
western shore, to prepare to tap the vast markets of Asia that lay just
across the Pacific.[39]

But when Crosby and others arrived in East Asia, they found the
demand for forest products in the native sector of the economy
extremely low. Little wood was used in native buildings: the less
pretentious were of mud-brick construction, the larger structures of
granite. Rice straw or tile was almost universally used for roofing.
What little wood was needed in buildings such as these could be
supplied from the forests of southern China.[40] Shippers sent the initial
cargoes to East Asia in order to supply foreign communities there and
to furnish goods and materials for provisioning and repairing vessels
calling in Oriental waters. The emergence of lumber markets in the
Far East was dependent not upon Asian consumers, but upon the developing foreign communities of the treaty ports and upon the business
enterprises their residents set in motion.

The treaty port communities on which this nascent trade depended
had evolved rapidly in the years since the First Treaty Settlement
(1842-44). The business environment on which the old China trade
had depended was destroyed by the treaties negotiated following the

38. Coy, *Humboldt Bay Region*, pp. 107–10; Melendy, "One Hundred Years of the Redwood
Industry," p. 20.
39. This interest is especially evident in State of California, *Legislative Journals, 1851* (n.p.,
1851), appendix LL, pp. 746–77. For discussions, see Joseph Schafer, "The Western Ocean as a
Determinant in Oregon History," in H. Morse Stephens and Herbert E. Bolton, eds., *The Pacific
Ocean in History* (New York: Macmillan, 1917), pp. 287–97; Henry Nash Smith, *Virgin Land:
The American West as Symbol and Myth* (Cambridge, Mass.: Harvard University Press, 1950),
pp. 16–51; Norman A. Graebner, *Empire on the Pacific* (New York: Ronald Press, 1955), pp.
1–9, 83–102; Sidney and Marjorie Greenbie, *The Gold of Ophir: The China Trade in the Making
of America* (New York: Wilson-Erickson, 1937), pp. 183–213.
40. Consular despatches, Hong Kong, 8 Oct. 1853; G. B. Endacott, *A History of Hong Kong*
(London: Oxford University Press, 1958), pp. 65–66; Rhoads Murphey, *Shanghai: Key to Modern
China* (Cambridge, Mass.: Harvard University Press, 1953), pp. 68–69; Norman Shaw, *Chinese
Forest Trees and Timber Supplies* (London: T. Fisher Unwin, 1914), *passim*. Information on the
forests of China during the nineteenth century is scattered; valuable information is to be found in
the Tientsin *Chinese Times*, 19 Feb., 11 June, 8, 22, 29 Oct., 5 Nov. 1887, 11, 25 Feb., 3 Mar., 14
Apr., 23 June 1888, 3 Aug. 1889, 4 Oct. 1890; Tientsin *Peking and Tientsin Times*, 15 Sept.
1894; China, Inspectorate General of Customs, *Decennial Reports on the Trade, Navigation,
Industries, etc., of the Ports Open to Foreign Commerce in China and Corea, . . . 1892–1901*
(Shanghai: Imperial Maritime Customs, 1904), 1:266–68, 2:99, 116.

Opium War. Canton was no longer the sole emporium for trade between Chinese and the Western barbarians: Amoy, Foochow, Ningpo, and Shanghai were now opened to foreign commerce. Tariffs were regularized. The protection of extraterritoriality was extended to foreigners, their goods, their property, and sometimes even to their Chinese servants and employees. Hong Kong was ceded to the British. Collectively these provided a new environment out of which developed the peculiar half-Oriental, half-Occidental world of the treaty ports.[41]

The treaty port economy did not emerge at once. In the spring of 1842, Sir Henry Pottinger, the British plenipotentiary in charge of negotiations with the Chinese, predicted that within "six months of Hong Kong being declared to have become a permanent Colony, it will be a vast emporium of commerce and wealth."[42] He was overconfident. Indeed, one cause of the renewal of strife between China and the Occidental powers, and of the Second Treaty Settlement (1858–60) that resulted, was the vexation of Westerners, especially the British, when their sanguine expectations of what would follow the First Treaty Settlement were not fulfilled. But change—and economic development—did come to the coast of China. When it did, it brought into being a world in which lumbermen of the West Coast found a continuing demand for forest products. The process was well under way by the time of the collapse of the market for lumber in San Francisco.

A number of forces were operating that encouraged the development of this new order. First, Chinese were migrating overseas both as contract laborers and as individual emigrants. Large numbers moved into Southeast Asia, North and South America, and various islands stretching from the West Indies to Mauritius and the Seychelles. Servicing the vessels engaged in the coolie trade was a boon to Hong Kong in particular. Moreover, the Chinese in these overseas communi-

---

41. Hallet Abend, *Treaty Ports* (Garden City, N.Y.: Doubleday, Doran & Co., 1944), *passim*; G. C. Allen and Audrey G. Donnithorne, *Western Enterprise in Far Eastern Economic Development: China and Japan* (London: Allen and Unwin, 1954), pp. 31–51; John King Fairbank, *Trade and Diplomacy on the China Coast*, 2 vols. (Cambridge, Mass.: Harvard University Press, 1953), 1:155–75, 200–225 and *passim*; Hosea Ballou Morse, *The Trade and Administration of China*, 3rd ed. (New York: Longmans, Green & Co., 1921), pp. 195–329. For a contemporary description, see Sir Hercules Robinson to Duke of Newcastle, 9 Mar. 1861, in G. B. Endacott, *An Eastern Entrepôt: A Collection of Documents Illustrating the History of Hong Kong* (London: Her Majesty's Stationery Office, 1964), pp. 201–12. See also Tientsin *Chinese Times*, 12 Mar. 1887; Tyler Dennett, *Americans in Eastern Asia* (New York: Macmillan, 1922), p. 597.
42. Quoted in Endacott, *History of Hong Kong*, p. 72.

ties clung tenaciously to their traditional way of life and were supplied with the Chinese products they demanded through Hong Kong and the treaty ports.[43]

Second, a large number of whaling vessels were present in the Pacific during the 1840s and 1850s. Once opened, ports along the coast of China were soon servicing these vessels, just as Honolulu and Lahaina did in Hawaii.[44]

Third, the stability that foreign dominance helped to insure in the treaty ports was in sharp contrast to conditions elsewhere in China. Official exactions, piracy, and corruption had long unsettled economic life. To these was added in 1850 the Taiping Rebellion, a bloody civil conflict that raged over much of South China for better than a decade and took perhaps twenty million lives. Seeking security for themselves and their wealth, large numbers of Chinese moved to Hong Kong and Shanghai. The influx of people, which led to the erection of new buildings and to an increase in the demand for foodstuffs and other necessities, stimulated the economy of the port cities. The money that the refugees brought with them and invested in various business enterprises also spurred the economy. In the long run, the investments of the refugees were probably of greater importance than their expenditures during the first months after their arrival on food, housing, and other necessities. As Rhoads Murphey has explained, for "the first time, safe and productive investments were available to capital which under the traditional system was almost exclusively, and unproductively, invested in land." [45]

A fourth factor at work was the growth of the opium trade. This not only attracted new men to the China trade, but also supplied a source

---

43. No thorough study of the coolie trade has been done; however, for contemporary recognition of its importance, see *Hong Kong China Overland Trade Report, Supplement*, 11 Sept., 12 Oct. 1859; Consular despatches, Hong Kong, 11 Apr. 1851; Senate Exec. Doc. 22: 35-2 (Washington, D.C., 1863), p. 622ff. See also Barth, *Bitter Strength*, pp. 9–31, 50–76; Endacott, *History of Hong Kong*, pp. 125–30; Fairburn, *Merchant Sail*, 4:2256–60; Endacott, *An Eastern Entrepôt*, pp. vii-xii, 131–42.

44. Whaling apparently began to decline in Far Eastern waters before it did in those serviced by the ports in Hawaii or Australia and New Zealand. Endacott dates the decline in the Far East from about 1854. See Endacott, *History of Hong Kong*, p. 126; Hong Kong *China Mail*, 29 May 1851 and other dates; Walter S. Tower, *A History of the American Whale Fishery* (Philadelphia: University of Pennsylvania, 1907), pp. 49, 51–62, 66–79; J. T. Jenkins, *A History of the Whale Fisheries* . . . (London: H. F. & G. Witherby, 1921), pp. 234–46; George Brown Goode, *The Fisheries and Fishery Industries of the United States*, Senate Misc. Doc. 124, 47–1 (Washington, D.C., 1887), 2:7–26, 67–71.

45. Murphey, *Shanghai*, p. 79. Cf. J. V. Davidson-Houston, *Yellow Creek: The Story of Shanghai* (London: Putnam, 1962), pp. 54–56.

of specie and foreign exchange that helped make possible expanded purchases, in Hong Kong or the treaty ports, of tea, silk, and other products of China.

Fifth, and finally, the opening and development of other areas around the Pacific Basin tended to increase the total mercantile traffic on the Pacific. It thus indirectly aided in the growth of Hong Kong and Shanghai, whose locations made them natural hubs of trade.[46]

The sum of these many developments was the rapid growth, both in population and economic activity, of Hong Kong and Shanghai. Amoy, Foochow, and Ningpo, with more limited hinterlands, lagged behind, but even there expansion was under way. Canton remained important but was gradually eclipsed by Shanghai, which tapped much of the hinterland Canton had once served, and by Hong Kong, which serviced the same area and, as a result of its insular location, superior anchorages, British rule, and freeport status, enjoyed advantages the Chinese city did not possess.[47]

Indicators of this growth are numerous. In 1848 some 700 vessels, totaling 228,818 tons registry, called at Hong Kong; by 1854 the number had risen to 1,100 vessels, and the tonnage had nearly doubled. The increase continued. By 1859 tonnage had passed the million mark; by 1864 it was over 2 million. In 1853 Hong Kong had 240 ship's chandlers and 12 rope manufactories. Twelve years later there were 427 chandlers and 20 rope works to supply visiting merchantmen. In 1843 the first locally built vessel of Western design, the 50-ton schooner *Celestial*, was launched in Hong Kong. The following year a small way capable of handling vessels of 300 tons was constructed. By 1860 a large and growing dock, financed by one of the leading mercantile houses in the colony, was in operation; other major facilities were available nearby at Whampoa. By 1865 a total of 93 boat builders were reportedly active in the colony.[48]

The development of Shanghai was less rapid until after the opening

46. Endacott, *An Eastern Entrepôt*, pp. vi–xiv, 131, 134–42; Endacott, *History of Hong Kong*, pp. 76, 125–32; Bernard Boxer, *Ocean Shipping in the Evolution of Hong Kong* (Chicago: University of Chicago Press, 1961), pp. 5, 14–15; Murphey, *Shanghai*, p. 67; Shanghai Mercury, *Shanghai: The Model Settlement* (Shanghai, 1894), p. 21 and *passim*.

47. Murphey, *Shanghai*, pp. 42–44, 47, 50, 52–53; Endacott, *History of Hong Kong*, pp. 8–10; Fairbank, *Trade and Diplomacy*, 1:123–29; *Hunt's Merchants' Magazine* 41 (1859):215.

48. Boxer, *Ocean Shipping*, p. 14; Endacott, *History of Hong Kong*, p. 132; Hong Kong *China Mail*, advertisements, various dates, 1851. In addition to these dockyards and those at Shanghai, discussed below, drydocks were also opened during this period at Amoy. See Hong Kong *China Mail*, 10 Feb. 1859 and other dates.

of the Yangtze River to trade in 1862. Even so, the city made significant progress during the 1840s and 1850s. Tea exported through Shanghai rose from 1 million pounds in 1844 to 80 million in 1855. During the same period silk exports rose from 6,433 bales to 92,000 bales. Moreover, with the establishment of the Shanghai Dock and Engineering Company in 1851, the city began to develop into a major center of shipbuilding and repair. It also became a center of commercial finance.[49] The first Western-style bank in the city, the Oriental Banking Corporation, appeared in 1848. This firm was soon joined by others—including the Mercantile Bank of India, London and China in 1854 and the Chartered Bank of India, Australia and China in 1857. By 1860 the foreign population of Shanghai had reached 1,000. In Hong Kong and the five treaty ports taken collectively, there were 219 foreign firms in operation by 1855.[50]

Lured by the potential of markets in China and driven by the collapse of demand in San Francisco, entrepreneurs dispatched six cargoes of lumber from Puget Sound to the Middle Kingdom in 1855.[51] Andrew Pope was among those heartened by developments in East Asia. In January 1855 he wrote, "I think in time we shall be able to market considerable lumber in China. . . ." Before the year was out, his vessel, the *Live Yankee*, earned $16,000 on a single run to South China.[52]

Among those who followed the example of Pope and Nathaniel Crosby in seeking to market lumber on the China coast were William J. Adams and his associates, who had just recently established the Washington Mill Company and erected a sawmill at Seabeck in Washington Territory. Like many firms during these years, the Washington Mill Company was short on operating capital.[53] As a

---

49. Murphey, *Shanghai*, pp. 108–9, 111; Davidson-Houston, *Yellow Creek*, pp. 31–43, 46–60; D. K. Lieu, *The Growth and Industrialization of Shanghai* (Shanghai: China Institute of Pacific Relations, 1936), p. 46. The first vessel built in Shanghai on Western lines was the *Pioneer*, built of native oak and camphor wood in 1857. Repair work with imported lumber had already been going on for some time, however. See Davidson-Houston, *Yellow Creek*, p. 59. Lieu's work contains a wealth of data on the growth of Shanghai during the nineteenth and early twentieth centuries, but his interpretations are debated by Chi-ming Hou, *Foreign Investments and Economic Development in China, 1840–1937* (Cambridge, Mass.: Harvard University Press, 1965).
50. F. C. Jones, *Shanghai and Tientsin* (London: Oxford University Press, 1940), pp. 75–76; Murphey, *Shanghai*, pp. 2, 21; Boxer, *Ocean Shipping*, p. 11.
51. U.S. Collector of Customs, Puget Sound, "Registers of Entrances and Clearances," vol. 1 *passim*.
52. Pope to Wm. Pope & Sons, 23 Jan. and 31 July 1855, PTP, Pope file.
53. For a discussion of the problems firms such as the Washington Mill Co. had in acquiring adequate working capital, see Porter and Livesay, *Merchants and Manufacturers*, pp. 62–63, 69–77.

result, Adams sought to secure sales in advance rather than tie up funds in cargoes dispatched as ventures carried on the company's account.

He found the owners of Pope and Talbot doing all they could to keep his firm from entering the trade. "It makes them feel awful to think that anyone else besides them should sell a cargo of lumber," Adams commented, and then added, "But they will get used to it before we get through with it the Lord being willing." Other times he was more vehement. "Look out for those d——d thieves," he warned his associates. When he learned that, using devious means, Pope had beaten him out of a coveted sale, Adams vowed, "We will get even as sure as there is a God in Israel." [54]

In spite of the efforts of Pope and Talbot, Adams was able to sell a cargo to a firm which, Adams explained to his partners, intended "doing quite a business in China" and would be "wanting cargoes of lumber quite often." [55] This was only a beginning. The Washington Mill Company sold numerous cargoes for shipment to the Far East in the years that followed. [56]

Not all the vessels that carried lumber to the Far East from Puget Sound or San Francisco during the 1850s were laden with materials cut at the sawmills of Pope and Talbot or the Washington Mill Company. [57] In addition, carriers were debarking from areas of production besides those where the two firms operated. The barks *Metropolis* and *C. E. Tilton* transported lumber to China from the Columbia River; the bark *Senator* loaded a cargo of forest products at Sooke Inlet for Shanghai; and a number of merchantmen—including the *Armitage, Ceres, Early Bird, Almatia,* and *Quatre Bras*—carried cargoes to the Far East from Humboldt County. [58]

54. Adams to Washington Mill Co., 12 Mar. 1858, WMCP (see also 8 Sept. 1858, and Adams to M. Blinn, 8 Dec. 1858, WMCP); Adams to Washington Mill Co., 21 Dec. 1858, WMCP; Adams to Washington Mill Co., 5 Nov. 1858 (see also 20 Oct. 1858), WMCP.

55. Adams to Washington Mill Co., 6 Apr. [1858], WMCP. See also 16 Oct. 1857, 4 Feb., 9, 12, 20 Mar. 1858, WMCP.

56. Thomas J. Prosch, scrapbook, WMCP, *passim;* Koopmanschap & Co. to Adams, Blinn & Co., 21 Feb. 1866, WMCP; memos of agreement between Adams, Blinn & Co. and Edwards & Balley, 24 Mar. 1860, and between Adams, Blinn & Co. and Koopmanschap & Co., 7 Feb. 1866, WMCP. For a discussion of the importance of such shipments, see below, chap. 6.

57. Commercial intelligence columns in the *China Mail* list numerous arrivals from Puget Sound and San Francisco that have no known connections with either firm. Some of these may have had such connections, of course, but it would hardly be likely that all of them did.

58. Astoria Customs House, "Marine List and Memoranda," vol. 1 *passim;* Astoria Customs House Papers, 1: F26, F122; Wright, ed., *Lewis and Dryden's Marine History,* pp. 53, 89; San Francisco *Herald,* 16 Nov. 1855, 4 May 1857; Arcata *Humboldt Times,* 28 Dec. 1856; Eureka *Humboldt Times,* 1 May 1858; Hong Kong *China Mail,* 7 Dec. 1854, 26 Aug. 1858; Hong Kong

Once in the Orient, these vessels did not seek cargoes of tea and silk to carry on around the Cape of Good Hope to ports on the Atlantic. Such employment was for the larger, faster clipper ships. Instead, these lumber droghers usually hastened back to the West Coast, either in ballast or with whatever cargoes of Chinese goods they could obtain. From these beginnings developed a trade that later led John Hittell to note proudly that San Francisco and New York were the only ports in the United States receiving regular shipments from China. Of the 246 vessels entering San Francisco harbor in 1856, 42 were from China.[59]

In spite of the growth of Western enterprise in the Far East and West Coast residents' continuing interest in shipping the products of their area to the Orient, the quantity of forest products marketed in China remained relatively limited through the 1850s. Reliable statistics are not available, but the annual totals appear to have been consistently less than those for Hawaii.[60] One limiting factor was the large amount of lumber dispatched to ports along the coast of China from areas other than the West Coast. Numerous cargoes arrived from Singapore, Borneo, Siam, Burma, and elsewhere in Southeast Asia.[61] Shipments also arrived from New Zealand, sometimes from Australia, and occasionally from the Baltic and Japan as well.[62] Timber from the

Friend of China and Hongkong Gazette, 26 Jan. 1856. According to *Lewis and Dryden's Marine History,* the cargo carried by the *Senator* consisted of piles, but advertisements in the *China Mail* indicate that it carried spars.

59. Gov. John Bigler, "Annual Message to the Legislature," in State of California, *Journal of the Senate* (Sacramento, 1855), pp. 44–45; Ernest Seyd, *California and Its Resources* (London, 1858), pp. 72–74; John S. Hittell, *The Commerce and Industries of the Pacific Coast of North America;* . . . (San Francisco, 1882), p. 207. For a discussion of the shipment of Chinese goods to the West Coast, see below, chap. 8. See also Hong Kong *China Overland Trade Report,* 14 Mar., 28 Oct. 1859.

60. Records are fragmentary on both sides of the Pacific. Hong Kong was a free port; even the Harbour Master did not know what items or quantities were entering. Extant records on the American side contain many omissions, errors, and contradictions. Nonetheless, based on the number of vessels sailing from the lumber ports to Hawaii in contrast to the number sailing to the China coast, the generalization that the former was the more important market appears irrefutable.

61. Hong Kong *China Mail,* 21 Feb. 1856, 24 June, 9 Sept., 7 Oct. 1858, 9 Mar., 19, 26 May, 16, 30 June, 7, 28 July, 25 Aug., 1, 15, 29 Sept. 1859; Shanghai *North China Herald,* 30 July 1859, and other dates. Endacott comments that trade in timber and other items from Manila virtually ceased after 1847, but occasional shipments were still arriving. See Endacott, *History of Hong Kong,* p. 75; Hong Kong *China Mail, Gazette Supplement,* 4 May 1854; Hong Kong *China Mail,* 21 Feb. 1856, 25 Nov. 1858. For further information on trade with Manila, see Fairburn, *Merchant Sail,* 4:2443–53.

62. Hong Kong *China Mail, Gazette Supplement,* 4 May 1854; Hong Kong *China Mail,* 1 Apr., 29 July 1852, 10 July 1856, 14, 21 Apr. 1859, 12 Jan. 1860; Evelyn Stokes, "Kauri and White Pine: A Comparison of New Zealand and American Lumbering," *Annals of the Association of American Geographers* 56 (1966):448.

mountains of southern China was also distributed through the minor ports to Shanghai and Hong Kong.[63]

The structure of Occidental business in the Far East was itself instrumental in limiting the quantity of lumber from the West Coast that could be marketed in China. The dominant American firms in the old China trade, such as Russell and Company and Olyphant and Company, had their business connections with the east, not the west, coast of the United States. When they dispatched lumber to the Far East, it came from the forests of the Northeast, not those of California or the Northwest.[64] Furthermore, by the 1850s these firms had begun to stagnate. Leadership in the mercantile communities of the Far East was passing to the British. Of the 219 foreign firms in Hong Kong and the treaty ports in 1855, 111 were British.[65] British entrepreneurs might purchase lumber from producers on the West Coast in order to fill the needs of their dockyards and other enterprises, but even greater profits could be earned if they could obtain forest products from enterprises they controlled themselves. It seems probable that this was the reason for the rapid increase in the importation of forest products from British-controlled areas of Southeast Asia during this period.

In contrast to the older American firms and the more important British enterprises, firms on the West Coast that were interested in capitalizing on the growing markets of the treaty ports were poorly financed and unknown in international business circles. Though the establishment of lumberyards in Hong Kong and Shanghai would have been a logical step, enterprises such as the Washington Mill Company and Pope and Talbot were in no position to do so. Their finances were limited, and they had neither experience nor connections in the Far East. Nathaniel Crosby, the one American to make a concerted effort in this direction in the 1850s, had almost no capital behind him. In spite of this handicap, Crosby appears to have made considerable headway. One contemporary observer referred to his home as the center of the American community in Hong Kong, and the American consul there

63. Hong Kong *China Mail, Gazette Supplement,* 4 May 1854. See also Murphey, *Shanghai,* pp. 93, 99.

64. For example, the famed clipper ship *Houqua* carried lumber to Canton on both her first and second voyages (1844 and 1846). As late as 1850, lumber was still being dispatched to the China coast from the eastern United States. See Helen Augur, *Tall Ships to Cathay* (Garden City, N.Y.: Doubleday, Doran & Co., 1951), pp. 152–54, 156; Consular despatches, Hong Kong, 1 July 1850.

65. Boxer, *Ocean Shipping,* p. 11; Dennett, *Americans in Eastern Asia,* p. 579.

described him as "one of our most prominent Merchants." [66] But Crosby died suddenly in December 1856. The loss of Crosby, coupled with the failure in the same year of two of the leading American firms in the Far East—Wetmore and Company and Nye Brothers and Company—made it all the more difficult to increase the West Coast's share of the lumber market of the China coast during the years that followed.[67]

China and Hawaii were not the only markets to which the lumbermen of the Far West turned their attention in the mid-fifties. Australia was a third. Lumber had long been produced in Australia and neighboring areas. Indeed, during the California Gold Rush, lumber had even been dispatched to San Francisco from down under. Following the discovery of gold in New South Wales and Victoria in 1851, however, Australia experienced a gold rush of its own.[68] By 1852, thousands of people, including many gold-seekers who had been unsuccessful in California, were swarming to this new source of sudden wealth. The influx created a demand for foodstuffs that soon reversed what had been a flow of flour to California and created a demand for building materials that soon outstripped the productive capacity of the lumber industry of the southwestern Pacific.

The southwest Pacific's timber industry had begun in New Zealand. During his visit to those islands in 1773, Captain James Cook made note of the potential value of their forest resources.[69] Utilization commenced in 1795 when a vessel called for a cargo of spars. The first shipment of spars from New Zealand to Australia occurred four years later. This shipment, carried on the *Hunter*, was followed by others in 1800, 1810, 1812, and 1814. By 1816, cargoes from New Zealand to Australia included sawn timber. The installation of sawmills transformed this early timber trade. Permanent settlements began to appear around the mills, such as that at Hokianga on North Island. In earlier

66. Cited in Griffin, *Clippers and Consuls*, p. 280; Consular despatches, Hong Kong, 14 Feb. 1857 (see also 29 Jan. 1857).
67. Consular despatches, Hong Kong, 29 Jan. 1857; Portland *Oregonian*, 28 Mar. 1857; Griffin, *Clippers and Consuls*, pp. 243–44.
68. Gold had been found in Australia off and on for a number of years, but not until after the California rush did it excite any wide circle. Authorities in the Australian colonies had apparently done what they could to stop the spread of news concerning the earlier discoveries for fear of the effect it might have on the populace, especially the convicts and ex-convicts. There was no hushing up the discoveries of 1851, however.
69. Captain James Cook, *The Journals of Captain Cook on His Voyages of Discovery*, ed. J. C. Beaglehole, 3 vols. (Cambridge: Cambridge University Press for the Hakluyt Society, 1955–67), 2:133–34.

times the crews of transient vessels that came to get cargoes of forest products provided the labor force in the woods. This practice now quickly passed. By the 1820s the timber trade, which together with whaling and flax production provided the backbone of the colony's economy, had become, as one authority has put it, a "fairly continuous" operation. By 1840, if not earlier, timber agents were being sent to New Zealand by Australians interested in the trade.[70]

Not all the forest products produced in the southwest Pacific came from New Zealand. The exportation of lumber from Australia began in 1803 when timber for the Royal Navy was dispatched to England. Many sawmills in Australia were abandoned during the peak of the gold rush excitement, just as they had been in California; but in time they were reopened. As has been mentioned, lumber was dispatched from Australia to China in the late fifties. By the 1870s and 1880s sawmills were common in Victoria and New South Wales; however, production, both actual and potential, was limited by the resource base on which the industry had to depend. Only about 4 percent of Australia was forested, a smaller portion than that of any continent except Antarctica. The wood of the leading timber species was hard and brittle; commercial softwoods were few and scattered prior to the introduction of nonindigenous species.[71]

American vessels, though present in the area, played no part in the early timber trade of the southwestern Pacific. The trade of the region was of two general types: a trade in luxury items, such as sandalwood, bêche-de-mer, pearl shell, and birds' nests; and a trade in bulk goods, such as pork from Fiji and Polynesia and flax and timber from New

70. C. Hartley Grattan, *The Southwest Pacific to 1900, a Modern History: Australia, New Zealand, the Islands, Antarctica* (Ann Arbor: University of Michigan Press, 1963), pp. 154, 157–58, 414. J. M. R. Young, "Australia's Pacific Frontier," *Historical Studies: Australia and New Zealand* 12 (1966):378–80. Stokes, "Kauri and White Pine," pp. 442–45. *Canada Lumberman* 1 (30 Nov. 1880): 1; 2 (1 Jan. 1882):1. For a revealing contemporary account of lumbering on North Island during the 1830s, see Edward Markham, *New Zealand, or Recollections of It*, ed. E. H. McCormick (Wellington, N.Z.: R. E. Owen, Gov't. Printer, 1963).

71. *Hunt's Merchants' Magazine* 34 (1856): 260; Grattan, *Southwest Pacific to 1900*, p. 282; Raphael Zon and William H. Sparhawk, *Forest Resources of the World* (New York: McGraw-Hill, 1923), pp. 11–13, 896–97; D. A. N. Cromer, "Australia," in Stephen Haden-Guest et al., eds., *A World Geography of Forest Resources* (New York: Ronald Press, 1956), pp. 573–90; Werner Levi, *American-Australian Relations* (Minneapolis: University of Minnesota Press, 1947), pp. 46–47, 57; Robert G. Albion, *Forests and Sea Power* (Cambridge, Mass.: Harvard University Press, 1926), pp. 364, 400–1; Thomas Dunbabin, *The Making of Australasia* (London: A. & C. Black, Ltd., 1922), p. 70. Four percent is a recent figure, but apparently Australian forests were not much more extensive in the nineteenth century than in the twentieth. Cf. *Canada Lumberman* 4 (2 June 1884):162.

Zealand. The trade in luxury goods often brought high profits; but prior to 1834 the principal market, Canton, while open to Americans, was closed to all British vessels save those of the East India Company. Trade between Australia and the islands of the southwest Pacific fell outside the monopoly of the East India Company, however. Thus, American vessels turned to the more profitable luxury trade, while Australian shipowners put their tonnage in the bulk trade as the only one open to them. The latter soon became a virtual monopoly of merchants from New South Wales and the basis of an Australian shipbuilding industry. The development of this domestic trade was encouraged by the Navigation Acts, which forbade foreign vessels from entering the competition. American vessels calling in Australia sometimes engaged in smuggling high-value goods ashore, but lumber was not such an item. Prior to the repeal of the Navigation Acts in 1849, Americans were effectively shut out of the Australian lumber market.[72]

With repeal of the Navigation Acts and the emergence of greatly enlarged markets as a result of Australia's gold rush, all of this quickly changed. Up to 1849, trade between the United States and Australia could reach $250,000 in value during an exceptionally good year. It now grew rapidly to the $3 million level. In 1850, eighty-six vessels carried goods from Australia to California. The flow of goods reversed when Australia's own gold rush commenced, but the volume remained high.[73]

The basic cause of the increase in shipments to Australia after 1852 was the rapid growth of the colonies there. Collectively they grew from a population of 400,000 in 1850 to 1,146,000 in 1860. Victoria, which in 1850 had a mere 76,000 residents, had 538,000 by 1860 and was the largest of the Australian colonies. The lure of gold had drawn people away from the farms and sawmills of the southwest Pacific. Importers soon began to purchase foodstuffs and building materials abroad in order to meet demand. Proximity made the west coast of the

72. Gordon Greenwood, *Early American-Australian Relations from the Arrival of the Spaniards in America to the Close of 1830* (Melbourne, Australia: Melbourne University Press, 1944), p. 167; Young, "Australia's Pacific Frontier," p. 376; C. Hartley Grattan, *The United States and the Southwest Pacific* (Cambridge, Mass.: Harvard University Press, 1961), pp. 83–95; Levi, *American-Australian Relations*, pp. 25–36; Fairburn, *Merchant Sail*, 2:1537–41.

73. Grattan, *United States and the Southwest Pacific*, p. 98; Levi, *American-Australian Relations*, p. 44; U.S. Dept. of State, Consular despatches, Sydney, National Archives, Record Group 59, microcopy 173, 13 Aug. 1852, 29 Aug. 1853; Bateson, *Gold Fleet for California*, pp. 142–44; Fairburn, *Merchant Sail*, 2:1537–41.

United States a logical source of supply. Sailing ships could reach Australia in sixty days or less under normal circumstances. To reach Australia from other major potential sources would take considerably longer, thus running up transportation costs and cutting into profits.[74]

From whatever source the imports came, it was Australian gold that paid for the bulk of them. This was especially true in the case of goods imported from the American West, for Australia produced nothing that was in sufficient demand in the Far West to balance the flow of lumber and foodstuffs from there. Exports of wool from Australia actually increased between 1851 and 1854 in spite of the gold rush, but Westerners had surplus wool of their own. Coal was plentiful down under and was occasionally shipped to San Francisco; but Californians had nearer sources of supply which, together with the coal brought in as ballast in many of the vessels that entered the Golden Gate, kept the market for coal in California relatively well stocked. Fortunately for those engaged in shipping goods to Australia, the production of gold was sufficient to more than make up the deficit in the balance of payments. From 1851 to 1860 Australia produced twenty-five million ounces of gold, 40 percent of the world's output.[75]

The American consul in Sydney noted an increase in trade with the United States as early as August 1852. One month later Andrew Pope reported that five vessels were loading lumber in San Francisco for Australia. Pope had high hopes for the market. Because of proximity and the possibility of return cargoes of coal, he felt Westerners could drive New York shippers out of the Australian lumber market.[76]

As the price of lumber in San Francisco dropped, shipments to Australia became more frequent and regular. Cargoes were shipped to the Australian colonies from the redwood coast, Puget Sound, and the Columbia River, as well as from San Francisco. Several cargoes were

---

74. Grattan, *The Southwest Pacific to 1900*, pp. 118, 245–46; Fairburn, *Merchant Sail*, 4:2297–2314; *Hunt's Merchants' Magazine* 34 (1856):416–27.

75. Grattan, *The Southwest Pacific to 1900*, pp. 246, 248, and *passim*; Grattan, *The United States and the Southwest Pacific*, pp. 98–99; Levi, *American-Australian Relations*, pp. 37–48; Fairburn, *Merchant Sail*, 4:2313–14; Consular despatches, Sydney, 2 July 1850; Bateson, *Gold Fleet for California*, p. 113. See also chap. 7, below. Not only the flow of commodities, but also the flow of "invisibles" tended to favor the United States, according to Grattan, *The United States and the Southwest Pacific*, p. 108.

76. Consular despatches, Sydney, 13 Aug. 1852; Pope to Wm. Pope & Sons, 30 Sept. 1852, PTP, Pope file. The first cargo of lumber carried to Australia directly may have been a cargo shipped as a venture on his own account by Capt. James Hall of the schooner *Spray*. See Astoria Customs House, "Marine List and Memoranda," vol. 1, entries for Dec. 1853; Astoria Customs House Papers, 2:F165; Wright, ed., *Lewis and Dryden's Marine History*, p. 48.

sent during 1854 from Duff and Chamberlain's mill on Humboldt Bay. J. R. Duff, searching for buyers for later shipments, himself sailed on the vessel carrying the first of these.[77]

The trade was not without its difficulties. E. C. Williams later claimed that the first cargo of redwood shipped to a foreign market was one hundred thousand feet sent to Sydney from the California Lumber Company mill at Big River. The shipment failed to return a profit. As Williams explained it: "The trouble with the lumber arose in the first place from the color, for the people there were very conservative; and in the next instance from the fact that all their nails came from England, and were pointed from the head down on all sides. Of course, these nails—no matter how they were driven relative to the grain of the wood—acted as a wedge and split the board, which confirmed the condemnation of this 'blasted stuff.' " The small quantities of redwood that remained on hand following the subsequent adoption of cut nails demonstrated that the initial impressions were unjustified. Regular sales commenced. As Williams concluded: "Our loss of $2,500 opened the door for the larger and more profitable market now enjoyed by shippers to that country." [78]

A second problem resulted from the undependability of Australian markets. Through the early months of 1858 William Adams watched news of the market in Australia. In August he sold a cargo from the Washington Mill Company for delivery there. However, by November he was writing to his associates that the bark *Glimpse* had arrived from Australia, "and her news in regard to the state of that lumber market is not flattering." Yet, only a month later he made arrangements to ship another cargo to Australia, half to be carried on the company's account, and reported that "from what we can learn from recent advices from Australia such a cargo will pay well." If the captain could find a return cargo for California, Adams noted, he would then take a third cargo of lumber to Australia.[79]

77. U.S. Collector of Customs, Puget Sound, "Registers of Entrances and Clearances," vol. 1 *passim*; Wright, ed., *Lewis and Dryden's Marine History*, p. 62; San Francisco *Herald*, 13, 28 Dec. 1856, 17 Jan. 1857, 18 Nov. 1858; Arcata *Humboldt Times*, 21 Oct. 1854; Eureka *Humboldt Times*, 1 May, 6, 9 Nov. 1858, 19 Nov. 1859; Adams to Blinn, 8 Dec. 1858, WMCP; Coman and Gibbs, *Time, Tide and Timber*, pp. 77, 79–80, 83; Coy, *Humboldt Bay Region*, p. 122. These vessels were usually considerably larger than those engaged in the coastwise trade; but since tugs were by then in use on the bars of the Columbia River and Humboldt Bay, no major problems resulted. As the Eureka *Humboldt Times* stated, 10 Apr. 1858, profits "fully justified the trade."
78. Williams, "First Redwood Operations," p. 13.
79. Adams to Washington Mill Co., 19 Apr., 11 June, 20 Aug., 8 Sept., 5 Nov., 8 Dec. 1858,

At times sea captains would load cargoes of lumber for Australian ports after having received news of favorable prices there, only to discover upon arrival down under that prices had broken and their venture netted them a loss. Uncertainty had plagued those who sought to market lumber in San Francisco during the California Gold Rush, too; but in the case of Australia the difficulties were compounded by the greater costs of transportation, the greater time lag between the dispatch of information regarding prices current and the arrival of shipments at the market, and the relative infrequency of contacts between Australia and the West Coast, as compared with those between San Francisco and the redwood coast or Northwest.[80]

The demand for building materials dwindled as the island continent's gold rush passed its peak. The hectic phases of the rush were over by 1855, gold production passed its peak in 1856, and employment in the mining areas began to drop after 1858. Wheat prices collapsed in 1856, and the commerce in provisions between the West Coast and Australia virtually ceased.[81] Lumber shipments reacted more slowly; but as the excitement of the gold rush subsided, lumber cargoes to Australia also became less common.[82]

Gold seekers were drawn to New Zealand following discoveries at Otago in 1861, but New Zealand was sufficiently endowed with timber resources to supply its own needs, even during a period of rapid development. New Zealand failed to supply a significant new market for lumbermen.[83]

Deep mining began to dominate in Australia after about 1860 as alluvial deposits were worked out, but lumbermen on the West Coast were unable to take advantage of the new demand for mining timbers because of the economic dislocations caused by the American Civil War and the presence of Confederate raiders in the Pacific.[84] After the

WMCP. See also A. Pope to Wm. Pope & Sons, Feb. 1855, PTP, Pope file.

80. W. Thorndike to Adams, Blinn & Co., 24 May 1859, WMCP; Younghouse & Co. to Washington Mill Co., 5 Apr. 1866, WMCP; Meany, "History of the Lumber Industry," pp. 106, 131; A. Pope to Wm. Pope & Sons, 27 June 1855, PTP, Pope file.

81. Fairburn, *Merchant Sail*, 2:1537–41; Levi, *American-Australian Relations*, pp. 43–47, 62–64; Grattan, *The Southwest Pacific to 1900*, pp. 247, 381–82; *Hunt's Merchants' Magazine* 34 (1856):416–27; Consular despatches, Sydney, 11 Oct. 1858, 2 July 1860.

82. Records are incomplete, but apparently lumber shipments to Australia passed their peak in 1857 or early 1858. See U.S. Collector of Customs, Puget Sound, "Registers of Entrances and Clearances," vol. 1 *passim*.

83. In later years, after its timber resources had been largely depleted, New Zealand did become a market for lumber from the United States, especially for larger dimensions.

84. Wright, ed., *Lewis and Dryden's Marine History*, p. 141; Levi, *American-Australian Relations*, p. 63. The first shipments of lumber from British Columbia to Australia occurred in 1864. That

war, Australia once again became a market for American lumber, but for the time being its importance had ceased.

In addition to Hawaii, China, and Australia, the Pacific coast of Latin America began to develop into a significant market for lumber from the West Coast during the 1850s. Occasional shipments had been sent there in earlier years, of course; but as development of the area accelerated in the 1850s, the quantities rose. ·

The forties and fifties were decades of growth for Chile, especially. Between 1845 and 1860 the exportation of minerals increased fourfold and agricultural exports slightly more than that. Under the autocratic rule of Manuel Montt (1851–61) railroad construction and public works were pushed forward. These factors, together with the growth of Chilean cities during the period, created unprecedented demands for timber products.[85]

Sufficient quantities to meet the demand were not available domestically. Though the southern third of Chile was heavily forested, the Araucanian Indians held sway there. Even if they had already been pacified, it probably would have proven easier to obtain the lumber and other forest products that were needed from the redwood coast or Pacific Northwest. Southern Chile is a rugged and wind-swept land with a dangerous, rock-bound coast. The obstacles to developing its resources are so great that even in the twentieth century they have remained but slightly tapped.[86]

Under these circumstances lumbermen shipped increasing quantities of Douglas fir and redwood to Chilean ports during the 1850s and early sixties. Among the most aggressive in penetrating this rising market was the California Lumber Company, which supplied large quantities from its mill on Big River for the construction of the railroad from Santiago to Valparaiso. The company's success in penetrating Chile may have been more than coincidental. Henry Meiggs, one of the firm's founders, had appeared in Chile and in a remarkable turn of

shipment from Canada occurred at a time when shipping to Australia under the American flag was risky may be more than a coincidence. See Lawrence, "Markets and Capital," pp. 8–9, 11–12, 15, 18.

85. Luis Galdames, *A History of Chile*, trans. Isaac J. Cox (Chapel Hill: University of North Carolina Press, 1941), pp. 290–92, 301–4. *Hunt's Merchants' Magazine* 34 (1856):607; 39 (1858):749; 41 (1859):98; 46 (1862): 301. A. Pope to Wm. Pope & Sons, 4 May 1857, PTP, Pope file.

86. Preston E. James, *Latin America*, 3rd ed. (New York: Odyssey Press, 1959), pp. 264–71; Zon and Sparhawk, *Forest Resources of the World*, 2:741–49; Haden-Guest et al., *World Geography of Forest Resources*, pp. 212–13.

events won the contract to construct the Santiago-Valparaiso line. There is no evidence that he gave preference to his old partners, but it seems likely.[87]

Peru was also a growing market during the period. The guano trade was developing rapidly by the 1850s and bringing new wealth to the nation. Government authorities speeded the process by borrowing from abroad against future earnings from the trade. Much of this money was passed on to private citizens in payment of old claims against the government. The result was an increase in the level of economic activity in the country. A class of entrepreneurs emerged in this land where the old aristocracy had for centuries looked upon commercial activity with disdain. In 1851 a railroad, the first in Latin America, was built to connect Lima with Callao. Somewhat later the line was extended to Chorrillos and another was built connecting Tacna and Arica. Railroad construction created a need for railroad ties. In time, this demand, in Peru and elsewhere, was to become a major factor in the lumber trade of the Pacific.[88]

During the mid-fifties, increasing quantities of salt were being imported into the Northwest from La Paz, the Carmen Islands, and other points along the west coast of Mexico. Apparently this activity failed to generate any demand for lumber in the areas from which the salt came, but the increasing frequency with which vessels were arriving from Latin America may have helped to turn the attention of lumbermen in that direction.[89]

Prior to 1854 most of the lumber available in the ports of Chile and Peru came from the northeastern United States. Vessels en route to California often stopped off to sell lumber that was aboard and to take on fresh provisions before proceeding on to San Francisco. However, in June 1854 a cargo was dispatched to Valparaiso from San Francisco. The quality of this lumber proved to be higher than that from the Northeast. It sold readily at good prices, and other shipments soon followed.[90]

87. Haight, "Journal," p. 34. Cf. U.S. Collector of Customs, Puget Sound, "Register of Entrances and Clearances," vol. 1 *passim*; Stewart, *Henry Meiggs*, pp. 23–32, 39–41.
88. Frederick B. Pike, *The Modern History of Peru* (London: Weidenfeld & Nicholson, 1967), pp. 91–96, 98–103; Jonathan V. Levin, *The Export Economies: Their Pattern of Development in Historical Perspective* (Cambridge, Mass.: Harvard University Press, 1960), pp. 58–85, 91–94, 100; *Hunt's Merchants' Magazine* 39 (1856):606.
89. Astoria Customs House Papers, 2:F104; 4:F213; 5:F220, F223a.
90. Meany, "History of the Lumber Industry," pp. 117–18; Samuel Eliot Morison, *The Maritime History of Massachusetts, 1783–1860*, 2nd ed. (Boston: Houghton Mifflin & Co., 1961), pp.

Not all the vessels that turned to this trade were American. In 1854 the British brig *Princess Louisa*, carrying goods from the Columbia River to England for the Hudson's Bay Company, filled out her cargo with 111,500 feet of lumber and eleven spars and then stopped off in Callao to market them before proceeding on to England. The Peruvian brig *Florencia* entered the trade between the Northwest and the west coast of South America, only to be wrecked in a storm while passing out of the Strait of Juan de Fuca.[91]

None of the new markets for lumber that emerged around the Pacific Basin in the mid-fifties was large enough by itself to have a major impact on the lumber industry of the West Coast. Collectively, they were of great value, nowever. By furnishing alternate outlets, they softened the immediate impact of the collapse in 1854 of the market for lumber in San Francisco. By providing a broader marketing base than had existed previously, they increased the potential for future expansion of the lumber industry and, at the same time, left lumbermen on the West Coast less at the mercy of the vagaries of the economic forces at work in San Francisco and its hinterlands. Mills that otherwise might not have weathered the collapse found in Hawaii, China, Australia, and South America outlets that made it possible for them to stay solvent until the return of better times. As it was, many failed to survive the crisis. Years were to pass before the lumber industry of Humboldt County again reached the level it had enjoyed just prior to the crash.[92] The recession had a similar effect on San Francisco's lumberyards. Several went out of business during the slump.[93]

But retrenchment, not catastrophe, followed the collapse of 1854. Many residents of the areas of lumber production continued to have faith in the future of the industry on which the economy of their locales depended. One experienced lumberman in Washington Terri-

---

269–71; Olympia *Pioneer and Democrat*, 3 June, 28 Oct. 1854, 3, 10 Mar., 26 Oct. 1855; San Francisco *Herald*, 28 June 1857; Coman and Gibbs, *Time, Tide and Timber*, pp. 80, 82–83.

91. Astoria Customs House Papers, 2:F147, F148; Astoria Customs House, "Marine List and Memoranda," vol. 1, entries for June 1854; Wright, ed., *Lewis and Dryden's Marine History*, pp. 96–97.

92. Coy, *Humboldt Bay Region*, pp. 120, 122–23; Melendy, "One Hundred Years of the Redwood Industry," pp. 278–80. Recovery was apparently somewhat more rapid around Puget Sound. See Meany, "History of the Lumber Industry," pp. 108–22.

93. A. Pope to Wm. Pope & Sons, 23 Jan. and 30 Apr. 1855, PTP, Pope file. Throughout the early years there was a rapid turnover of firms engaged in selling lumber in San Francisco. In 1859 there were thirty-two lumber dealers in the city. During the next year nine of these closed and ten new ones opened. See *Hunt's Merchants' Magazine* 43 (1860):647. Cf. Hubert Howe Bancroft, *California Inter Pocula* (San Francisco, 1888), p. 341.

tory decried the "cowardly croakings of despair" of those who believed California would never again provide a favorable market for the fir of Puget Sound. "San Francisco," he said, "has now the *chills and shakes*—her case is rather bilious, but she will recuperate and yet depend upon us for lumber." [94] The editor of the Olympia *Pioneer and Democrat* urged lumbermen in the area of Puget Sound to seek out actively new markets around the Pacific Basin where they could sell their cut, rather than simply wait for a change of the situation in San Francisco.[95]

This confidence in the face of setbacks was shared by residents of Humboldt County. Convinced that the future prosperity of their region was assured and that it was not tied to San Francisco alone, they repeatedly sought to have a customs house established on Humboldt Bay. Markets for lumber had emerged around the Pacific Basin; given a customs house, they felt, Humboldt County could compete for these markets on even terms with other areas of production and its future prosperity would be assured.[96] Though the residents of Humboldt County failed in their efforts to get a customs house, they were correct in their assessment of the future importance of the lumber markets of the Pacific Basin. In the years that were to come, these markets and the coastal lumber industry of the Far West, which turned increasingly to supplying them, were destined to prosper, and occasionally to suffer, together.

San Francisco was also to be affected. The city had become a port from which lumber was exported.[97] Moreover, as the San Francisco *Daily Times* observed, "as any portion of the country up or down the coast advances in population and wealth, so must it add to our already generous store." [98] The editor of *Hunt's Merchants' Magazine* believed the trade that had sprung up "with the other countries of the Pacific [shore of the Americas], and with China, Australia, and the islands of the ocean" had "such extended ramifications" that San Francisco would continue to prosper even if gold should cease to be produced in California.[99] It was an accurate forecast.

94. Quoted in Meany, "History of the Lumber Industry," p. 115.
95. Olympia *Pioneer and Democrat*, 18 Nov. 1854, 3, 10 Mar. 1855.
96. Coy, *Humboldt Bay Region*, pp. 131–32, 224–25; Arcata *Humboldt Times*, 4 Oct., 1 Nov. 1856.
97. In addition to sources cited previously, see *Hunt's Merchants' Magazine* 41 (1859):52.
98. Quoted ibid., 42 (1860):616.
99. Ibid., 41 (1859):43.

CHAPTER VI

Rise of
the Giants

THE lumber industry of the Pacific Coast continued to grow through the 1860s and 1870s. However, some mills and some areas prospered more than others. By the 1880s the industry had come to be dominated by a few large firms whose manufactories were mainly located in the vicinities of Puget Sound and Humboldt Bay. Small mills existed, both in these locations and elsewhere, but they catered to a different sort of market. The cargo trade, as the maritime lumber trade of the Far West was generally known, was dominated by the giants.

The mills that rose to prominence in the cargo trade did so largely because of advantages derived from their locations. Their relative success might thus have been anticipated. Yet, even in the most favored locations, the survival and expansion of firms established during the Gold Rush and nurtured through the mid- and late fifties by the timely development of markets around the Pacific Basin were not assured. These firms had been established, for the most part, with capital from California, but the amount available to them was often inadequate for either immediate or long-range needs. Some of the mills suffered from poor management, others from faulty equipment and ill fortune. Moreover, even after the most speculative aspects of the lumber business disappeared with the passing of the Gold Rush, the industry continued to be one of high risks. Millmen could never be certain their enterprises would flourish. Not all did.

Henry Yesler's mill in Seattle, the first steam sawmill in what was to become the state of Washington, is an excellent case in point. The building of the mill was hailed by the editor of the Olympia *Columbian*: "It would be folly to suppose that the mill will not prove as good as a gold mine to Mr. Yesler. . . ." [1] Yet, a gold mine it was not. Yesler had liquidated his holdings in Massillon, Ohio, his former home, and taken into partnership Arthur Denny and George F. Frye in order to raise money to construct a mill in Seattle. Though these measures gave him enough capital to build the mill in 1852, they apparently failed to furnish enough to cover operating expenses as well. In 1856 Yesler and his partners sold a one-third interest in their operation to Adams, Blinn and Company of San Francisco for $5,000. The firm from San Francisco—and perhaps Yesler himself—anticipated that, with this additional money, the mill would prosper. For their part, Adams, Blinn and Company apparently expected soon to be receiving regular shipments from the mill in Seattle and, on this expectation, entered into contracts to furnish lumber to buyers in California. [2]

They were overly optimistic. The equipment in Yesler's sawmill was faulty, and he himself had no experience in the lumber business. The mill still did not prosper. In February 1858, while Adams, Blinn and Company were in need of millstuff to meet contract obligations, Yesler's plant was shut down because Yesler had failed to acquire an adequate supply of sawlogs to carry the mill through the winter. One of the partners in the San Francisco firm commented, "Yesler has disappointed us in nearly everything . . . but he has probably done the best he could under the circumstances." [3]

By 1861, unable to keep his mill running with the regularity that the cargo trade demanded, Yesler began to direct most of his efforts at the domestic market of the Seattle area, which could be supplied with lumber on a more irregular basis than the cargo outlets and required less operating capital of the millowner catering to it. Adams, Blinn and Company divested itself of its share of the business soon thereafter. In 1863 Yesler again tried to compete in the cargo trade. Adams, Blinn

1. Olympia *Columbian*, 30 Oct. 1852. For an account of Yesler's mill, see Finger, "Seattle's First Sawmill."
2. J. McClane to Yesler, 12 Aug. 1851, Henry Yesler Papers, University of Washington library, Seattle; Clarence B. Bagley, *History of Seattle from the Earliest Settlement to the Present Time*, 3 vols. (Chicago: S. J. Clarke Publishing Co., 1916), 2:723; Thomas Gedosch, "Seabeck: The Story of a Company Town, 1856–1886" (master's thesis, University of Washington, 1967), pp. 2, 16.
3. Adams, Blinn & Co. to Washington Mill Co., 9 Mar. 1858, WMCP. See also Adams, Blinn & Co. to Washington Mill Co., 7 Dec. 1857, 24 Feb. 1858; Adams Blinn & Co. to Williamson & Co., 16 July 1857, WMCP.

and Company's share had passed to Captain George H. Plummer, who had both lumber interests in San Francisco and vessels of his own; but, once again, Yesler's attempt to gain a permanent place in the lumber trade of the Pacific failed.[4]

Much of the trouble stemmed from his plant. Yesler once complained to Plummer: "I think it is rediculas for any set of men to run a saw mill at so great a disadvantage as ours. The same Expence it takes to run this mill, would if there was more Power, do one half more and in that would consist the Proffit." [5] Yesler was soon talking of remodeling the mill and adding new boilers. There "is no sense in fixing up, unless we increase the power," he wrote. Yesler tried to reassure Plummer, to whom he was already in debt: "Just Imagine the same amount of men and expence to saw 4 or 5 M feet more each day, and you can form some Ide[a] of the advantage [to be] gained. . . ." [6] At other times Yesler talked of building an entirely new mill.[7] Remodeling commenced in November 1864 and was not completed until mid-February. "We have had the D——st bother fixing up the mill," Yesler wrote. He estimated that it took three times longer than it should have.[8] With the improvements completed, the mill was capable of sawing fifteen thousand feet per day.[9]

Yet the improvements failed to better Yesler's circumstances. He had been short on operating capital all along and was now more deeply in debt than ever. To cut expenses he tried to have goods shipped to the mill only on the vessels Plummer had brought to the partnership and, even more important, to avoid purchasing goods on the sound. Unfortunately, Yesler was not always able to do so. Having to send "to other stores for our supplies is taking the *life* right out of us," he once complained, adding that "the *clear proffit* on our lumber [that this leaves] is a mear nothing." [10] In addition, Yesler found it difficult to get tonnage to handle his mill's cut. Captains apparently preferred to take

4. Yesler, Denny & Co. to M. Blinn, 10 Oct. 1861, WMCP; Gedosch, "Seabeck," p. 16; Yesler, Denny & Co., merchandise lists, Yesler Papers. The latter show that Yesler dispatched at least one shipload of lumber during this second attempt at penetrating the cargo trade, this in March 1863 aboard the *Iconium*.
5. Yesler, Denny & Co. to G. Plummer, 31 Oct. 1863, Yesler Papers.
6. Ibid., 22, 29 Aug. 1864. M is used to indicate one thousand; MBF, indicating one thousand board feet, is widely used in the lumber industry. See also 5, 19 Sept., 17 Oct. 1864.
7. Ibid., 15 May, 1 June, 30 Aug. 1863.
8. Ibid., 13 Feb. 1865. See also 28 Nov. 1864, 6 Mar. 1865.
9. Ibid., 17 Oct. 1864, 6 Mar. 1865.
10. Ibid., 10 Oct. 1865. See also 15 May 1865.

their craft to mills that gave them better terms or faster dispatch, or that were more apt to give them full cargoes than was Yesler's small plant. At one point, with his wharf full of lumber, Yesler wrote in desperation: "If you do not send a vessel soon I fear the wharf will break down." [11] There are also indications that Yesler failed to keep adequate records and was careless in managing the business.[12] When added to the other handicaps under which his mill was operating, this was enough to insure that Yesler's enterprise would not prosper.

Perhaps because of these financial problems, the partnership between Henry Yesler and George Plummer was a stormy one. By 1866 Plummer was seeking to buy out Yesler. The latter thought Plummer's offer too low, but he was in a poor position to bargain. He owned only a third of the mill and lacked both vessels of his own and outlets in San Francisco. However, Yesler managed to obtain sufficient funds to buy out Denny and Frye and thus, with the controlling interest in the business, was able to counter Plummer's offer.

But Yesler's indebtedness to the captain could not be repaid if the mill did not prosper. Only by keeping his vessels running to the mill could Plummer have any assurance that Yesler would make enough money to be able to repay him. Once Yesler had done so, he would be in a position to buy out the San Franciscan's share. When the debt was finally paid off, Yesler, still short of capital, offered to buy Plummer's third of the business, with payments in lumber.[13] By this time, Yesler had built a second, larger mill and was deeply in debt again.[14] To salvage his investment before disaster might strike his chronically overextended partner, Captain Plummer accepted Yesler's offer. Once his partnership with Plummer was terminated, Yesler abandoned the Pacific lumber trade and concentrated on sales in Seattle and its environs. By 1868 Henry Yesler had given up his attempts at finding a regular place in the cargo trade, though occasionally a vessel bound for San Francisco still loaded at his mill.[15]

11. Ibid., 10 Oct. 1865. See also 31 Oct., 6 Dec. 1863, 25 Apr., 20 June 1864; Yesler to G. Prescott [May 1868].
12. Yesler to Plummer, 15 Feb. 1864, 15 June 1866, Yesler Papers.
13. Ibid., 28 Jan. 1866; Yesler to A. Denny, 17 Aug. 1867; Yesler to Prescott, 19 Jan., 3 Feb., 13 June, 10 Aug. 1868; Yesler and Plummer, agreement dated 19 Aug. 1868, Yesler Papers. See also Yesler to Plummer, 15 May 1863, 1 Aug. 1864, ibid.
14. In 1876 Yesler put up this second mill as grand prize in a lottery designed to bring in money for the building of a railroad to Seattle. See John R. Finger, "Henry Yesler's 'Grand Lottery of Washington Territory,'" *PNQ* 60 (1969):121–26.
15. See, for example, Seattle *Weekly Intelligencer*, 7 Mar. 1870, 17 July 1871.

The sawmill erected by Henry Yesler had failed to live up to its initial promise. In the process it undoubtedly helped direct San Francisco merchants with investment capital away from existing mills and toward the erection of new plants that they would themselves control.

Yesler's was not the only pioneer sawmill that failed to win a permanent share of the lumber trade of the Pacific. J. J. Felt had a mill across the sound on Appletree Cove that commenced operation only a few days after Yesler's mill began sawing.[16] Both it and the mill that Captain William Renton and C. C. Terry erected on Alki Point the next year soon closed, however, and the equipment was moved to other sites. Neither location had been well suited for a sawmill. Arthur Denny later commented: "It now seems strange that men of such marked intelligence and experience as they possessed could have overlooked and passed by such superior locations as Madison and Blakely. I suppose it was on the theory that Puget Sound is all a harbor and it was not necessary to be particular. . . ." [17] Thus, the mills on Appletree Cove and Alki Point failed to gain a permanent place in the cargo trade, but for a reason quite different from that which precluded Yesler's success.[18]

The most spectacular failure of all was that of the mill at Port Orchard. After a winter of great difficulties, Captain William Renton moved his sawmill from Alki Point to Port Orchard. At first the operation was fairly successful in its new location, keeping busy both with the sawing of lumber and shipbuilding; but in 1857 Renton was seriously injured in a boiler explosion and returned to San Francisco to recover his health. In 1862 the firm of Coleman and Falk bought the mill. Two years later J. M. Coleman took over the superintendency of

16. There is some disagreement in early accounts as to whether the mill was built by Felt or someone else. Most of the more reliable sources give the credit to Felt. For a summary of the evidence, see Buchanan, "Economic History of Kitsap County," pp. 83–84. Felt, who had an interest in several vessels plying between Puget Sound and San Francisco, had the mill in operation by 4 Apr. 1853. See Olympia *Columbian*, 9 Apr. 1853.

17. Arthur A. Denny, *Pioneer Days on Puget Sound*, ed. Alice Harriman (Seattle: A. Harriman Co., 1908), pp. 57–58. The failure of Felt's mill may not have been entirely due to its location. The wife of an employee rejoiced when her husband left Felt's employ, for "Mr Felt was so important, overbearing & terribly lazy withal. . . ." L. Smith to H. Huntting, 29 Sept., 4, 5 Oct. 1856, Smith Papers.

18. Meany, "History of the Lumber Industry," pp. 101–2; Snowden, *History of Washington*, 4:351–53; Olympia *Columbian*, 9 Apr., 21 May 1853; Hubert Howe Bancroft, *Chronicles of the Builders of the Commonwealth: Historical Character Study*, 7 vols. (San Francisco, 1891–92), 4:616, 621–22.

the mill. In 1868 Coleman and A. K. P. Glidden, to whom Falk's interest had passed, completely rebuilt the mill. They went deeply into debt to finance the improvements, but the result seemed to justify the expenditure. The Port Orchard mill was now considered the equal of any on the coast. It had three large boilers, two engines, circular saws, a planer, an edger, and a lath-making machine. The plant began sawing in late 1868 and by early 1869 was running at full capacity.[19]

Unfortunately, the market for lumber was down and Coleman and Glidden, heavily in debt, were unable to keep the mill running. It shut down in June, and the firm filed notice of bankruptcy soon thereafter. In an attempt to salvage something from the debacle, Glidden apparently resorted to chicanery. The plant, which had cost $40,000 to build, was leased to one of Glidden's clerks for $100 a month. Logs belonging to the company were sold to the same clerk for about half their value. The lease was quickly invalidated when this deception became known, and arrangements were then made to sell the mill and property at public auction to satisfy the claims of the creditors.

Glidden opposed the idea of selling at auction. Times were not good in California, he argued, and the mill was not apt to sell for anything near its worth. However, creditors pushed for the sale, arguing that, if it were postponed, bills would continue to mount and there would soon be little left for anyone. In December someone attempted to blow up the mill, but the damage was not extensive. Then, on the morning before the auction was to be held, the mill was destroyed by fire. An attempt had been made to drug the watchmen at the mill, probably by those who set the blaze. The county sheriff quickly seized the ruins for back taxes. When sold, they brought only $520. Port Orchard's days as a lumber port were over.[20]

Other pioneer operations were more successful. From these many of the giant firms that were to dominate the cargo trade for decades emerged. These included the Port Madison mill, the Port Blakely mill,

19. Bancroft, *Chronicles of the Builders*, 4:622–24; Seattle *Weekly Intelligencer*, 7 Sept., 21 Dec. 1868, 11 Jan. 1869; Bagley, *History of Seattle*, 1:228; Buchanan, "Economic History of Kitsap County," pp. 89, 237–38; W. B. Seymore, "Port Orchard Fifty Years Ago," *WHQ* 8 (1914):258. One assumes the plant was run efficiently, for *West Shore* later described Coleman as "the best saw-mill man on the coast, if there is one better than another." *West Shore* 7 (1881):59. Glidden's reputation was less favorable. See J. Smith to H. Huntting, 25 Sept. 1865, Smith Papers; L. Smith to H. Huntting, 2 Oct. 1865, ibid.
20. Buchanan, "Economic History of Kitsap County," pp. 238–39, 279–80; Seattle *Weekly Intelligencer*, 28 June, 6 Sept., 27 Dec. 1869, 7, 14, 21 Mar. 1870; Seymore, "Port Orchard Fifty Years Ago," p. 259.

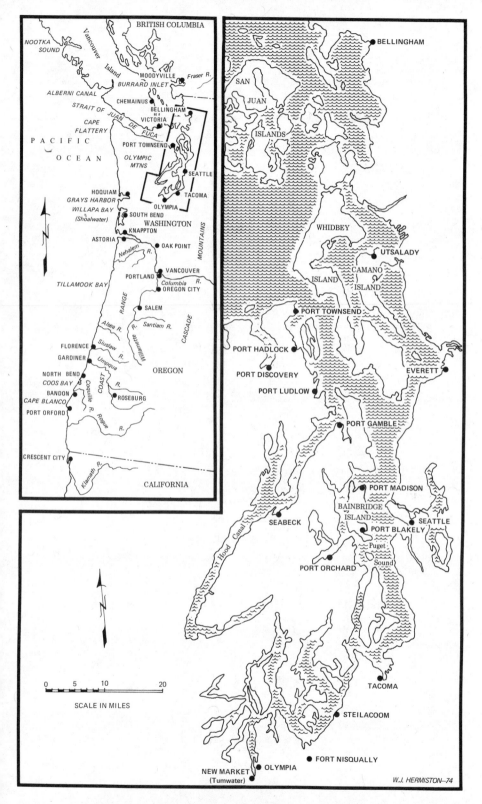

Map 2. Lumber Centers of the Pacific Northwest prior to 1900

the Seabeck mill, and Pope and Talbot's mill located at Port Gamble.

The mills at Port Madison and at Seabeck both suffered through financial crises; however, unlike Yesler's mill at Seattle and the mill at Port Orchard, they were not forced out of the cargo trade. When J. J. Felt's mill on Appletree Cove proved to be poorly located, its owners had it dismantled and moved to Port Madison on Bainbridge Island. Shortly afterward, George A. Meigs bought, enlarged, and improved the plant. Meigs spent most of his time at Port Madison, leaving his lumberyard in San Francisco in care of William H. Gawley. Both enterprises prospered. Gawley turned out to be a hard worker and was eventually rewarded with a partnership.[21] By 1858 Meigs had developed a capacity of fifteen thousand feet per day at the mill at Port Madison and had begun making at least occasional shipments to foreign markets. By February 1859 he had a foundry in operation in conjunction with the mill. According to Meigs, the foundry lost money but was kept in operation in order to protect against the greater losses from long stoppages when the mill broke down, for the nearest facilities for repairs were in San Francisco.[22] A shipyard was soon added. Meigs continued to add to and improve his plant, so that by 1860 it was reportedly the largest gang mill on the sound.[23]

Then came a severe setback. On 18 February 1861 a boiler explosion made a complete wreck of the mill. Meigs had the debris cleared away, and within a short time the mill was back in production; however, in his own words, he was "damned hard up" as a result of the accident and the cost of rebuilding.[24] In 1864 fire destroyed the mill. The loss was set at $100,000. Once more, Meigs had the mill rebuilt and its capacity enlarged. As after the disaster of 1861, Meigs found himself in tight financial circumstances, but once again he was able to survive them.[25]

21. L. Smith to H. Huntting, 29 Sept., 4, 5 Oct. 1856, 2 Oct. 1865, Smith Papers; J. Smith to H. Huntting, 25 Sept. 1865, ibid.
22. G. Meigs to J. Williamson, 24 Jan. 1860, WMCP; Buchanan, "Economic History of Kitsap County," pp. 84–85. The foundry did work for other mills around the sound, too.
23. Wright, ed., *Lewis and Dryden's Marine History*, pp. 27–28; Olympia *Pioneer and Democrat*, 25 Feb. 1859. For the early history of this mill, see Buchanan, "Economic History of Kitsap County," pp. 84–85, 172–75, 259–60.
24. Meigs to M. Blinn, 28 Nov. 1861, WMCP. See also Meigs to M. Blinn, 18 Mar. 1861, WMCP; Olympia *Pioneer and Democrat*, 22 Mar. 1861; Bancroft, *History of Washington, Idaho, and Montana*, p. 338; Hunt and Kaylor, *Washington West of the Cascades*, 1:184; Buchanan, "Economic History of Kitsap County," pp. 260–61.
25. Buchanan, "Economic History of Kitsap County," p. 261; G. Haller to M. Blinn, 19 Oct.

Serious though they were, these setbacks did not stop the mill's growth. During the first six months of 1870 the mill, running day and night, cut and shipped 11,872,000 feet of lumber, in addition to such other items as lath and pilings. The company's fleet, also expanded, now totaled some forty-five thousand tons. Additional vessels were chartered. During September alone eight vessels loaded over four million feet of lumber at Port Madison. In 1871 the plant was improved by the addition of four new boilers. Though the lumber market was down, the mill at Port Madison kept busy filling large contracts with the Central Pacific Railroad.[26]

The company might have been able to weather the depression of the seventies without difficulty had Gawley, by now a partner, not speculated with funds taken from the company. The depression created circumstances from which he was unable to extricate himself. The company itself was nearly bankrupted by Gawley's pilfering and inability to repay. However, Meigs made arrangements to carry on the business as the Meigs Lumber and Shipbuilding Company. Because of the depression, the mill did not run at full capacity through the mid-seventies, but gradually Meigs was able to get the firm back on its financial feet. By the fall of 1877 one newspaper was describing the sawmill as probably the strongest on the coast.

Once again, expansion was undertaken. The additions were expensive, however, and the company was soon back in financial difficulty. Conditions grew worse rather than better, and the mill was finally sold at a sheriff's auction in 1881. The mill, under its new owner, William Sayward, continued to operate during the eighties. At the end of the decade it was still one of the largest on the sound, in spite of all the reverses it had suffered. With its fleet of lumber carriers, the mill was busily engaged in shipping forest products to markets throughout the Pacific Basin.[27]

1866, and D. Finch to M. Blinn, 8 Nov. 1866, WMCP. The capacity of Meigs's mill at this point has been reported as 125 M per twenty-four hours and 80 M per day [twelve hours?]. Seattle *Weekly Gazette*, cited in Buchanan, "Economic History of Kitsap County," p. 261; Wright, ed., *Lewis and Dryden's Marine History*, p. 142.

26. Seattle *Weekly Intelligencer*, 29 Aug., 19 Sept. 1870, 10, 24 Apr., 7 Aug. 1871; Buchanan, "Economic History of Kitsap County," pp. 261–62; Clark, "Analysis of Forest Utilization," pp. 39, 68–69. Shipyards became a major adjunct of the lumber industry at an early date. See Buchanan, "Economic History of Kitsap County," pp. 95, 173–89; Wright, ed., *Lewis and Dryden's Marine History*, p. 62; Bagley, *History of Seattle*, 2:617; Coy, *Humboldt Bay Region*, pp. 219–20; Bancroft, *History of Washington, Idaho, and Montana*, p. 328.

27. Bancroft, *History of Washington, Idaho, and Montana*, p. 338; Hunt and Kaylor, *Washington*

In 1856 four residents of San Francisco joined forces to form the lumber marketing firm of Adams, Blinn and Company.[28] That same year the new firm purchased a one-third interest in Henry Yesler's sawmill in Seattle; but when Yesler failed to meet their needs, the owners of Adams, Blinn and Company moved to establish a sawmill of their own. In 1857 they formed the Washington Mill Company and began construction of a plant at Seabeck on the arm of Puget Sound known as Hood Canal. William Adams stayed in San Francisco to manage the parent firm, which was to serve as both sales and purchasing agent for the Washington Mill Company. Marshall Blinn departed for Puget Sound to direct the construction and operation of the sawmill.[29] For the next three decades Adams, Blinn and Company and the Washington Mill Company were to be major factors in the cargo trade. However, as at Port Madison, the operations were plagued by recurring financial problems. These commenced almost as soon as the mill was completed.

The Seabeck mill began cutting lumber in the fall of 1857, and the first reports forwarded to San Francisco were encouraging. Adams welcomed the news and, since the market for lumber was good for the moment, urged Blinn to run the plant night and day.[30] These halcyon days were not to continue. Difficulties with the sixty-six-inch circular saw that served as the mill's head rig, as well as other problems, kept the plant from furnishing an adequate supply of timbers to keep construction moving on a bridge that Adams, Blinn and Company had contracted to build near Sacramento. Yesler, too, was unable to supply the needed timbers. Adams found it necessary to purchase construction materials on the retail market and from manufacturers on the redwood coast—both costly alternatives.

West of the Cascades, 1:184; Bagley, *History of Seattle*, 1:228; Buchanan, "Economic History of Kitsap County," pp. 262–69; Seattle *Weekly Intelligencer*, 26 Apr. 1873, 16 Mar. 1878; *West Shore* 8 (1882):182; Portland *Oregonian*, 1 Jan. 1886. See also Buchanan, "Economic History of Kitsap County," pp. 280–84. Later Sayward engaged in the sawmill business in Victoria. See *Canada Lumberman* 13 (Apr. 1892):8.

28. The four were William J. Adams, Samuel P. Blinn, Marshall Blinn, and James N. Prescott. Marshall Blinn, as master of the *Brontus*, had hauled lumber from Puget Sound in 1855. Adams was apparently the William J. Adams who had headed the ill-fated banking firm of Adams & Co. which had failed following the flight of Henry Meiggs from San Francisco. One wonders if this new firm was established with some of the money that vanished when Adams & Co. failed. See Bancroft, *History of California*, 7:176ff.

29. Gedosch, "Seabeck," pp. 1–4.

30. Adams, Blinn & Co. to Washington Mill Co., 22 Oct., 6 Nov. 1857, WMCP.

Like many early firms, Adams, Blinn and Company was hampered by a shortage of working capital. Much of what it did have was tied up in the bridge project, and there was no prospect of return on the investment until the bridge was completed. Adams hoped that sales from the firm's lumberyard would bring in enough to meet day-to-day expenses, but cash sales were few. In addition, Blinn had difficulty keeping the yard stocked with the dimensions in demand in San Francisco. Even when the yard was full, Adams sometimes found it necessary to buy lumber on the retail market in order to fill orders for dimensions not in stock. The rush to the newly discovered gold fields on the Fraser River also hurt the fledgling firm: much of its business had been in Sacramento, and, as thousands left that area bound for Canada, sales fell off. As one carpenter in Sacramento concluded, "There is no use staying here; there is little or no business of any kind owing to the large number leaving for the new diggings. . . ." Having heard that carpenters were earning from forty to one hundred dollars a day on the Fraser River, he joined the rush northward.[31]

Short on cash and with his credit nearly used up, Adams found it necessary to stop sending supplies to Seabeck and to stall off creditors. With supplies no longer arriving from the south, Blinn had to buy needed goods at retail prices from the store of Pope and Talbot at Port Gamble. This alleviated immediate problems, but it also worsened the firm's financial situation. In January 1858 Adams had to resort to paying freight bills in interest-bearing notes rather than cash. The following month, Adams was forced to suspend bridge construction because he could no longer afford to purchase building materials in California.[32]

Gradually Adams worked his way out of this financial crisis. He obtained contracts to furnish material to the naval yard at Mare Island and to the Pacific Mail Steamship Company, as well as for construction of a railroad bridge across the American River near Folsom. He sold lumber for shipment to Valparaiso, Melbourne, Hong Kong, and Amoy. Cargoes were also dispatched to Victoria to capitalize on the demand that had developed there as a result of the rush to the Fraser River mines.[33] In addition, a revival of lumber sales in California during

31. McCulloch to Pumyea, 12 June 1858, Samuel McCulloch Papers, Henry E. Huntington Library, San Marino, Calif. See also San Francisco *Alta California*, supplement, 1 Jan. 1859.
32. Adams, Blinn & Co. to Washington Mill Co., 16 Oct., 6 Nov. 1857, 18, 21 Jan., 6 Feb., 26 Apr., 17 June, 20 Oct., 30 Dec. 1858, WMCP.
33. The impact of the Fraser River gold rush on the lumber industry of Puget Sound is partially

the last half of 1858 aided Adams in his efforts to extricate Adams, Blinn and Company and its affiliate from their financial difficulties.[34] Indeed, as early as June 1858 Adams had begun to exude confidence. He wrote Blinn: "We are crying . . . an awful dull prospect ahead for lumber business to such men as Balch and others, but they do not affect our trade much, although they sell a dollar or two less than we do our lumber has a little the best reputation of any that is brought here and we have already got the best trade of any one in the business, Meigs not excepted." [35] Adams' confidence was not unwarranted. His firms, which only a short time before had appeared in danger of bankruptcy, had by the end of the 1850s clearly weathered the crisis.[36]

Difficulties other than financial plagued the companies. Andrew Pope was unrelenting in his efforts to keep Adams and his associates from cutting into the business of Pope and Talbot.[37] Sea captains complained of the poor dispatch that they had received at Seabeck. Logging operations on Hood Canal were "behind the times" and sometimes failed to keep the mill adequately supplied. Also, according to one contemporary observer, there was disagreement among the partners. Perhaps because of this, or perhaps only to free himself for his campaign for the position of territorial delegate to Congress, Marshall Blinn retired from management of the mill in 1869. Samuel Blinn took over many of the duties that had been handled by his brother, but he was in ill health, and the responsibility soon shifted to Richard Holyoke, the mill's first hired manager.[38]

---

revealed by the fact that at its height few cargoes were dispatched anywhere but to Victoria. See U.S. Collector of Customs, Puget Sound, "Register of Entrances and Clearances," vol. 1 *passim*.

34. Adams, Blinn & Co. to Washington Mill Co., 6, 7 Nov. 1857, 7 June, 10, 14 July, 20 Aug. 1858, 18 Mar. 1859, WMCP; Thomas Prosch scrapbook, WMCP, *passim*; McCulloch to Pumyea, 2 Oct. 1858, McCulloch Papers. Blinn wrote in connection with the contract with the Mare Island Naval Yard: "We consider ourselves fortunate in getting the contract . . . although it has cost us quite a little sum to get it as we had to buy off a man whose bid was below ours. This however you had better say nothing about." Adams, Blinn & Co. to Williamson & Co., 16 July 1857, WMCP.

35. Adams, Blinn & Co. to Washington Mill Co., 11 June 1858, WMCP.

36. For an account of the history of the two firms during this period, see Gedosch, "Seabeck," pp. 6–14.

37. See above, chap. 5.

38. Adams, Blinn & Co. to Washington Mill Co., 2 Mar., 18 Apr. 1859, WMCP; Gedosch, "Seabeck," pp. 16–17, 23–24; Edward Clayson, Sr., *Historical Narratives of Puget Sound: Hood's Canal, 1865–1885: The Experience of an Only Free Man in a Penal Colony* (Seattle: R. L. Davis Printing Co., 1911), p. 8. Clayson claimed Holyoke ran the company town as if it were a penal colony, exploiting the men and forcing them to struggle against one another to make a bare living. It is unclear just what Clayson meant by these charges, but Buchanan tends to dismiss them as biased, arguing that the mill was apparently no worse than others insofar as such things as wages

In spite of all their problems, the two enterprises prospered. They had started out owning only two vessels, *Brontus* and *Blunt*, and had found it necessary to ship much of the cut of the mill in chartered bottoms. By 1869 six sailing vessels and the tug *Colfax* had been added to the company fleet. Moreover, the mill was enlarged from its original capacity of 20,000 board feet per day to 33,000 board feet in 1859, 40,000 board feet in 1863, and 47,000 board feet in 1867. It is not clear what the earnings of either company were during these years, but they apparently were considerable.[39]

In spite of Adams' attempts to sell lumber abroad almost from the beginning, foreign sales were neither regular nor numerous. Adams was loath to dispatch cargoes as ventures carried on the company's account, though this was the likeliest way of developing new outlets. Perhaps such ventures seemed too risky, or perhaps Adams hesitated simply because they would tie up capital for too long a time with no certainty of profits. To be sure, Adams sold cargoes abroad in the 1860s. In 1862 and 1863 he dispatched lumber to Shanghai. The Yangtze River had just been opened to foreign trade and many Europeans and Americans foresaw a great commercial future for Shanghai, which was the logical entrepôt to the entire river basin. Old firms in Shanghai expanded, and new ones hastened to the scene in an effort to reap as large a share as possible of the anticipated business.[40] The result was a building boom that attracted shipments of lumber from almost every quarter of the Pacific. Ports from which lumber had not been exported hitherto, such as Nagasaki and Hakodate in Japan, joined the established lumber ports in dispatching building materials to Shanghai.[41] Adams' shipments were but a small part of the total. When

and charges at the company store went (p. 253). In actuality, of course, the life of the common laborer was grim throughout the industry.

39. Gedosch, "Seabeck," pp. 17–18 and *passim*; Kitsap County Treasurer to S. Blinn, 11 Nov. 1869, WMCP. Cf. Wright, ed., *Lewis and Dryden's Marine History*, pp. 77–78, 142. The new vessels were the ship *Isaac Jean*, the barks *Carlotta* and *Fremont*, and the brigs *Tanner*, *Florence*, and *Jennie Pitts*. *Brontus* and *Blunt* were still company vessels.

40. Prosch scrapbook, WMCP, *passim*; Shanghai *North China Herald*, 12 Jan. 1861, 31 Jan., 21 Feb. 1863; Shanghai Mercury, *Shanghai: The Model Settlement*, p. 23. The new firms that began to appear were a mixed blessing. For critical comments on them, see Dennett, *Americans in Eastern Asia*, pp. 579–80, 597; Tientsin *Chinese Times*, 12 Mar. 1887.

41. Shanghai *North China Herald*, 14 Sept., 19 Oct. 1861, 24 May, 12, 19 July 1862, 18 Apr., 18, 25 July 1863, 28 May, 23, 30 July, 3, 10, 17 Sept., 12 Nov. 1864. Shipping reports in the *North China Herald* clearly reveal the increase in building that attended the opening of the Yangtze. Cargoes of forest products entering the port each year numbered as follows: 1851, 1; 1852, 2; 1853, 0; 1854, 0; 1855, 6; 1856, 2; 1857, 0; 1858, 12½; 1859, 15; 1860, 5; 1861, 23; 1862, 35; 1863, 47; 1864, 97; 1865, 50; 1866, 97½; 1867, 87½; 1868, 61.

the brief boom in Shanghai ended, the shipment of lumber there from Seabeck quickly ceased. Lumber from the Washington Mill Company was also sent to Peru, Chile, Australia, Hong Kong, and even Tahiti during the 1860s, but shipments were intermittent and of far less importance to the Washington Mill Company than those to San Francisco.[42]

The primacy of San Francisco is hardly surprising. Adams lived there and had established customers in the area. Beyond that, San Francisco was prospering during the 1860s. The discovery of huge deposits of silver ore on the slopes of Sun Mountain in 1859 set off a new mining boom which, coupled with San Francisco's increasing commerce and industry, brought the economic activity in the city to a level that had not been seen since the Gold Rush. In 1864, when the production of the silver mines topped $16,000,000 for the first time, over one thousand new buildings were erected in San Francisco.[43] Between 1 October 1868 and 9 August 1869, the sole portion of the decade itemized in the surviving records of the Port Townsend Customs House, twenty-two cargoes of lumber went to San Francisco from Seabeck. All other markets combined took only twenty-seven, even though this was when the demand in Shanghai was near its peak.[44]

By the 1870s, however, much of the capital of San Franciscans was being drawn off into speculation in mining stocks. As investment in capital improvements declined, the lumber business dropped off. When the prices of mining stocks tumbled in June 1872, wiping out millions of dollars of paper profits, the lumber industry had already been in the doldrums for months. Though there was a brief recovery in mid-decade, the lumber market remained dull and prices low through most of the seventies.[45]

42. Prosch scrapbook, WMCP, *passim.* Much of the lumber shipped to San Francisco passed on through that port to Sacramento and other interior points.
43. Z. S. Eldredge, ed., *History of California*, 5 vols. (New York: Century History Co., [1915]), 4:223–52, esp. 251–52; William H. Brewer, *Up and Down California in 1860–1864: The Journal of William H. Brewer* . . . , ed. Francis P. Farquhar, 3rd ed. (Berkeley: University of California Press, 1966), pp. 496–500; Oscar Lewis, *San Francisco: Mission to Metropolis* (Berkeley, Calif.: Howell-North Books, 1966), pp. 110–15; Bancroft, *History of California*, 7:649–52, 687.
44. U.S. Collector of Customs, Puget Sound, "Register of Clearances," vol. 5 *passim.* Four-fifths of the lumber shipped to San Francisco at this time was rough. San Francisco *Alta California*, 12 July 1866.
45. Eldredge, ed., *History of California*, 4:491–99; Lewis, *San Francisco*, pp. 147–54, 160; Bancroft, *History of California*, 7:161–63, 667–80, 688–89; Seattle *Daily Intelligencer*, 23 Apr. 1877.

In spite of their restricted market base, Adams, Blinn and Company and its affiliate on Puget Sound easily weathered the slump. Renewed efforts were made to penetrate foreign markets and were met with a degree of success. Shipments were also made to ports along the coast of southern California. Though no permanently important outlets were developed, sufficient quantities were sold to prevent having to close down or completely glut the San Francisco market. In 1872 nearly 50 percent of the lumber shipped from Seabeck went to ports outside the Bay Area.[46] Results may not have been all that the mill's owners wished, but they were sufficient to keep the enterprise growing through the decade. Improvements expanded the daily capacity of the mill to over fifty thousand board feet, while such capacious lumber carriers as the bark *Cassandra Adams* and the ship *Olympus* went down the ways of the shipyard at Seabeck to join the company fleet. The enterprises regularly returned profits above and beyond those plowed back into the business.[47]

In spite of its shaky beginnings, the Washington Mill Company had emerged by the end of the 1870s as an established and apparently financially sound member of that small circle of firms that dominated the cargo trade of Puget Sound and, together with a handful of mills elsewhere, of the whole Pacific Coast. The company played an active role in the cooperative attempts of the owners of large firms to bring stability into the risky, highly competitive lumber business when these efforts were undertaken during the late seventies and early eighties.[48] This and other surface manifestations must have made it appear to contemporaries that the mill at Seabeck was destined to continue to play a major role in the Pacific lumber trade for years to come. Such was not the case.

During 1883–84 a new mill with a capacity of one hundred thousand board feet per day was added at Seabeck. The company went deeply into debt to finance the new plant. Unfortunately for the owners, many other mills also expanded during these years. By the time

46. Prosch scrapbook, WMCP, *passim.* Gedosch, "Seabeck," p. 19, estimates that 30 percent of the mill's cut was marketed abroad, but he merely counted the number of cargoes to each destination and made no allowance for the fact that the vessels going foreign were generally larger than those in the coastwise trade.
47. Gedosch, "Seabeck," *passim.* See also E. Ferry to Washington Mill Co., 30 July 1870, 7 Aug. 1875, WMCP; *Marine Digest* 19 (19 July 1941):2. According to *Lewis and Dryden's Marine History,* p. 272, the *Olympus* "could sail like a yacht carrying an immense cargo."
48. See chap. 11, below.

demand returned to normal in 1885 and 1886, the productive capacity of the lumber industry was so great that few mills were able to run full time. Under the circumstances, the Washington Mill Company found it impossible to work off its indebtedness. The loss of some of the company-owned vessels, only partly insured, added to the difficulties. When the Seabeck mill burned in 1886, it was not rebuilt. Instead, the owners chose to buy the mill at Port Hadlock, a recently constructed plant of one hundred thirty thousand feet capacity that failed to achieve a position of lasting importance in the Pacific lumber trade. The days of Seabeck as a lumber port were over.[49]

Of all the pioneer lumber manufactories on Puget Sound, by far the most successful was the Puget Mill Company, formed by Pope, W. C. Talbot, and their associates in 1852. The sawmill that they established at Port Gamble enjoyed a history of remarkably steady growth and expansion.[50] The firm got an early start in the lumber trade and repeatedly plowed profits back into the business to keep the advantage "we have always had," as Pope put it.[51] By 1862 the Puget Mill Company owned a fleet of ten vessels and was shipping a total of nearly nineteen million board feet of lumber a year to foreign and domestic ports. The company's trade was the largest on the sound. To accommodate the business, a second mill and new docking facilities had been added in 1857. Other improvements followed in regular succession. In 1869–70 the old mill erected in 1853 was torn down and replaced by an enlarged and improved plant that more than doubled the company's capacity, bringing it to one hundred sixty thousand feet per day. By 1875 the company was shipping forty-three million feet annually, and the quantities were continuing to rise. Seven years later the company owned a fleet of at least sixteen sailing vessels and four tugs, between one hundred and one hundred fifty thousand acres of timberland, and, in addition to their two plants at Port Gamble, mills at Utsalady and Port Ludlow. The combined capacity of the four mills was three hundred thirty-five thousand board feet per day.[52]

49. *West Shore* 12 (1886):44; W. Chase to Adams, 1 Aug. 1883; Washington Mill Co. to Adams, 11 Aug. 1883, 22 Jan., 3 Feb. 1884; R. Kendrick to R. Holyoke, 14 Feb., 6 Mar. 1884; Kendrick to Adams, 5, 8 Apr. 1884, WMCP.

50. Indeed, according to local tradition, the slab fire started in 1855 never went out until more modern methods of using waste were adopted in 1925. See E. L. Riddell, "History of Port Gamble," Poulsbo *Kitsap County Herald*, 17 July 1931.

51. A. Pope to Wm. Pope & Sons, 18 Nov. 1857, PTP, Pope file.

52. *West Shore* 7 (1882):183; Coman and Gibbs, *Time, Tide and Timber*, pp. 103–28, 361–63 and *passim*; Buchanan, "Economic History of Kitsap County," pp. 85–89, 248, 250.

The Utsalady and Port Ludlow mills had both been present for years before they were acquired by the Puget Mill Company. Thomas Cranney had been operating at Utsalady, on Camano Island, at least as early as 1855. Spars were his specialty, and sales were made to such exacting customers as the English, French, and Spanish navies. In 1862 sales included ten cargoes that were shipped abroad. But in spite of such sales, the mill was forced to close down in 1875. Attracted by Utsalady's fine harbor, its pivotal location on the sound, and its closeness to the timber stands along the Skagit River, Pope and Talbot decided to bid on the plant at the impending bankruptcy sale. They got the mill and some timber on Camano Island for $32,000—about half their actual value according to contemporary accounts.[53]

The mill at Port Ludlow was also a pioneer operation, having been built in 1853 by William Sayward. In 1858 the firm of Amos and Phinney of San Francisco rented the plant and increased its capacity to thirty-five thousand feet per day. In 1866 control passed to Phinney, who conducted the business as the Port Ludlow Mill Company. In 1877, while he was in the midst of having the sawmill rebuilt, Phinney died, and the operation was put up for sale. "I think we have mills enough," Pope argued; but Talbot, who was in a more expansive mood, prevailed. On 7 November 1878 the Puget Mill Company acquired the Port Ludlow mill for $64,850.[54]

More than chance was responsible for the vitality that enabled the Puget Mill Company not only to expand but to absorb others. One factor was the caution that Pope and Talbot applied in the company's financial affairs. They tried to conduct business on a cash basis and to keep the firm out of debt. Unlike many of their competitors, they were not wont to go so deeply into debt for mill improvements or other purposes that the entire operation might be endangered by an untimely fire or unforeseen collapse of the market. Moreover, funds from the enterprises of the Pope and Talbot families in Maine were always available to the Puget Mill Company and affiliated firms if the need should arise.[55] These turned out to be of little importance, however; in

53. Coman and Gibbs, *Time, Tide and Timber*, pp. 110–11, 364–65; Olympia *Puget Sound Weekly Courier*, 5 May 1876, cited in Buchanan, "Economic History of Kitsap County," p. 248; Wright, ed., *Lewis and Dryden's Marine History*, p. 113.

54. Pope to Talbot, 30 Sept. 1878, PTP, letter book 1. See also Coman and Gibbs, *Time, Tide and Timber*, pp. 110–11, 365–67; Buchanan, "Economic History of Kitsap County," p. 250; Portland *Oregonian*, 5 Jan. 1880.

55. A. Pope to Wm. Pope & Sons, 4 June, 2 July, 5 Oct., 18 Nov. 1857, PTP, Pope file; A. Pope to "brother" [Edwin?], 19 July 1877, PTP, letter book 1; Coman and Gibbs, *Time, Tide*

fact, funds earned on the Pacific Coast were transferred eastward to buttress the faltering family businesses in Maine and an ill-fated milling operation that relatives purchased in Quebec. The enterprises on the West Coast supported those in the East, not the reverse. The greatest problems faced by Pope and W. C. Talbot during the formative years of their empire in lumber came not from a lack of capital but from fragmented ownership and encumbering ties with family operations elsewhere. The steady progress of the Puget Mill Company continued for years after these problems were removed as a result of reorganization in the early 1860s.[56]

More than the soundness of the financial underpinnings of the Puget Mill Company needs to be taken into account, however, if one is to understand how the firm achieved its position of leadership. Both Pope and Talbot were shrewd and capable businessmen. In Cyrus Walker, who became superintendent of the operations at Port Gamble in 1862, they had a dedicated and effective ally. Each man was valuable in himself; moreover, the talents of each complemented those of the others. No firm on the sound had a better management team than that of the Puget Mill Company. By the time of Pope's death in 1878 and Talbot's in 1881, the Puget Mill Company was in a position of dominance which other members of the families were long able to maintain.[57]

Another major factor in the firm's success was the broad market base that it serviced. As has been seen, during its first year of operation over a third of the cut of the mill at Port Gamble was shipped to foreign ports. The percentage of sales abroad continued to be high in the years that followed. In 1857 some 5.5 million board feet out of a total cut of 7.9 million feet were dispatched to foreign ports. This represented more than half of all the lumber exported from the entire West Coast that year.[58]

---

and Timber, pp. 132–36 and *passim*. The main affiliates during the period covered by this chapter included Pope & Talbot and the Puget Sound Commercial Co. Several other affiliates were formed during the 1890s. For a corporate history, see Coman and Gibbs, *Time, Tide and Timber*, pp. 368–429.

56. Coman and Gibbs, *Time, Tide and Timber,* pp. 90, 97–101; A. Pope to Wm. Pope & Sons, 29 July 1850, 4 Nov. 1857; A. Pope to E. Pope, 13 Dec. 1850; A. Pope to W. Pope, 3, 28 July, 13 Sept., 25 Nov. 1875, and other dates, PTP, Pope file; A. Pope to E. Pope, 14 Dec. 1876, PTP, letter book 1; Coman and Gibbs, *Time, Tide and Timber*, pp. 92–102.

57. Coman and Gibbs, *Time, Tide and Timber*, pp. 141ff. For a revealing contemporary assessment of Pope, see Oscar T. Shuck, *Sketches of Leading and Representative Men of San Francisco* ([London], 1875), p. 897.

58. A. Pope to Wm. Pope & Sons, 4 Nov. 1857, PTP, Pope file; Coman and Gibbs, *Time, Tide and Timber*, p. 437; *Hunt's Merchants' Magazine* 38 (1858):476–77. The latter source indicates

Pope, who was in charge of lumber sales, aggressively pursued new outlets for lumber. Unlike William Adams of the Washington Mill Company, he did not hesitate to dispatch ventures to unfamiliar markets. When such a cargo was sent, the captain of the vessel was instructed to sell to an agent if possible. By this and other means the captains of company-owned vessels helped to lay the groundwork for continued sales.[59] Similarly, when others discovered areas where lumber could be sold, Pope quickly took advantage of the knowledge. Soon after John Kentfield began making regular shipments to southern California from the mill at Port Discovery, Pope moved to win the trade from Kentfield and others who entered the market.[60] In time southern California was to become a major outlet for the Puget Mill Company.

Pope and his partners established W. C. Talbot and Company in 1855 in order to provide a vehicle for sales, as well as such related activities as arranging for charters and purchasing supplies. When they resurrected the name of Pope and Talbot in renaming the company in 1862, its general functions, and Pope's role, remained the same.

In the penetration of new markets abroad, Captain William C. Talbot also played a key role. As master of one of the company's vessels, Talbot often accompanied venture cargoes dispatched abroad. The captain not only worked harder to assure the financial success of such voyages than the captains of chartered vessels might have, but also, as a partner in the firm he represented, could carry on negotiations with would-be customers with a degree of authority and freedom others would have lacked.

One such effort that was particularly effective involved the Hawaiian Islands. In 1855 the first cargo of Puget Mill Company lumber was dispatched to H. Hackfeld and Company of Honolulu. Hackfeld became a regular customer, and Pope and Talbot lumber was still being shipped to Hackfeld's successors, American Factors, in the mid-twentieth century.[61] Pope shipped to other buyers in Hawaii in addition to Hackfeld. So successful were the efforts of Pope and Talbot that in 1878 the Tacoma *Herald* reported that the Puget Mill Company

that the Pope and Talbot enterprises shipped 2,755,000 feet to Australia, 1,350,000 to the Hawaiian Islands, 875,000 feet to Hong Kong, and 530,000 feet to Chile during 1857. On 1858, see Wright, ed., *Lewis and Dryden's Marine History*, p. 77.

59. For example, see A. Pope to Wm. Pope & Sons, 2 July 1857, PTP, Pope file.
60. McIntyre to Kentfield, 11 Feb. 1875, JKCP, 1:2.
61. Coman and Gibbs, *Time, Tide and Timber*, pp. 74–77, 79–80, 82–85.

"enjoyed a monopoly of the Sandwich Island trade." The statement was an exaggeration, but the position of Pope and Talbot was indeed formidable.[62] When C. Brewer and Company, one of the handful of firms dominating Hawaiian business, ordered lumber for one of their customers from the mill at Port Discovery, they asked that it be kept quiet "as it might make trouble with Pope & Talbot who have formerly supplied them." [63]

Pope attached great importance to these foreign markets. He strove to keep other firms from entering them and undermining the Puget Mill Company's position. His efforts must have rankled other lumbermen besides Adams, who complained that Pope hated to see any firm other than his own sell lumber through foreign outlets. Unlike firms that paid little attention to foreign sales unless demand was lagging on the domestic market, Pope attempted to keep his foreign contacts in order and foreign demand supplied even when they gave little promise of immediate gain. Pope cultivated these markets knowing that someday they would be vital to the company.[64] The policy was continued after he and William C. Talbot passed from the scene. In 1887, Pope's successor, William H. Talbot, wrote to Cyrus Walker noting that he was refusing orders every day, but that foreign demand should "fall off considerably soon and then we can do more for the coast business." [65] The broad market base that such a policy helped to make possible, together with the Puget Mill Company's sound financial position and capable executive leadership, lay behind the firm's rise to a position in the lumber trade of the Pacific that was to remain unsurpassed through the rest of the nineteenth century.

Some of the mills established on Puget Sound during the mid-fifties failed to prosper. Others, such as those at Seabeck, Port Madison, and Port Gamble, survived and grew. However, not all the mills destined

62. Tacoma *Herald*, 17 May 1878, quoted in Buchanan, "Economic History of Kitsap County," p. 249. Shipping reports in the *Pacific Commercial Advertiser* for 1874–79 indicate lumber cargoes from the Columbia River and Humboldt Bay, as well as the mills of Pope & Talbot, were arriving in Honolulu regularly.

63. C. Brewer & Co. to J. Kentfield & Co., 13 Oct. 1877, JKCP, 2:7. See also J. Emerson to Kentfield, 12, 17 Nov. 1877, ibid.

64. Adams to Washington Mill Co., 12 Mar., 8 Sept. 1858, WMCP; Adams to M. Blinn, 8 Dec. 1858, WMCP; Coman and Gibbs, *Time, Tide and Timber*, pp. 82–83.

65. W. H. Talbot to C. Walker, 3 June 1887, AC, Pope & Talbot, incoming corres. See also 18 June 1887, and Pope & Talbot to C. Walker, 22 Dec. 1888, AC, Walker, incoming corres.; McIntyre to Kentfield, 18 Jan. 1876, JKCP, 2:3; Comer to Kentfield, 5 Aug. 1876, ibid. William H. Talbot was the older of William C. Talbot's two sons.

to become members of the small circle of giants that dominated the lumber industry of Puget Sound had their beginnings in the fifties. Some did not put in their initial appearance until the following decade. The most notable of these were the Port Blakely Mill Company and the Tacoma Mill Company.

The sawmill operation at Port Blakely followed a pattern of growth remarkably similar to that of the mill at Port Gamble. In 1863 Captain William Renton, having recovered from the injuries suffered when the boiler of his old mill at Port Orchard exploded, returned to Puget Sound to re-enter the lumber business. By April of the following year Renton and his partner, Daniel S. Howard, had a mill in operation at Port Blakely near the southern end of Bainbridge Island. Initially, the plant had a daily capacity of fifty thousand board feet. It expanded steadily. In 1866 Renton and Howard acquired three vessels to aid in hauling the mill's cut. The fleet, too, grew steadily, especially after the Hall brothers moved their shipyard to Port Blakely from Port Ludlow. Reorganization took place in 1870, following Howard's death, and again in 1876, but in neither case did the changes seem to slow the growth of the operation.[66] By 1881 the mill had a daily capacity of two hundred thousand feet, and growth was continuing. The following March, *West Shore* described it as the largest single mill on the sound.[67]

The growth of the enterprise at Port Blakely resulted from factors similar to those behind the rise of the Puget Mill Company. First, the firm was soundly financed; unlike many early mills, the Port Blakely Mill Company had access to enough capital to be able to expand facilities without serious risk of bankrupting the firm. Moreover, the financial policies of the company's management were remarkably conservative for years. Even the firm's first logging railroads were financed on a cash basis.[68]

Second, as was the case with the Puget Mill Company, the Port Blakely Mill Company appears to have had a management team that was better than most. A reporter for the Seattle *Daily Intelligencer* noted, "Capt. Renton, the guiding spirit and good genius of the place, is always on hand early and late, directing every department. . . ."

66. In 1876 the mill was incorporated as the Port Blakely Mill Co. Sales, purchases, and other transactions were handled through Renton, Holmes & Co. in San Francisco.
67. Buchanan, "Economic History of Kitsap County," pp. 269–74; *West Shore* 8 (1882):55, 183; Seattle *Weekly Intelligencer*, 6 Apr. 1878. Cf. John Hittell, *Commerce and Industries of the Pacific Coast*, p. 593; Bancroft, *History of Washington, Idaho and Montana*, p. 339.
68. Richard C. Berner, "The Port Blakely Mill Company, 1876–1889," *PNQ* 57 (1966):168.

Everything about the place suggested "solidity, system, thrift, and economy." [69] While keeping the Port Blakely Mill Company in sound financial condition, Renton and his associates also managed to obtain the necessary orders to keep the plant busy. More than once, the mill at Port Blakely was running near capacity, and night shifts were even being added, while others were curtailing their cut.[70] The management also displayed a high degree of foresight. Realizing that the gradual disappearance of waterside timber stands would soon make necessary fundamental changes in the lumber industry around Puget Sound, Renton and his partners moved at an early date to acquire extensive stands away from the water that could be tapped by railroad. They were among the vanguard in building logging railroads. This early adaptation to changing conditions paid rich dividends in the long run.[71]

Third, the Port Blakely Mill Company serviced a broad marketing base, though it was somewhat slower than the Puget Mill Company in moving in this direction. Renton concentrated his initial efforts on gaining a secure place in the San Francisco trade. With this won, Renton and his new partner, Charles S. Holmes, who played a major role in marketing activities, began during the mid-seventies to seek out new outlets for the cut of their mill. In 1875 large quantities of railroad ties were sent to South America. Renton and Holmes moved soon thereafter to try to wrest some of the markets in the interior of California from Pope and Talbot, which had been dominant there, and opened yards at Vallejo and Alameda to further this endeavor.[72] At about the same time, they began to dispatch cargoes to Honolulu in an attempt to crack the near-monopoly the Puget Mill Company had on the island trade. Both efforts were successful. The Port Blakely Mill Company was too large and too prosperous to be blocked out of these markets the way Adams and his fledgling Washington Mill Company had been kept out of the foreign trade in the late fifties.[73] As further

69. Seattle *Daily Intelligencer*, 28 Nov. 1877. See also 13 July 1877.
70. Ibid., 29 Aug. 1870, 17 July 1871, 14, 21 Apr. 1877. The firm's aggressive merchandising practices earned it the disapprobation of E. G. Ames. He wrote, "I will do the best I can to secure . . . orders [for the Puget Mill Co.], but will make no dirty marks such as Hadlock, Discovery, and Blakely are constantly." See Ames to C. Walker, 16 Apr. 1890, AC, Walker, personal corres. Cf. Renton, Holmes & Co. to Pt. Blakely Mill Co., 16 Sept. 1901, PBMCP, 65:2.
71. Berner, "Port Blakely Mill Company," pp. 161–64.
72. Ibid., p. 164. For a description of the Alameda yard see Munro-Fraser, *History of Alameda County*, p. 410. As demand in southern California expanded during the next few years, yards were also established at Santa Barbara, San Buenaventura, and San Pedro.
73. Once in the Hawaiian market Renton, Holmes & Co. joined with Pope & Talbot in trying to exclude interlopers. Their task was made easier by Hawaiian business practices. A few large firms dominated business in the islands and each refused to compete for the others' recognized

steps in their efforts at broadening the market base of their mill, Renton and Holmes also shipped to southern California, South America, and Australia. In an attempt to develop a market in Buenos Aires, Holmes agreed to a charter to that port at less than market rates.[74]

Collectively these new outlets reduced the fluctuation in demand that had previously been a problem, albeit surmountable, for the company. The new markets not only made it possible to keep the mill busy while demand was low in San Francisco, but also provided a means of keeping the company fleet active during slack periods. With its broad market base, as well as its sound financial position and capable management, the Port Blakely Mill Company had developed along lines quite similar to those of the Puget Mill Company.[75]

The Tacoma Mill Company followed a rather different pattern. As a boy, Charles Hanson, the principal owner of the firm, had run away from his home in Denmark to go to sea. Still a young man, by the mid-fifties he had acquired a sloop of his own and was engaged in ferrying lumber and shingles from Redwood City to San Francisco. It was there, not in the Northwest, that he first attracted notice. During the rush to the Fraser River gold fields, Hanson cornered the shingle supply of the San Francisco Peninsula, and a good share of its lumber supply as well, by staying in California and buying up the stocks of operators who were joining in the rush to the new diggings. When prices went up with declining production, Hanson earned a substantial profit.

---

customers. Similarly, once a mill had the business of one of these firms, it could expect to retain it. In 1894 an outsider tried to open Hawaii to competition by dispatching a cargo to Honolulu on his own account. After the lumber had lain unsold for several weeks, dealers there "took pity on the 'skirmisher'" and purchased his cargo at $5 a thousand, which was less than the cost of shipment. See *Canada Lumberman* 16 (May 1895):13; T. Comer to Kentfield, 19 Dec. 1876, JKCP, 2:3; C. Brewer & Co. to J. Kentfield & Co., 8 Oct. 1877, JKCP, 2:7; Castle & Cooke to J. Kentfield & Co., 9 June 1879, JKCP, 3:7; Renton, Holmes & Co. to Pt. Blakely Mill Co., 30 Oct. 1901, PBMCP, 65:4. See also JKCP, boxes 16–22 *passim*, for extensive records of the trade with Hawaii.

74. Buyers in southern California and Latin America generally bought many railroad ties and construction timbers. Such orders used the whole log and thus did not leave large quantities of difficult-to-market, low-grade stock in the mill yard. Until later this was not the case with orders from Australia, but this market too was worth cultivating for, while vessels normally returned in ballast from South America and southern California, vessels returning from down under could bring return cargoes of coal.

75. Buchanan, "Economic History of Kitsap County," pp. 269–76; Berner, "Port Blakely Mill Company," pp. 164, 166–67; Seattle *Pacific Tribune*, 13 Aug. 1875; Olympia *Puget Sound Weekly Courier*, 27 Nov. 1875; Bancroft, *Chronicles of the Builders*, 4:624–25. Other mills were also shipping abroad. According to one authority, seventy-two cargoes of lumber were carried to foreign ports during 1877 on sixty-five different vessels. See Wright, ed., *Lewis and Dryden's Marine History*, p. 255.

He continued to increase his lumber interests in the years that followed. By the mid-sixties, his firm of Hanson, Ackerson and Company was dominant in the production and shipping of lumber in the area of Redwood City. The productive capacity of the peninsula was limited, however, so Hanson began to look outside the Bay Area for new opportunities. In 1864 he began receiving on consignment cargoes of lumber from a newly constructed mill at Gardiner, near the mouth of the Umpqua River in Oregon. Four years later he purchased the plant outright, only to sell it three years later, perhaps because the possibility of expansion on the Umpqua seemed too limited. Hanson then concentrated his efforts on a new, larger plant that he had built shortly before on Commencement Bay near the southern end of Puget Sound.[76]

Hanson's Puget Sound mill shipped its first cargo of lumber on 8 December 1869.[77] During 1870, nineteen more followed this initial shipment. Most went to San Francisco, but the Tacoma Mill Company soon began to service foreign markets as well. By March 1883, 425 cargoes had been carried from the mill. Of those whose destination can be determined, eighty-one—approximately 20 percent—went to foreign ports (not including Victoria, British Columbia). If one assumes that the vessels going abroad averaged 50 percent larger than those going to San Francisco, then some 33 percent of the mill's cut was marketed abroad. The percentage may well have been even higher. Only two foreign markets were of major significance to the Tacoma Mill Company: Central and South America, which received fifty cargoes; and Australia, which received twenty-five.[78]

When the Tacoma Mill Company began cutting, it had a daily capacity of forty thousand board feet, large for a new operation. This was increased frequently in the years that followed. Prior to a major expansion of the plant in 1874, Hanson's chief partner, John W. Ackerson, visited several of the largest mills in the country. Ideas gleaned during the observation of these mills were applied to the mill in

76. Stanger, *Sawmills in the Redwoods*, pp. 14–19; *West Shore* 4 (1878):38; Roseburg (Ore.) *Plaindealer*, 11 June 1899. For more on the Umpqua River operation, see below, chap. 8.

77. Stanger, *Sawmills in the Redwoods*, p. 19. Just five days before, a plat of the town of Tacoma had been filed in the Pierce County auditor's office. The new city was, as *West Shore* put it, "a small town of few inhabitants, which had grown up around and was chiefly dependent on the saw mill of Hanson, Ackerson & Co." *West Shore* 9 (1883):1.

78. Stanger, *Sawmills in the Redwoods*, p. 19; Tacoma Mill Co. lumber account book, MS, University of Washington library, Seattle.

Tacoma. It quickly became one of the most modern on the sound and in the estimation of one Seattle observer, "the best worked." [79] By 1882 the company owned thirty thousand acres of timberland, four large sailing vessels, the largest tug in the Northwest, and a plant that was "almost constantly" being expanded.[80] Improvements installed in 1883 brought the capacity to two hundred twenty-five thousand feet per day, making it one of the largest anywhere. Hanson continued to live in the Bay Area and to cut lumber on the San Francisco Peninsula; but the Tacoma Mill Company had a capacity greater than all the mills of San Mateo County combined. In time, Hanson was actually shipping lumber *to* Redwood City from his operations in the north.[81]

The advantages of the modern plant of the Tacoma Mill Company were for a time partially offset by its location, which was difficult and expensive to reach by sea. As one competitor put it, "If you should blind fold a man and put him in a boat . . . whare ever he landed on the shore . . . would be as well located for a mill as Tacoma is." [82] At the time the assessment had some truth in it, but years later the mill's location at the terminus of the Northern Pacific Railroad was to prove a boon to the firm.

For all the superficial differences among the mills that had come to dominate the cargo trade of Puget Sound by the beginning of the 1880s, these operations shared certain basic characteristics. Not a one had been built with capital generated in the Northwest: all were erected with money from the Bay Area and were controlled from there in the years that followed. As Richard Berner has observed, from its beginnings the Puget Sound lumber industry was "dominated by San Francisco marketing and shipping interests." [83] All the successful mills had offices in San Francisco to market their cut and take care of other business that could not readily be conducted from Puget Sound. All opened their own lumberyards in the bay city (and sometimes

79. Seattle *Daily Intelligencer*, 2 Feb. 1878. See also Seattle *Weekly Intelligencer*, 21 Mar. 1870; *West Shore* 4 (1878):38.
80. *West Shore* 7 (1882):183.
81. Ibid., 9 (1883):2; Stanger, *Sawmills in the Redwoods*, pp. 19–23.
82. G. Stetson to W. H. Tolbert [sic], 1 May 1888, PTP, Talbot file. See also W. Schelay to J. Kentfield & Co., 28 Oct. 1875, JKCP, 1:7.
83. Berner, "Port Blakely Mill Company," p. 158. See also *West Shore* 8 (1881):53; Portland *Oregonian*, 1 Jan. 1882; W. H. Hutchinson, "California's Economic Imperialism: An Historical Iceberg," in *Reflections of Western Historians*, ed. John A. Carroll (Tucson: University of Arizona Press, 1969), pp. 67–83; J. C. Lawrence, "California's Influence on the Industrial and Commercial Development of British Columbia, 1858–1885," paper presented to annual meeting of Pacific Coast Branch, American Historical Association, Santa Clara, Calif., 29 Aug. 1968.

elsewhere, as well), acquired their own fleets of lumber carriers and tugs, and sought to remain competitive by enlarging and modernizing their mills whenever circumstances would permit. As the size of their fixed investments grew, the need for market stability increased. In time the cargo mills sought to reduce fluctuation in demand by broadening their market base. Though not all began as participants in the international lumber trade, all ended up that way.

Production facilities were also remarkably similar. All were steam-powered plants with very nearly the latest equipment. All, if not located in company-owned towns, at least were serviced by company-owned stores and other enterprises affiliated with the parent operation.[84] All began by depending upon independent loggers, but moved in time toward acquiring their own timber stands and toward running their own logging operations.[85] Mills that were unable to adapt their operations to this overall pattern disappeared or, at least, were forced to abandon the cargo trade and concentrate on local sales. Those that did not had, by the 1880s, made "Oregon pine" known in the lumber markets of the entire Pacific Basin.

Not all the giant sawmills that had appeared on the West Coast by 1880 were located on Puget Sound. Along the redwood coast, and especially around Humboldt Bay, a handful of large enterprises had also risen to dominance. In spite of the very different topography and resource base with which these mills had to deal, the pattern of their operations was strikingly similar to that which developed on the sound.

Natural obstacles made development of giant firms in the redwoods slower than on Puget Sound. Though logs from many acres of timberland could be floated to the mills surrounding it, Humboldt Bay was a pale shadow of the sound. Once the timber along the shores of the bay had been cut, it became necessary to tap inland stands and float the logs out to the coastal mills by stream or, later, haul them out by railroad. The necessity of adopting these more expensive alternatives arose sooner in California than in Washington Territory. Floating logs out of the woods depended on freshets. When they did not come, mills frequently had to close for lack of logs; and when they did, it

---

84. For descriptions, see James B. Allen, *The Company Town in the American West* (Norman: University of Oklahoma Press, 1966), pp. 14–32.
85. One of the best descriptions of such an operation is Berner, "Port Blakely Mill Company." See also Portland *Oregonian*, 1 Jan. 1881, 1 Jan. 1882; *West Shore* 10 (1884):124; Coman and Gibbs, *Time, Tide and Timber*, pp. 361–67; Clark, "Analysis of Forest Utilization," pp. 52–58.

sometimes rained so heavily as to halt all work. Even with the advent of railroads, none of the redwood mills was completely free from concern over its log supply.[86]

As a result of such factors, redwood lumber tended to be more expensive to produce than Douglas fir. Manufacturers had to ask higher prices than their northern competitors and thus had difficulty penetrating markets for low-cost construction materials. The specialty markets for which redwood was peculiarly suited were as yet too small to provide much relief. In addition, whenever there was a dry winter, the small mills of Santa Cruz County and along the Russian River were able to run during months they were normally shut down.[87] They then flooded the market in San Francisco with their cut. Being closer to the Bay Area, their transportation costs were less and the larger mills on the northern redwood coast were further pressed. Thus, resistance at both the supply and demand levels caused the development of firms engaged in the manufacture of redwood lumber to proceed more slowly than did the development of the Puget Mill Company and other giant firms to the north.[88]

Nonetheless, large firms did develop along the redwood coast. By 1865 the eight largest mills in Humboldt County were producing 12.6 million board feet of lumber annually. The biggest was the firm of Kentfield, Buhne and Jones, thanks largely to the aggressive marketing practices of John Kentfield.[89] Production increased rapidly in the years

86. D. Jones to J. Kentfield, 2 and 11 Feb. 1873, JKCP, 1:1; 14 Oct. 1882, JKCP, vol. 26; C. Nelson to J. Kentfield, 12 Feb. 1873, JKCP, 1:1; J. Henderson to Kentfield & Co., 8 June 1876, JKCP, 2:1; C. Nelson to Kentfield & Co., 12 Dec. 1878, JKCP, 3:6; D. R. Jones & Co. to Kentfield & Co., 24 Oct. 1852, 16 and 21 Feb., 12 May 1883, JKCP, 1:26; Weidberg, "History of John Kentfield & Co.," pp. 20–21.

87. For a description of one such small operation, see Phyllis Bertorelli Patten, *Oh, That Reminds Me,* ed. Elizabeth Spedding Calciano (Felton, Calif.: Big Trees Press, 1969), pp. 1–11.

88. *West Shore* 6 (1880):98–99; 8 (1882):182. Portland *Oregonian,* 1 Jan. 1886. California State Mineralogist, *Second Biennial Report,* 1880–82 (Sacramento, 1882), pp. 210–16. California State Board of Forestry, *Fourth Biennial Report,* 1891–92 (Sacramento, 1892), p. 51. Eureka *Humboldt Times,* 4 Nov. 1865, 8 Jan. 1876, 16 Jan., 4 June 1878, 5 Apr. 1888. Coy, *Humboldt Bay Region,* pp. 218, 260–62. Melendy, "One Hundred Years of the Redwood Industry," pp. 148, 159–60, 177–85, 223, 237–48, 281. Melendy has stated (p. 160) that from "1875 on the redwood industry commenced to become big business, causing the formation of joint stock companies for sufficient capital to insure success." The process began at least a decade earlier on the sound.

89. Weidberg, "History of John Kentfield & Co.," pp. 5–29, 71; C. Nelson to J. Kentfield, 12 Feb. 1873, JKCP, 1:1; Coy, *Humboldt Bay Region,* p. 218; Wilde, "Chronology of the Pacific Lumber Company," p. 11; Palais and Roberts, "History of the Lumber Industry in Humboldt County," pp. 8–9. Both the size of the largest mills and the total production consistently lagged behind those on Puget Sound. For statistics on production on the sound, see Coman and Gibbs, *Time, Tide and Timber,* pp. 436–37.

that followed. In 1869 the Kentfield operation alone cut a total of 11.6 million board feet, while that of John Vance cut 8 million and Dolbeer and Carson 7.6 million. Three years later sagging demand kept the mills from running at full capacity, yet Kentfield's still cut 10.5 million feet and the Vance and Dolbeer and Carson mills 8.6 million feet apiece. Only one other mill in the county cut over 3 million feet that year, but by 1875 other major producers had developed. Buhne, Jones and Company (successors to Kentfield, Buhne and Jones) led production with a cut of 16 million feet, and Dolbeer and Carson held second place with 12.7 million feet; but the Occidental mill of Evans and Company and the mill of Russ, Pickard and Company had both passed John Vance's mill in production. It was an exceptional year for the lumbermen of Humboldt County. Seventy-seven million board feet were sawn, 60 percent more than in any previous year. By 1880 demand had eased and less than 40 million board feet of lumber were cut. As in previous years, the vast bulk came from a few large mills.[90]

Though demand slumped in 1881, John Kentfield and Company (the marketing agent for Buhne, Jones and Company) found itself unable to fill all its orders. To avoid becoming overstocked, John Kentfield regularly took more orders than could be filled by the mills of his associates and then he purchased lumber from outside mills to make up any deficits. In 1881, heavy rains kept woodsmen from getting out enough logs to keep the mill busy, and an absence of vessels kept much of what lumber was cut from reaching San Francisco. The firm weathered the crisis, however, and, like Pope and Talbot, profited from the sort of managerial caution revealed in this episode.[91]

In Mendocino and Sonoma counties the pattern was the same. A few larger mills turned out most of the lumber.[92]

By the seventies a pattern of business had emerged in the redwood industry that was quite similar to that on the sound. The success of the handful of companies that dominated production was in large part a

90. Coy, *Humboldt Bay Region*, pp. 260–62; Melendy, "One Hundred Years of the Redwood Industry," pp. 160, 237–48; California State Mineralogist, *Second Biennial Report*, p. 205; Wilde, "Chronology of the Pacific Lumber Company," pp. 11–13; California State Agricultural Society, *Transactions, 1877* (Sacramento, 1878), p. 146. Available evidence does not seem to support Melendy's statement that by 1875 the Occidental mill was the largest in the county. See Melendy, "One Hundred Years of the Redwood Industry," p. 160.
91. Weidberg, "History of John Kentfield & Co.," pp. 6–10.
92. Robert A. Thompson, *Historical and Descriptive Sketch of Sonoma County* . . . (Philadelphia, 1877), pp. 30–32; Menefee, *Historical and Descriptive Sketch Book*, pp. 276, 333; Ryder, *Memories of the Mendocino Coast*, p. 16.

result of their vertical integration, which meant they did more than produce lumber: they had their own ships, their own offices and yards in San Francisco, and their own stores to supply their crews. The records are not as clear as those for Puget Sound, but like the mills to the north, those on the redwood coast appear to have been developed not with capital generated locally, but with funds from San Francisco. Most were joint stock companies; the days of small, trial-and-error proprietorships were over. Apparently the largest on the redwood coast was the California Redwood Company, which by 1884 had two mills at Eureka and one at Trinidad Bay, two logging railroads, nine logging camps, and a complete machine shop. It employed no less than a thousand men.[93]

As was the case with the cargo mills of Washington Territory, San Francisco was the primary market of the larger firms on the redwood coast. They sold less there than did mills in the Northwest, but had a good portion of the market, nonetheless. During the decade of 1872 through 1881 these mills regularly shipped around 35 to 40 percent of the lumber that entered San Francisco Bay. However, though the percentage of the total market remained rather constant, the amount shipped varied widely. In 1873 only 70.6 million feet of redwood entered the bay; in 1875 the quantity jumped to 116.1 million feet. Of course, shipments from Puget Sound varied with fluctuations in demand, too.[94]

Important as San Francisco was, it was not the only market to which the burgeoning mills of the redwood coast sent their cut. Shipments to other ports in California began early. In 1861 San Pedro, San Luis Obispo, and Sacramento all received cargoes. Santa Barbara and San Buenaventura also received shipments during the decade. John Kentfield pursued the southern California market assiduously during the 1870s, dispatching not only numerous cargoes of redwood from the saws of Buhne, Jones and Company but Oregon pine from Port Discovery in Washington Territory as well.[95] Other redwood produc-

93. Edgar Cherry & Co., *Redwood and Lumbering in California Forests* . . . (San Francisco, 1884), pp. 69–72 and *passim*; Melendy, "One Hundred Years of the Redwood Industry," p. 160; California State Mineralogist, *Second Biennial Report*, p. 205; Palmer, *History of Mendocino County*, pp. 132–46, 246, 262, 381–84, 429–36, 470–71.

94. Redwood Manufacturers' Association, *Redwood Lumber: Statistics for the Port of San Francisco from 1862* (n.p., n.d.); California State Mineralogist, *Second Biennial Report*, pp. 210–16.

95. Coy, *Humboldt Bay Region*, pp. 220–21, 269; J. McIntyre to J. Kentfield, 11 Feb. 1875, JKCP, 1:1; 8 and 16 Apr. 1875, JKCP, 1:5.

ers competed for the trade, as the complaint of Perry Woodworth, a
leading lumber dealer in Los Angeles, makes clear. Woodworth wrote
that a cargo from Kentfield's associates at Buhne, Jones and Company
was "the worst of refuse being full of Rotten nots, not Holes and
Shakes. . . . [We] Cannot Sell Such trash against Such lumber as
Dober [*sic*] & Carson Sends here." [96]

Disgruntled customers were not the only problem Kentfield faced as
he sought to penetrate the markets of southern California. The ports of
the area were poor, facilities for landing lumber inadequate. The
captain of one of Kentfield's vessels wrote from Hueneme on the
Ventura County coast, "This is the last Place God Made for a Vessel
to kom." [97] Other ports presented problems, too. At Santa Barbara
insufficient dock space slowed the landing of lumber,[98] while at San
Pedro and Wilmington, gateways to the Los Angeles area, lumber
long had to be ferried ashore via lighters, which were too few to give
quick dispatch either in discharging the cargo or in taking on ballast for
the return north. Coastal steamers, which had first call on the use of the
lighters, added further complications. "It is not rush it out and in like
San Francisco Cargoes," Captain James McIntyre noted.[99] All the
southern California ports were open to storms and thus dangerous in
winter.[100] Yet, for all these complications, the area was not an
inconsiderable market for the growing mills of the redwood coast.

Redwood was also shipped abroad. The foreign markets to which
northern redwood had first been dispatched in the 1850s continued to
receive it in the sixties and seventies. Redwood was shipped to Tahiti,
the Hawaiian Islands, Chile, Peru, and perhaps Australia during the
sixties. These markets continued to receive shipments during the
seventies and, in addition, cargoes were dispatched to other markets
including New Zealand, Asiatic Russia, Shanghai, New York, and
England.[101]

96. P. Woodworth & Co. to J. Kentfield & Co., 15 Dec. 1874, JKCP, 1:3. See also C. Reel to J.
Kentfield & Co., 23 Dec. 1874; W. Allen to J. Kentfield & Co., n.d., JKCP, 1:3; P. Woodworth
& Co. to J. Kentfield & Co., 11 June 1875, JKCP, 1:6.
97. J. Emerson to [Kentfield], 28 Jan. 1875, JKCP, 1:4.
98. Emerson to [J. Kentfield & Co.], 21 June 1874, JKCP, 1:3; C. Pierce to J. Kentfield & Co.,
24 Jan. 1876, JKCP, 2:2. See also Pierce to J. Kentfield & Co., 5 Oct. 1876, JKCP, 2:5.
99. McIntyre to Kentfield, 12 Nov. 1874, JKCP, 1:2.
100. McIntyre to Kentfield, 12, 13, and 17 Nov. 1874, 18 Oct. 1875, JKCP, 1:2; McIntyre to
Kentfield, 16 Apr. 1875, JKCP, 1:5; P. Woodworth & Co. to J. Kentfield & Co., 18 Nov. 1874,
JKCP, 1:2.
101. California State Agricultural Society, *Transactions*, 1864–65 (Sacramento, 1866), pp.
383–86; *Transactions*, 1879 (Sacramento, 1880), p. 220; Coy, *Humboldt Bay Region*, pp. 220–21,

In all of these markets redwood faced stiff competition from Douglas fir and other woods. Even in Tahiti, where the rot- and insect-resisting qualities of redwood gave it a distinct advantage, penetration of the market was slow. A lumber dealer in Papeete complained in 1874 that redwood "lies unsaleable in our yard"; it could not be sold, he explained, so long as good Oregon pine was available at the same price.[102]

Yet sales continued to be made. California's state mineralogist reported in 1882 that during the first two years of the new decade redwood had been sent to England, Mexico, the Hawaiian Islands, and Australia, and that, of the forty million feet cut in 1881, over eighteen million were sent abroad. He predicted that from the redwood region "will continue to come a large portion of that shipped to foreign markets." Much of this must have been reshipped from San Francisco, for the records of Hans H. Buhne show that 534 of the 555 lumber cargoes that left Humboldt Bay were destined for San Francisco Bay.[103]

John Vance was one of the most active in cultivating the foreign trade. The *Humboldt Times* reported both that he nearly monopolized the Oriental markets for redwood and that his main foreign outlets were in Australia and Hawaii.[104] However, other large firms of the redwood coast were also shipping abroad. In 1873 and 1874, and possibly in other years as well, the amount some of them dispatched exceeded that shipped by Vance, so perhaps his claim to primacy should not be taken too seriously. Vance; Evans and Company; Dolbeer and Carson; Kentfield, Buhne and Jones (and its successors); and the larger mills on the Mendocino coast all shipped lumber to foreign markets.[105]

269; Melendy, "One Hundred Years of the Redwood Industry," p. 153; D. J. Flanigan, "California Lumber" (typescript account, Bancroft Library, Berkeley, Calif.).

102. Meuel, Smith & Co. to C. Nelson, 28 Sept. 1874, JKCP, 1:3. Cf. *Canada Lumberman* 16 (1895):13.

103. California State Mineralogist, *Second Biennial Report*, p. 205; Hans H. Buhne, Miscellaneous Papers (microfilm copies, Bancroft Library, Berkeley, Calif.), shipping records, 1887. See also Seattle *Daily Intelligencer*, 17 Jan. 1878.

104. Eureka *Humboldt Times*, 27 Jan. 1892, 5 Aug. 1904. For a brief sketch of Vance's career, see Melendy, "One Hundred Years of the Redwood Lumber Industry," pp. 169–70.

105. Eureka *Humboldt Times*, 23 Jan. 1875; Coy, *Humboldt Bay Region*, pp. 269–70. The increased marketing of lumber in the Hawaiian Islands during the last half of the 1870s came largely as a result of the economic boom experienced there following the treaty of reciprocity with the United States. See Honolulu *Pacific Commercial Advertiser*, 7 July, 19 Sept., 21 Nov. 1874, 17 Apr., 29 May 1875, 1 Apr., 26 Aug. 1876, 19, 26 May, 14, 21 July, 11 Aug., 20 Oct., 29 Dec. 1877, 5 July, 5 Oct., 14 Dec. 1878, 18 Jan., 8 Feb., 28 Apr. 1879.

In the 1870s even pine from the Sierra entered the Pacific lumber trade. Earlier high transportation costs had not only kept it out of the cargo trade, but had also kept it from capturing a significant portion of the business of San Francisco and Sacramento. According to one historian, of the ten million feet of lumber consumed in Sacramento in 1858, 95 percent was Douglas fir or redwood.[106] But by the 1870s this was changing. The completion of the California and Oregon Railroad to Chico in 1870 and to Redding in 1872, coupled with the adoption of V-flumes, made it possible to get lumber out from mills in the mountains more cheaply than ever before. Sierra pine was soon a common sight in the yards of Sacramento and even San Francisco.

With the advent of better transportation facilities and, through them, of wider markets, consolidation of the many small mills that had dominated production in the Sierra began. In November 1875 the Sierra Flume and Lumber Company was incorporated. This firm brought under its control 10 mills, 23 miles of logging tramways, 156 miles of flume, 250 miles of telegraph line, 3 planing mills, 2 sash and blind factories, 3 lumberyards, plus a main office and exporting agency in San Francisco. Twelve hundred men were employed in the firm's various operations. During its first year of operation the Sierra Flume and Lumber Company sold forty-three million feet of lumber. Its products were soon being shipped to markets around the Pacific Basin. The firm developed "Australian clear," a grade too pitchy for domestic use but welcome down under because it was less apt to fall victim to wood-eating insects than the ordinary clear.[107]

For all its early successes, the Sierra Flume and Lumber Company was not destined to play a permanent role in the lumber trade of the Pacific. By September 1877 the firm was in financial trouble. It had expanded too rapidly and, when two years of drought in the Sacramento Valley caused a sharp drop in demand, was soon in trouble with its creditors. In 1878 it was succeeded by the Sierra Lumber Company. Unlike its predecessor, the Sierra Lumber Company played no major role in the shipment of lumber by sea. It and other producers of pine lumber in the mountains of California continued to ship their cut to domestic markets to compete with Douglas fir and redwood.

106. Hutchinson, *California Heritage*, pp. 7–8.
107. Ibid., pp. 9–20; California State Agricultural Society, *Transactions*, 1877, pp. 143–44, 146–47; W. H. Hutchinson, "The Sierra Flume & Lumber Company of California, 1875–1878," *Forest History* 17 (1973):14–20.

Much went eastward to meet the burgeoning demands of the Comstock Lode and other mining centers, but for the rest of the nineteenth century Sierra pine was seldom seen in transit across the Pacific.[108]

That lumbermen all along the West Coast were shipping to foreign ports ought not obscure the fact that through most of the nineteenth century California, and especially San Francisco, remained their primary market. As new foreign outlets appeared and old ones grew larger, they often drew off grades of lumber that were in oversupply on the West Coast. They also brought an increased degree of stability to this high-risk industry by broadening the market base of the cargo mills. But, in spite of the importance of complementary demand and a broader market base, foreign markets never reached a point where their importance to the cargo mills outweighed that of markets in the area of San Francisco. Not only did the larger mills maintain lumberyards in the Bay Area, they also sold to independent yards, such as that of Z. B. Heywood and Company in West Berkeley. Collectively the independent yards serviced a large portion of the area's market.[109] Indeed, foreign markets alone could hardly have supported the cargo mills at anywhere near the level they reached during the sixties and seventies.

The troubled histories of the mills located on Alberni Canal and Burrard Inlet serve to illustrate San Francisco's importance. Being located north of the international boundary, mills on these anchorages had to set their prices low enough to remain competitive even after import duties were paid. Log prices and labor costs tended to be lower north of the border, but the differential was insufficient to make up for the tariff charges. By May 1868 an associate of Captain Edward Stamp, owner of the main mill on Burrard Inlet, admitted that earlier hopes of being able to compete in San Francisco had not been fulfilled. As

108. Seattle *Daily Intelligencer*, 7 Jan. 1878; *Lumberman's Gazette*, reprinted in *Canada Lumberman* 3 (15 June 1883):183; Hutchinson, *California Heritage*, pp. 21–26; Hutchinson, "The Sierra Flume & Lumber Company," p. 20. Most available statistics do not break down the lumber being exported from San Francisco Bay by species. Therefore, it is possible that more sugar pine was actually exported than available evidence seems to indicate. Clearly, Robert Dollar shipped sugar pine to Pacific markets during the 1890s. See *Canada Lumberman* 15 (Mar. 1894):7.

109. On Z. B. Heywood & Co., see Mary Johnson, "California History and Biography: Mr. Z. B. Heywood and Family" (MS, Bancroft Library, Berkeley, Calif.); Joseph Eugene Baker, ed., *Past and Present of Alameda County* . . . , 2 vols. (Chicago: S. J. Clarke Publishing Co., 1914), 2:422ff.

Henry Yesler reported, "Now he sais Stamp can't stand it—He is loosing money all the time. . . ." [110]

The Fraser River gold rush stimulated lumber production in British Columbia, but it was not until the 1860s that significant attempts were made to participate in the cargo trade. In 1857 Edward Stamp, an employee of one of England's largest timber brokers, visited Alberni Canal while his vessel was being loaded at Puget Sound. The captain was impressed by the timber on the canal and told his employers of it upon his return to England. The outbreak of the Civil War opened the possibility that England might be cut off from sources on the East Coast upon which it had been drawing. To guarantee a continuing supply, Stamp's employers dispatched him to Vancouver Island to begin exploitation of the Alberni stands.

The Alberni Mill Company quickly gained a position of prominence in West Coast lumber circles. The London office got orders for the mill from dockyards of the French, Spanish, and Sardinian governments. The mill also sent cargoes to Australia, Chile, the Hawaiian Islands, China, and, occasionally, to San Francisco. At its peak the operation employed seven hundred men. However, when the abnormally high demand in Shanghai vanished in the mid-sixties (the mill had shipped several cargoes to that port) and the East Coast once again began to furnish lumber and spars to European dockyards as it had before the war, the mill on Alberni Canal was deeply in trouble. Gilbert M. Sproat, who replaced Stamp as manager of the operation, complained that the decision to build on the canal had been a mistake, for the stands there were inadequate to support a large mill. The success of later operations on the canal would seem to contradict this claim. The real problem appears not to have been a shortage of sawlogs but the tariff barriers that generally kept the firm's lumber out of San Francisco and left it without markets large and dependable enough to keep it running during slack times as well as flush. Before the decade was over, the mill closed, never to reopen. [111]

---

110. Yesler & Co. to G. Prescott [May 1868], Yesler Papers. See also Carrothers, "Forest Industries of British Columbia," p. 256. *Canada Lumberman* 3 (1 Oct. 1883):290; 9 (July 1889):1.

111. Carrothers, "Forest Industries of British Columbia," pp. 256–59, 265; Lawrence, "Markets and Capital," pp. 7–8, 19–23; Robert Brown, *The Countries of the World: Being a Popular Description of the Various Continents, Islands, Rivers, Seas, and Peoples of the Globe*, 6 vols. (London, 1876), 1:262; Wright, ed., *Lewis and Dryden's Marine History*, p. 110; Lamb, "Early Lumbering on Vancouver Island," pp. 31–53, 95–121; Jackman, *Vancouver Island*, pp. 63–64. While the flush days continued, the profits must have been remarkable. One visitor reported that

The mills on Burrard Inlet faced the same problem. William Sayward erected the first permanent sawmill there during the Fraser River gold rush and profited handsomely from the venture. However, Burrard Inlet did not become an important port in the cargo trade until the 1860s when Stamp, having resigned his position with the Alberni Mill Company, arrived to construct a mill of his own. The connections that Stamp had made during his long career in the international lumber trade were an invaluable asset. During the first five weeks that his Hastings Mill Company was in operation, he cut seven hundred thousand feet of lumber for shipment to Australia. He had more orders than his mill could handle during its first months and passed some of them on to the Pioneer Mill across Burrard Inlet in Moodyville which, lacking Stamp's overseas connections, had been unable to get orders of its own and had been idle. In the next five months the mill at Moodyville reportedly cleared $40,000. In the years that followed, Stamp shipped to Java, New Zealand, Australia, Chile, China, Honolulu, London, and San Francisco. The latter, however, never became a mainstay of the mill's trade.

As it became evident that it was possible to have successful sawmill operations north of the international border, investment capital became available. Both the Moodyville and Hastings mills expanded with the help of money from San Francisco.[112] In underwriting this expansion, Bay Area merchant capitalists were investing in going concerns in an established area, rather than in new enterprises in virtually untapped territory, as they had at Puget Sound and Humboldt Bay. Investments of the former sort seem to have been more typical than the latter. American merchants appear to have done more in financing industrial expansion than in pioneering in new fields.[113]

In 1870 the enlarged mills exported nearly 9 million feet of lumber, plus shingles, spars, and other forest products. Peru, the leading market, consumed 3.2 million feet of lumber, Australia 1.9 million,

a captain of a British vessel claimed that he had purchased lumber on Vancouver Island and sold it in Foochow after a two months' voyage for a gain of approximately 700 percent. Matthew Macfie, *Vancouver Island and British Columbia* (London, 1865), cited in Carrothers, "Forest Industries of British Columbia," p. 258.

112. Eric Nicol, *Vancouver* (Toronto: Doubleday Canada, Ltd., 1970), pp. 29–32; Carrothers, "Forest Industries of British Columbia," pp. 260–68; Lawrence, "Markets and Capital," pp. 7–9, 24–29; Hubert Howe Bancroft, *History of British Columbia, 1792–1887* (San Francisco, 1887), pp. 359, 708; Victoria *Colonist*, 15 May 1865, 8 Aug. 1868; Seattle *Weekly Intelligencer*, 27 July, 7 Sept., 14 Dec. 1868, 11 Jan. 1869, 5 Feb. 1872.

113. Porter and Livesay, *Merchants and Manufacturers*, p. 8 and *passim*.

China 1.5 million, and Hawaii 1.1 million.[114] Though Yesler had reported in 1868 that Stamp was said to be "loosing money all the time," aggressive salesmanship helped Stamp overcome the disadvantages inherent in having his mill located in British Columbia. He traveled extensively, seeking new customers and attempting to hold old ones. Hugh Nelson, who took over as manager of the operation after Stamp's death, continued Stamp's policies for some time.[115]

The success of the Hastings Mill in the cargo trade was hard-earned. Indeed, it seems that only Stamp's unbounded energy and many contacts made it possible. After his death, the Hastings Mill gradually moved away from the trade. In the late 1880s and 1890s more and more of the mill's cut was sent to the interior by rail, rather than out to the markets of the Pacific Basin.[116] Though some lumbermen on the West Coast had Stamp's drive, no others had world-wide connections of the sort that contributed so substantially to the success of the Hastings Mill. Stamp proved that it was possible to have a successful cargo mill without depending on the markets of San Francisco, but he also demonstrated how difficult it was to do so. Moreover, even Stamp was not entirely independent of San Francisco's influence. Though he marketed little of his cut there, San Francisco was the source of investment capital upon which he drew to finance the expansion of his operation. Millmen in British Columbia were tied to San Francisco in other ways. Supplies often came from there, and, when it became necessary for millmen to charter vessels, they obtained these too in the city by the Golden Gate. There is no reason to assume that Stamp's practices differed from the general pattern in this regard.[117]

The economic influence of San Francisco was felt all along the Pacific slope. It was felt as keenly in the lumber industry as in any other, and Stamp's operations could not escape it. To the vast majority

114. H. L. Langevin, *British Columbia Report of the Hon. H. L. Langevin* (Ottawa, 1872), p. 150. See also Carrothers, "Forest Industries of British Columbia," pp. 265–68; Lawrence, "Markets and Capital," pp. 26–27. After the United States and Hawaii began tariff reciprocity in 1876, lumber from British Columbia had difficulty maintaining its place in the islands. See *Canada Lumberman* 12 (July 1891):6.

115. Lawrence, "Markets and Capital," pp. 27–29, 66. *West Shore* 15 (1899):5. *Canada Lumberman* 1 (15 Oct. 1880): 8; 1 (30 Nov. 1880):1. Lumbermen in eastern Canada seem to have been less aggressive in seeking foreign markets. See *Canada Lumberman* 2 (1882):231.

116. Lawrence, "Markets and Capital," pp. 66–67. *Canada Lumberman* 9 (Mar. 1889):6; 16 (May 1895):3. As late as 1895, however, R. H. Alexander, then the mill's manager, traveled abroad in quest of customers. See *Canada Lumberman* 16 (Aug. 1895): 10.

117. Lawrence, "Markets and Capital," pp. 28–29, 111–12. Lawrence, "California's Influence," *passim.* *Canada Lumberman* 8 (Apr. 1888):10; 16 (Apr. 1895): 12.

of the emerging giants of the Pacific lumber trade, San Francisco was even more important. From Burrard Inlet and Alberni Canal in the north to Humboldt Bay, the Mendocino coast, and Santa Cruz in the south, millmen were never far removed from the influence of the California city. In spite of setbacks, San Francisco grew during the sixties and seventies. The cargo trade, aided by broadening foreign markets, grew with it. By the beginning of the eighties the coastal lumber industry had not only evolved into one of the most important industries in the Far West, but had assumed what was to prove to be its permanent character. So long as a separate, identifiable industry of cargo mills remained, it was to be dominated by a handful of giants who stood with one foot in San Francisco and the other at the point of production.

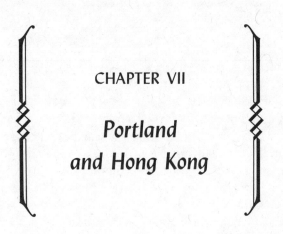

CHAPTER VII

Portland and Hong Kong

WHILE a few producers controlled and financed from San Francisco were coming to dominate the lumber trade of Puget Sound and the redwood coast, development was following a different pattern along the Columbia River. The latter area had fallen far behind the other two by the 1860s. Accessible timber stands along the Columbia and its tributaries were inferior to those in much of the area around the sound and in California's northern redwoods. The rocky bluffs that fronted much of the lower Columbia restricted the area that could be logged and pushed the cost of woods operations upward, while the river's dangerous bar and shoal-strewn watercourse made shipment from the river more expensive than from competing regions.[1] As one correspondent wrote to the *Alta California*, "From 1850 to 1855 the sawmills of Oregon exported great amounts of sawn lumber principally to San Francisco, but the sawmills of California and Puget Sound have now driven the Oregonian mills from the market, and the consequence is, that some of the latter are abandoned, and the others have no aspiration beyond supplying the home demand."[2] Andrew Pope agreed: "The Columbia River folks can't compete with us and they

1. Olympia *Columbian*, 11 Sept. 1852. Portland Board of Trade, *Annual Reports* (microfilm copies, University of Oregon library, Eugene), 1879, pp. 10–11; 1881, pp. 14–15. Hittell, *Commerce and Industries*, pp. 196–97, 399–400. *West Shore* 7 (1881):16–17; 10 (1884):8. The *West Shore* reported that a cargo of wheat shipped from Tacoma paid port, loading, and pilotage charges totaling $2,059.25, whereas the same cargo shipped from Portland would have paid charges totaling $6,075. The differential was apparently about the same for shipments of lumber.
2. San Francisco *Alta California*, 7 Sept. 1858.

dont do much as its vessels are coming down [to San Francisco] in balast for want of freights."[3] The records of the Astoria Customs House confirm these observations. During the sixties they show that exportation of lumber from the Columbia virtually ceased. Lumber continued to be produced in the area, but through the decade most of it was consumed locally. The sawmills themselves remained small.

Then, during the 1870s, the lumber industry along the Columbia River experienced a revival.[4] The resurgence was triggered by the advent of railroad building in Oregon and the growth of economic activity that accompanied it.

In 1868 construction started on two competing railroad lines up the Willamette Valley. The first to complete twenty miles of track was to win the land grant the federal government had offered to induce construction of a rail link between the Pacific Northwest and California. Soon a number of vessels new to Pacific waters entered the Columbia River with rails and equipment for the rival enterprises. Once discharged of their cargoes, many took on wheat to carry eastward around Cape Horn. They thus became pioneers in the then-emerging Northwestern grain trade. Not all were employed in this manner. The fleet of coasting vessels that had grown up in preceding decades kept the San Francisco market rather amply supplied, but a few of the agents of vessels lying at anchor off Portland managed to find consignments to carry to the Bay Area. Others followed the example of Nathaniel Crosby and the pioneer captains of the fifties who had sailed their vessels westward from the Columbia to Asiatic ports.

It was not just the possibility of finding markets for the products of the Pacific Northwest that led to this revival of interest in the Far East. In 1868 the French bark *Jennie Alice* brought 432 Chinese laborers to Portland to help build the railroads. Although they were later to decry the presence of Chinese, at this stage Northwesterners welcomed the newcomers as useful additions to the labor supply. Others followed the lead of the French vessel, and a regular traffic in Chinese passengers soon developed.[5]

3. Pope to Wm. Pope & Sons, 15 Sept. 1857, PTP, Pope file. See also San Francisco *Alta California*, 4 July 1861, 7 Jan. 1862, 9 July 1867.
4. However, the river continued to plague shippers, as the letters of one captain who called at Asa Simpson's Knappton mill show. See J. McIntyre to Kentfield, 8 and 19 Nov. 1875, JKCP, 1:2.
5. Wright, ed., *Lewis and Dryden's Marine History*, pp. 167, 179; Astoria Customs House Records, 1:F81, 2:*passim*; Portland *Oregonian*, 22 Apr. 1868, 20 Feb., 2, 7 Aug. 1869. Cf. San

Increasing numbers of emigrants were also leaving South China for other areas. During 1867 only 4,283 left Hong Kong; in 1875 over ten times that number departed.[6] Shipowners welcomed this passenger traffic, for there was far more tonnage in the world's sailing fleets than existing demand justified. Increasingly, cargoes on the world's main trade routes were being dispatched in ships powered by steam. Windjammers would often lie idle for months at a time only in the end to sail in ballast to another port where, it was hoped, a cargo might be found. The coolie traffic offered welcome employment, and more and more owners of sailing ships ordered their vessels to Hong Kong to engage in it.[7]

The servicing of vessels was a major activity at Hong Kong almost from the beginning of the colony. The port supplied provisions and refitting not only for ships taking on cargoes in the British colony but for nearly all the rest of those operating along the coast of South China. These ships represented considerable tonnage.[8] As the flow of Chinese passengers increased in the seventies, the vessels engaged in the coolie trade helped swell the increasing demand at Hong Kong for spars, ship planking, and provisions. Cargoes began to move westward across the Pacific to service these needs.

It quickly became apparent that a two-way trade between Oregon and South China was possible: lumber, spars, and provisions on the outbound leg and coolies on the return. Such possibilities even attracted some who did not own tonnage. In 1869 three Portlanders with capital to invest—James B. Stephens, A. M. Loryea, and Walter Moffitt—purchased the bark *Edward James* and put it in the China trade. They sent the vessel to Puget Sound to take on lumber and then to Hong Kong. It returned to Portland in 1872 with 380 Chinese passengers.[9] Other investors followed. In 1876 the Portland commission house of Corbitt

Francisco *Alta California*, 12, 16, 31 May 1851, for early reactions in California to Chinese immigrants.

6. From *Historical and Statistical Abstracts of the Colony of Hong Kong, 1841–1930*, 3rd ed. (Hong Kong, 1932), reprinted in Endacott, *An Eastern Entrepôt*, pp. 132–33.

7. The problems faced by merchant sail at this time are discussed in Cutler, *Greyhounds of the Sea*, pp. 263–376; William Armstrong Fairburn, *Merchant Sail*, 2:1534–84; Walter MacArthur, *Last Days of Sail on the West Coast: San Francisco* (San Francisco: James H. Barry Co., 1929), pp. 19–27. Revealing materials on the coolie trade are found in United States, 34th Cong., 1st Sess., House Exec. Doc. No. 105, and 36th Cong., 1st Sess., House Report No. 443.

8. Endacott, *An Eastern Entrepôt*, pp. xii–xiii; Bernard Boxer, *Ocean Shipping*, pp. 5, 11; Edward Szczepanik, *The Economic Growth of Hong Kong* (London: Oxford University Press, 1958), pp. 3, 133; Endacott, *History of Hong Kong*, pp. 132, 194, 253–54.

9. Wright, ed., *Lewis and Dryden's Marine History*, pp. 178–79.

This V-flume, which during the 1880s carried lumber from the mills at Bridal Veil, Oregon, to the banks of the Columbia River some five miles away, could handle up to 75,000 feet per day. In the background a conveyor carries sawdust to the burner. Disposal of waste was a continuing problem for most mills, but especially for redwood mills, whose sawdust was most often too wet to burn. Courtesy of the Forest History Society, Weyerhaeuser Timber Company Collection

Unlike at the bar harbors and outports, sizable vessels could call at the mills on Burrard Inlet. In this etching made in 1872, several wait to take on cargoes. Courtesy of the Forest History Society

Horse teams were used to pull lumber from the Mendocino mill to the nearby ocean-front bluffs, where it was shot down apron chutes to waiting vessels. Horses continued in use for yard work around mills until well into the twentieth century, but steam engines were widely used in other aspects of lumbering from the 1880s. Courtesy of the San Francisco Maritime Museum, Union Lumber Company Collection

William Kyle's ill-fated schooner *Bella*, which had a retractable centerboard to aid her in entering and leaving the treacherous bar harbors of the area. Courtesy of the San Francisco Maritime Museum, Mrs. C. Jacobsen

and Macleay, one of the largest firms in the city, purchased a pioneer of the trade, the bark *Garibaldi*.[10] Then in 1879 John C. Ainsworth, one of the greatest entrepreneurs of the Northwest, purchased the barks *Coloma* and *Alden Besse*. Two years later he added the full-rigged ship *Kate Davenport*. Ainsworth dispatched all three to Hong Kong, the *Coloma* and *Alden Besse* directly and the *Davenport* by way of Melbourne.[11]

Ainsworth's two barks were the most important vessels engaged on the run from the Northwest to Hong Kong. Year after year they plied back and forth hauling produce, Northwestern forest products, and passengers on the outward voyage and Chinese laborers, merchandise, and provisions on the return. Together, they and a number of less regular visitors made the market in Hong Kong a near-monopoly of Portland interests.

Portland's ascendancy was no accident. Railroad construction began in Oregon before it did north of the Columbia. Though lumber prices and port costs tended to be higher in Portland than on Puget Sound, the need for coolies, which made the two-way trade possible, existed only in the Portland area. By the time railroad construction began on a large scale north of the Columbia, Portlanders already had their connections secured. Moreover, by then the end of sail was rapidly approaching, steamers were hauling much of the coolie traffic, and the need for spars and ship planking was declining even in so busy a harbor as Hong Kong's.[12]

Had the demand remained firm after railroad construction commenced north of the Columbia, the larger firms that dominated the

10. Ibid., pp. 187, 198, 202, 245–46; Seattle *Daily Intelligencer*, 12 Dec. 1877. Just how regularly Corbitt and Macleay dispatched the vessel to Hong Kong is not clear, but passages clearly continued to be made. See, for example, Hong Kong *Times*, 17 Mar. 1876.

11. Bills of sale, marine surveys, and other papers relevant to the purchase of these vessels are in APSA, box 1. See also J. McCracken to Ainsworth, 3 June 1879, and Noyes & Fairweather to Ainsworth, 10 July 1879, John C. Ainsworth Papers, University of Oregon library, Eugene, General, box 1. These show that Ainsworth first sought to charter the *Coloma*, but that in order to do so he would also have had to charter *Stonewall Jackson*. Rather than do this he purchased the *Coloma*. Both the *Coloma* and *Alden Besse* had carried lumber on the run to Hong Kong before Ainsworth purchased them. See U.S. Bureau of Customs, series 19, vol. 3, "Lumber Exported in U.S. Vessels," entries for 3 Nov. 1870 and 2 Sept. 1873, MS, Federal Records Center, Seattle, Wash.; Seattle *Weekly Intelligencer*, 21 June 1873.

12. From an early date Pacific Mail and other steamers were actively engaged in the coolie trade. However, their role increased in importance with time. See Ta Chen, *Chinese Migrations with Special Reference to Labor Conditions* (Washington, D.C.: U.S. Government Printing Office, 1923), pp. 12–16; Russell H. Conwell, *Why and How: Why the Chinese Emigrate and the Means They Adopt for the Purpose of Reaching America* (Boston, 1871), pp. 212–23.

cargo trade would still have had difficulty securing entry into the Hong Kong trade. Their fleets of lumber carriers, as well as many of the charter vessels operating on the Northwest coast, were designed for hauling millstuff.[13] Their single decks and yawning holds were well adapted to carrying large cargoes of lumber, but they lacked the versatility of down-Easters such as the *Edward James, Garibaldi, Coloma,* and *Alden Besse* which, with their two decks and clean lines, could carry cargo and passengers on the same crossing and at speeds most of the lumber droghers could not match.[14] While entrepreneurs from Portland serviced the demand in Hong Kong, the large lumber-producing firms on the sound and redwood coast busied themselves elsewhere.

John C. Ainsworth was the largest single investor in the trade. Though he had not pioneered in this commerce and never journeyed to the Far East himself, this self-made millionaire was in large part responsible for Portland's ascendancy in this portion of the Pacific lumber trade. Fortunately, extensive records of his role have been preserved. They are records that tell much, not just of the Portland–Hong Kong trade in which Ainsworth's vessels participated, but of the problems facing all the independently owned vessels that sought to engage in the lumber trade. Indeed, even vessels owned by mill companies faced many of the challenges so graphically depicted in the records of Ainsworth's little fleet.

Ainsworth, who first came to Oregon in 1851, had been active in both mercantile and steamboat enterprises before he turned his attention to deep-water trade. He was influential in forming the Oregon Steam Navigation Company, which monopolized trade on the Willamette and Columbia rivers through the sixties and seventies.[15] When he finally sold his steamboat holdings, Ainsworth found himself with capital to reinvest. The sale "made me worth more than a million," he recorded. "To carefully invest this money, so that it will

13. See below, chap. 10.
14. Basil Lubbock, *The Down Easters: American Deep-Water Sailing Ships, 1869–1929* (Boston: C. E. Lauriat Co., 1929), *passim.* Rowe, *Maritime History of Maine,* pp. 207–36. Fairburn, *Merchant Sail,* 2:1577–80; 3:1733–52, 1755–59; 4:2284–87. Basil Lubbock, *Last of the Windjammers,* 2 vols. (Glasgow: J. Brown & Son, 1927–29), 2:28. The *Coloma* was built in Warren, R.I., but nonetheless fits the type that has come to be called down-Easter.
15. *National Cyclopedia of American Biography* (New York: James T. White & Co., 1891–1971), 25:204; Joseph Gaston, *Portland: Its History and Builders* . . . (Chicago: S. J. Clarke Publishing Co., 1911), 3:8; Dorothy O. Johansen, "The Oregon Steam Navigation Company: An Example of Capitalism on the Frontier," *PHR* 10 (1941):179–88.

be of the greatest advantage to my children is now my daily study and care." The most important of his new enterprises were the Ainsworth National Bank, established in Portland in 1883, and the Central Bank of Oakland (California), established in 1892 after he had moved to that city.[16] His interest in ships was not dead, however; even before his connections with the Oregon Steam Navigation Company had been fully terminated he was arranging for the purchase of *Coloma* and *Alden Besse.*

The two barks were much alike. Both were built in the down-Easter mold that became popular in the years of the decline of clipper ships. Both measured some 850 tons, and, though there were slight differences between them, the master of the *Besse* reported that "the vessels sail about alike bye and large." [17] Apparently they sailed well. The master of the *Coloma*, who had taken a one-eighth share in the bark, wrote from Hong Kong on his first voyage aboard: "I suppose you would like to know how I like the 'Coloma' Very much. . . . She is all I expected." [18] Though the two barks were much alike, *Coloma* was a better purchase. Ainsworth paid $18,414.21 for his seven-eighths interest in her, some four thousand dollars less than he paid for the same interest in *Alden Besse.*[19]

As masters of his new vessels Ainsworth signed on the brothers Allen and Cyrus Noyes. The two brought a wealth of experience to their jobs. Raised in Searsport, Maine, famed for its associations with Cape Horners and blue-water captains, they had known the sea since boyhood.[20] They had come to the Northwest in 1870 aboard the *Garibaldi*, which they sailed to Hong Kong with Allen Noyes as master, Cyrus Noyes as mate, and had been running in the China trade ever since. When Corbitt and Macleay bought *Garibaldi*, Cyrus moved over to assume command of *Alden Besse*, on which he continued to serve until Ainsworth acquired her. He then assumed command of the *Coloma* while his brother took over "the Bessie." [21]

16. John C. Ainsworth, "Autobiography of John C. Ainsworth, Oregon Capitalist," MS, Ainsworth Papers, box 1, entry dated 9 Feb. 1881; *National Cyclopedia of American Biography*, 15:204; Gaston, *Portland*, 3:8.

17. Allen Noyes to J. Ainsworth, 1 Apr. 1881, APSA, 1:*Besse* 2.

18. Cyrus Noyes to J. Ainsworth, 1 Feb. 1880, APSA, 1:*Besse* and *Coloma* 1.

19. Ibid., 15 Apr. 1880.

20. On Searsport, see Lubbock, *The Down Easters*, pp. 24–25, 260–62; Rowe, *Maritime History of Maine*, pp. 15, 286–87.

21. Portland *Oregonian*, 24 Jan. 1903; Wright, ed., *Lewis and Dryden's Marine History*, p. 198; Lubbock, *The Down Easters*, p. 262; Hong Kong *Times*, 3, 19 Jan., 16 Feb. 1876. Apparently

The choice of captains was important. The master of a vessel not only had to sail his ship, but was, more often than not, also responsible for selling its cargo and selecting one for the return that would yield a good profit. So it was with Cyrus and Allen Noyes. Ainsworth worked out an elaborate code so his captains could communicate by telegraph with him or with L. L. Hawkins, Ainsworth's agent in Portland after Ainsworth moved to California; but the extensive records that have been preserved make it clear the majority of the decisions were made by the two captains, especially as Ainsworth's confidence in them grew after the first voyage or two.[22]

Ainsworth chose well. The brothers proved canny traders in the best Yankee tradition. Included in the cargo of the *Coloma* on the first voyage were several barrels of beef and pork that the owners of the bark hoped to sell for ship's provisions. "The beef and Pork was such poor stuff we shall not make much profit on it," Cyrus Noyes reported from Hong Kong. "The Pork I have left with a good man to sell. I thought it better to do so than to sell it cheap." On his return to Hong Kong a year later, he reported:

When I arrived here I found the Pork I left all unsold and it had not improved any while I was gone. The man I left it with thought he had sold it but when the man that was going to take it saw it he said it would not do. I took it all on board ship as soon as I got the cargo out and went to work on it and scraped it all over and washed it And got it off for over one thousand dollars. I done much better with it than I expected when I saw it on my arrival[;] some of the bbls had spoiled after lying here one year.[23]

Ainsworth could see his interests were being carefully attended to; from that point on his direct participation in management of the vessels declined sharply.

The first voyage that the Noyes brothers made to Hong Kong in the employ of John Ainsworth was in part experimental. Seeking to determine which items would bring the greatest return, they filled the lower holds with diversified cargoes, which they carried on the ships' accounts. On board *Alden Besse* were lumber, spars, fresh beef, 250

Allen Noyes had sometimes served as captain of *Besse* even before Ainsworth entered the picture. See Hong Kong *Times*, 16 Mar. 1876, supplement.

22. The code did not prove foolproof; the messages sometimes came through in garbled form. See A. Noyes to J. Ainsworth, 12 Jan. 1881, and A. Noyes to L. L. Hawkins, 13 Jan. 1881, APSA, 1:*Besse* 2.

23. C. Noyes to J. Ainsworth, 15 Apr. 1880, APSA, 1:*Coloma* 1; Noyes to Ainsworth, 2 Apr. 1881, APSA, 2:*Coloma* 2. For another example of Noyes's sharp practices in business, see C. Walker to W. H. Talbot, 13 and 16 July 1885, PTP, Walker file. Cf. J. Winding to J. Kentfield, 9 Sept. 1878, JKCP, 3:4.

barrels of preserved beef and pork, and over sixty tons of scrap iron. Eight tons of coal sold by the master of the *Coloma* in Hong Kong do not appear on the ship's manifest; they must have been stowed as ballast. Both vessels left Portland with the upper holds and deck spaces chartered to Chinese businessmen of Portland for use in transporting Chinese passengers and freight.[24]

The first crossing of the two vessels proved profitable for Ainsworth and his captains. The *Coloma* had left Portland with its upper hold and deck space already under charter for the return; Allen Noyes easily obtained a return charter for his vessel once he had arrived in the British colony. The charters alone more than covered the expenses of the vessels, in spite of the fact that repairs and improvements costing over $6,000 were put into the *Coloma* while in Hong Kong. Everything earned on the cargoes carried on ship's account was clear profit.[25]

Selling the cargoes was not easy, however. We have already seen the difficulties Cyrus Noyes faced in trying to dispose of his preserved meats. The brothers faced a different sort of problem in selling their lumber: competition. Corbitt and Macleay had chartered the *W. F. Holcomb* to carry an order of lumber to the Whampoa Dock Company. The *Garibaldi* was due shortly with an additional cargo of lumber, while a third vessel, similarly laden, had arrived somewhat earlier.[26]

Undaunted, Allen Noyes—whose vessel had reached port ahead of the *Coloma*, *W. F. Holcomb*, and *Garibaldi*—set about to dispose of his cargo. Through Ainsworth's agent in Hong Kong, Rozario and Company, Noyes received an offer of $25 per thousand for his entire lot. Noyes considered the offer low, "but after looking about I thought if I went to selling in small lots [even] if I did get a little more I should spoil the market for my brother's cargo. . . ." The offer had come from the owner of the Spratt dock, one of two dry docks in the colony.

24. Manifest of *Alden Besse*, dated 16 Oct. 1879, APSA, 1:*Besse* 1; manifest and receipts of *Coloma*, various dates, APSA, 1:*Coloma* 1; undated passenger list, *Alden Besse* out of Portland for Hong Kong, APSA, 1:*Besse* 1; receipt for lumber from Willamette Steam Mills shipped by Tong Duck Chong, n.d.; C. Noyes to J. Ainsworth, 15 Apr. 1880, APSA, 1:*Coloma* 1. Unable to engage enough passengers to fill the space they had chartered, the Chinese also took on some lumber. Apparently they lost money on the venture.

25. C. Noyes to J. Ainsworth, 1 Feb. 1880, APSA, 1:*Coloma* 1; general statement, *Coloma* and *Alden Besse*, voyage 1, APSA, 1:*Besse* 1.

26. C. Noyes to J. Ainsworth, 1 Feb. 1880, APSA, 1:*Coloma* 1; A. Noyes to J. Ainsworth, 23 Dec. 1879, APSA, 1:*Besse* 1.

The other, the Whampoa Dock Company, was controlled by Jardine, Matheson and Company. Inasmuch as Jardine served as Corbitt and Macleay's agents in Hong Kong, Noyes considered the chances of a sale there almost nil. Moreover, the *Garibaldi* was in need of repairs. If he did not close before it arrived, Noyes feared his own customer might take the cargo of the *Garibaldi* in order to get its repair work. On the other hand, if Noyes sold to the Spratt dock, which could not possibly take two cargoes, the larger Whampoa Dock Company might take the cargo of the *Garibaldi* as well as that of the *W. F. Holcomb* in order to get the repair work on Corbitt and Macleay's vessel. Noyes closed.[27]

When the younger Noyes arrived some three weeks later, he found the way well prepared for him. The small-lot market, thanks to his brother's foresight, was still undisturbed. By 1 February 1880 Cyrus Noyes was able to write Ainsworth, "The lumber is sold—at $30 per M."[28]

The lumber carried to Hong Kong on the *Coloma* and *Alden Besse* was well received. It had been produced at George W. Weidler's Willamette Steam Mills Lumbering and Manufacturing Company, a firm noted for up-to-date equipment and quality products. Allen Noyes reported, "Mr. Spratt who purchased my lumber said . . . [it] cannot be beaten." By contrast, the cargo aboard the *Holcomb* was a disappointment to its purchasers. "They are very dissatisfied with the lumber sent them by Mess. Corbet-McClay, And it so happens that Mr. Rozario landed his lumber that my broth[er] brought on their dock and when they saw that they were still more dissatisfied."[29] Cyrus Noyes was optimistic: "My brother and I will get the business from the [Whampoa] Dock Company after this for they are very much dissatisfied with the cargo Macleay sent them. They have seen our Lumber and say they wish they had sent the order to us. . . . I know we can do this buisness [*sic*] better than any one else and make money out of it."[30]

In spite of this, securing future orders did not prove easy. Spratt formed his dock into a joint-stock company with Russell and Company

27. A. Noyes to J. Ainsworth, 23 Dec. 1879, APSA, 1:*Besse* 1.
28. A. Noyes to J. Ainsworth, 6 Feb. 1880, ibid.; C. Noyes to J. Ainsworth, 1 Feb. 1880, APSA, 1:*Coloma* 1.
29. A. Noyes to J. Ainsworth, 6 Feb. 1880, APSA, 1:*Besse* 1. See also C. Noyes to J. Ainsworth, 1 Feb. 1880, APSA, 1:*Coloma* 1.
30. C. Noyes to J. Ainsworth, 1 Feb. 1880, APSA, 1:*Coloma* 1.

as managing agents. To work the stock up above par, Russell wanted to keep overhead down. Allen Noyes offered to bring them a full cargo the following year, and Spratt gave him a list of specifications. "I told him that $12,000 would be my price to lay the lot down in Hong Kong," Noyes wrote, adding later that they "thought it rather high but I did not like to say less as every thing has risen so much lately." Russell and Company countered by asking for a quotation on a list that would have constituted a partial cargo. Again, they considered Noyes's price high.[31] In the end, they contracted for a full load from Burrard Inlet and offered the charter to the *Alden Besse*. John Ainsworth's son, George, who took over negotiations after Noyes's return to Portland, was chagrined. He wanted the profit from the sale of the cargo as well as the freight. Rather than settle for the freight alone, the owners dispatched the *Alden Besse* as they had the year before, laden with lumber but without a buyer under contract.[32]

The Whampoa Dock Company's business also proved difficult to obtain, though Allen Noyes suggested a way it might be won:

I fear Corbet and McClay have got a strong hold of Jardine Matheson so they will get the order next year. If such is the case they will have to charter a ship to bring the cargo[.] Nou [*sic*] if we could manage to charter the AB to them . . . we should have the business all right—and I think we could hold it[;] . . . it will be rather hard to get a hold there unless we can work in Portland by chartering with them and get hold of the business that way.[33]

Such a stratagem would have meant, at least for the time being, freights without the additional benefits of profits from the sale of the cargo.

This was a course that did not have to be taken, however. When examiners in the British colony declared the *Garibaldi* unfit for use as an emigrant transport, she went into service on the China coast. Allen Noyes commented laconically, "She will not be liable to trouble us much more." [34]

Disqualification of the *Garibaldi* put Corbitt and Macleay in a difficult position. Although some eleven sail were due at Hong Kong, Allen Noyes suspected none of them could meet the standards of the Chinese Passengers Act of 1855. If the Portland firm could not obtain

31. A. Noyes to J. Ainsworth, 22 Mar., 17 Apr. 1880, APSA, 1:*Besse* 1.
32. For a résumé of subsequent negotiations see Macondray & Co. to George J. Ainsworth, 4, 5 Aug. 1880, ibid.
33. A. Noyes to J. Ainsworth, 22 Mar. 1880, ibid.
34. A. Noyes to J. Ainsworth, 6 Feb. 1880, ibid.

a passenger vessel, the only way they could continue to import rice and Chinese provisions to Oregon would be to charter a vessel for that purpose or send their goods by steamer via San Francisco. Either would be expensive, for steamer rates had recently been raised, and, at the moment, charters were scarce and therefore dear in Hong Kong. Further, without their own captain present to superintend purchases in the colony, Corbitt and Macleay would have to pay additional commissions on what they bought. Moreover, as Allen Noyes observed, "I doubt if they will get as good quality rice as we do." The dilemma must have been only one of many faced by the firm, for its fortunes appear to have waned rapidly. The company had dispatched fifteen vessels in the European grain trade in 1876; by 1880 they had dropped out of the trade altogether. Unable to find a satisfactory solution to the problems confronting them in Hong Kong, they dropped out of that trade as well.[35]

Corbitt and Macleay was not the only competitor that disappeared from the scene at this juncture. Stephens, Loryea, and Moffitt sold the *Edward James* in 1879, and the vessel left the China trade. The reasons are not clear, but it would appear that it, too, was no longer able to meet standards set by the Chinese Passengers Act.[36] Local competitors had dwindled; if they could only hold off outsiders, Ainsworth and his associates were in a good position to rise quickly to dominance in the Portland–Hong Kong trade.

Once the Noyes brothers had sold the goods on board their vessels, they not only had to try to arrange for the sale of future cargoes, they also had to select and load goods for the return passage. They settled on rice for the bulk of the cargo to be carried on the vessels' accounts on their return, but also purchased eighteen tons of fine-punched granite, a boiler and condenser for the *Alden Besse*, and rattan furniture and baskets for the *Coloma*.[37]

In addition to cargo carried on the ships' accounts, much of the unchartered space was filled with consignments of freight. Again, the largest single item was rice, but a wide variety of Chinese merchandise and produce was taken aboard.[38] Both captains advanced funds that

35. A. Noyes to J. Ainsworth, 22 Mar. 1880, ibid.; Wright, ed., *Lewis and Dryden's Marine History*, pp. 246, 272, 279; Portland *Oregonian*, 1 Jan. 1886.
36. Wright, ed., *Lewis and Dryden's Marine History*, p. 272.
37. A. Noyes to J. Ainsworth, 17 Apr. 1880, APSA, 1:*Besse* 1; Manifests and receipts, various dates, APSA, 1:*Coloma* 1.
38. Bills of lading, various dates, APSA, 1:*Coloma* 1 and *Besse* 1.

they did not wish to invest in merchandise to Chinese merchants so that the latter might make purchases to send as freight. The goods shipped served as collateral on the loans, which were payable in gold in Portland. It was a profitable means of returning money. The vessel would collect freight on the goods shipped; to have purchased bills of exchange to return the funds to the West Coast would have cost some 9 percent.[39] The chartered space in the upper hold and on deck was, once more, used to transport Chinese passengers. The *Coloma* carried 346.[40]

Thus the *Coloma* and *Alden Besse* completed their first round trip under Ainsworth's colors. They proved a profitable pair of voyages. Though the two captains had set sail with mixed cargoes, uncertain of just who their buyers would be, and various developments worked to keep down the rate of return, between them the vessels cleared $13,080.85 on the run out and back. From an investment of some $40,000 in the two barks, Ainsworth earned a profit on the first voyage of $9,810.63—not a high return by the standards of the Gold Rush years, but a rate being earned by few others in the seventies and eighties. Charters and freights were the main source of income, but the cargoes carried in the lower hold also brought large profits.[41]

Subsequent voyages followed the general pattern of the first. On the second voyage, scrap iron, lumber, and spars—all of which yielded good returns on the first voyage—were dispatched to Hong Kong on the ship's accounts, as were bar iron, horseshoes, an old boiler, and forty-nine tons of bones, hoofs, and horns. Since the preserved meat carried on the first voyage was still unsold, none was sent on the second.[42]

Upon their arrival in Hong Kong the two captains found the price of old iron "very lou" and spars "a complete drug" on the market. Allen Noyes found a potential buyer for the spars, a Mr. Duncan, but he "got into some trouble and had to leave in the night." In the end, the elder

39. Rozario & Co. to C. Noyes, 12 Apr. 1880, APSA, 1:*Coloma* 1; A. Noyes to J. Ainsworth, 6 Feb., 22 Mar. 1880, APSA, 1:*Besse* 1. See also C. Noyes to J. Ainsworth, 22 Mar. 1880, APSA, 1:*Besse* 1; C. Noyes to J. Ainsworth, 12 Jan. 1881, APSA, 2:*Coloma* 2.
40. C. Noyes to J. Ainsworth, 15 Apr. 1880, APSA, 1:*Coloma* 1.
41. General statement, *Coloma* and *Alden Besse*, voyage 1, APSA, 1:*Besse* 1; vessels' journal and day book, 28 Oct. 1879 to 11 Nov. 1884, APSA, vol. 1; vessels' ledger, trial balances, APSA, vol. 2; A. Noyes to J. Ainsworth, 23 Dec. 1879, APSA, 1:*Besse* 1; A. Noyes to J. Ainsworth, 7 Mar. 1881, APSA, 1:*Besse* 2; C. Noyes to J. Ainsworth, 1 Feb. 1880, APSA, 1:*Coloma* 2. Cf. Financial records, statement of ship earnings [sailing vessels], OICP.
42. Manifests, bills of lading, and receipts, various dates, APSA, 1:*Besse* 2; 2:*Coloma* 2.

Noyes reported, "I fairly peddled them out with my brothers assistance."[43]

Prices of lumber, by contrast, were up. Moreover, marketing conditions had been greatly stabilized when Russell and Company sold the Spratt dock to the Whampoa Dock Company. Part of the lumber of the Noyes brothers went to the Jardine, Matheson firm and, true to Allen Noyes's estimate, once they had obtained them as a customer, the Portlanders were able to hold them.[44] When the master of the *Alden Besse* said in reference to his outbound cargo, "I wish mine had been all lumber. . . . ," he was pointing the way toward future developments.[45] As time passed, lumber came to constitute a larger and larger share of the goods shipped on the ships' accounts.

The owners of the *Coloma* and *Alden Besse* must have been pleased by the course of events. Their vessels had turned good profits on the first voyage and seemed well on their way toward doing even better on the second. The main competitors had been removed from the scene, and the combined dry docks had been obtained as a customer. Nonetheless, there were still problems. New competition was threatening to enter the trade. The bark *W. H. Besse*, another Bath-built down-Easter, appeared in Hong Kong and proceeded to take on cargo for Portland. Cyrus Noyes suspected that her captain intended putting his vessel on the run on a permanent basis.[46] Whatever the intentions of the captain of the *W. H. Besse*, the threat failed to materialize. The vessel remained an irregular transient.[47]

Nor was the *W. H. Besse* the only apparent problem. Hop Kee and Company, Chinese merchants of Hong Kong, chartered both vessels for the return trip before they were dispatched from Portland. Once again, advance chartering worked against Ainsworth and his associates. Upon his arrival in Hong Kong, Cyrus Noyes was offered $11,000 for the charter of his bark, but already under charter to Hop Kee for $9,400, he had to reject the offer. "I never saw Hong Kong so bare of

43. A. Noyes to J. Ainsworth, 12, 31 Jan., 23 Mar., 1 Apr. 1881, APSA, 1:*Besse* 2.
44. A. Noyes to J. Ainsworth, 12 Jan. 1881, ibid.; C. Noyes to Hawkins, 30 Dec. 1880, 18 Feb. 1881, APSA, 2:*Coloma* 2; C. Noyes to J. Ainsworth, 7 Mar. 1881, ibid.; C. Noyes to J. Ainsworth, 4 Apr. 1884, APSA, 4:*Coloma* 5.
45. A. Noyes to J. Ainsworth, 12 Jan. 1881, APSA, 1:*Besse* 2.
46. C. Noyes to J. Ainsworth, 2 Apr. 1881, APSA, 2:*Coloma* 2. *W. H. Besse* had engaged in the trade between Hong Kong and the Northwest before. See Hong Kong *Times*, 17 Jan. 1876; Seattle *Daily Intelligencer*, 5 June 1876.
47. E. B. Mallett to G. Ainsworth, 23 Jan. 1882, APSA, 3:*Kate Davenport* 3; Laine & Co. to Hawkins, 15 Dec. 1884, APSA, 6:general accounts 6.

ships," he wrote Ainsworth. By the time he sailed for Portland, Cyrus Noyes was estimating that the two vessels might have brought $6,000 to $7,000 more than they did, had they let the charter in the British colony.[48]

Talk of prohibiting Chinese immigration into the United States worried John Ainsworth, but the Noyes brothers reassured him. They doubted that such a prohibition would go into effect soon. At any rate, the charterers were "amply able to pay," and business in Hong Kong was good. If immigration were stopped, Hop Kee and his associates could easily find freight to fill the space they had acquired. Besides, Cyrus pointed out, "If they stop the Chinamen from going to the United States we can go to Victoria [British Columbia]."[49]

The charter of Hop Kee created other problems. Unlike the charterers on the first voyage, Hop Kee had taken the entire tonnage of both vessels. Only a small space on the poop deck was reserved for stowing goods carried on the owners' accounts. The sale of the outbound cargo left the Noyes brothers with considerable money credited to their accounts in the British colony. If they returned the money to the West Coast by bills of exchange, it would cost 10 percent, whereas if they purchased rice with the money, they would have to ship it in chartered space and pay the charterers $6.00 per ton freight. The latter alternative was tempting: though they had lost on rice the year before, the grain was now down to $1.05 per bag; and although reports indicated the price of rice was also low in San Francisco, Cyrus Noyes thought it would rise. At worst, it was not apt to go much lower, for there were so few vessels at Hong Kong that any rice shipped to the West Coast would have to go by steamer, the cost of which would tend to make it more expensive. When the rate on bills of exchange climbed to 12 percent, the brothers decided to buy rice. They managed to squeeze some fifty tons on the poop deck, but had to pay freight on two hundred fifty tons more. In the end, the decision to buy proved a wise one, for they turned a 4.3 percent profit on the grain and, at the same time, avoided paying for bills of exchange to transfer the credits on their account back to the West Coast.[50]

48. A. Noyes to J. Ainsworth, 12 Jan. 1881, APSA, 1:*Besse* 2; C. Noyes to Hawkins, 12 Jan. 1881, APSA, 2:*Coloma* 2; C. Noyes to J. Ainsworth, 30 Jan., 7 Mar. 1881, APSA, 2:*Coloma* 2.
49. C. Noyes to J. Ainsworth, 12 Jan. 1881, APSA, 2:*Coloma* 2. See also A. Noyes to J. Ainsworth, APSA, 1:*Besse* 2.
50. A. Noyes to J. Ainsworth, 7, 23 Mar., 1 Apr. 1881, 1:*Besse* 2; C. Noyes to J. Ainsworth, 12, 30 Jan. 1881, APSA, 2:*Coloma* 2; vessels' journal and day book, APSA, substatement for both vessels, voyage 2, in front of book.

When the accounts were closed, the second voyage proved to have been even more profitable than the first. Of the various items carried on ships' accounts, lumber and spars gave the best returns. Lumber earned a 253 percent profit, and, in spite of Allen Noyes's complaints about the poor market for spars, in the end spars did even better, showing a gain of 342 percent. Between the two, lumber and spars brought in over $11,000 above their cost in Oregon. Other items carried on ships' accounts also showed profits, but none did anywhere near so well as the wood products. Granite showed the third best rate of return, 74.3 percent. Once again, however, it was not the cargoes of the owners but charter fees that proved to be the greatest source of income.[51]

The profits earned on the first two voyages of the *Coloma* and *Alden Besse* apparently lived up to John Ainsworth's expectations, for in August 1881 he purchased a third deep-water vessel, the 1,249-ton *Kate Davenport*, a full-rigged medium clipper.[52] Ainsworth appears also to have begun to supplement his own vessels by chartering others, such as the *W. H. Besse* and *Edward Kidder*, which he also put into the Hong Kong trade.[53]

From the first Ainsworth utilized the *Kate Davenport* differently than the smaller vessels. The latter were able to take care of the lumber needs of Hong Kong; laden with 804,000 feet of lumber from Weidler's mill, *Kate Davenport* sailed under charter for Melbourne. In Australia, it was hoped, a second charter could be obtained to haul coal to Hong Kong, where the ship would then be made available to the Chinese merchants who had chartered it for the Hong Kong to Portland leg of the trip.[54]

There were difficulties almost from the moment the ship cleared the bar of the Columbia. The winds proved contrary. Worse, the vessel was too lightly laden in the lower hold and thus lacked stability at sea. Captain E. B. Mallett could not hoist enough sail to take full advantage of the few favorable winds he did receive without running the risk of

51. The largest gain was shown by lumber: $7,952.92 for the two vessels. The charter of the *Coloma* alone brought in more than this. See vessels' journal and day book, APSA, for a breakdown.
52. See above, n. 12. See also J. Ainsworth to Hawkins, 1 Sept. 1881, APSA, 1:general; Lubbock, *The Down Easters*, p. 59. Hawkins and Ainsworth's son George each owned one-sixteenth of the ship.
53. Laine & Co. to Hawkins, 15 Dec. 1884, APSA, 4:general accounts 6; Laine & Co. to Hawkins, 21 Nov. 1886, APSA, 24:*Coloma* 8.
54. J. Ainsworth to Hawkins, 17 Sept. 1881, APSA, 3:*Davenport* 3.

capsizing the vessel. As the *Davenport* worked slowly southward, the mounting wages of the crew ate into the would-be profits of the voyage.[55]

Ainsworth may have had hopes of entering his new acquisition in the Australian lumber trade on his own account. If so, Captain Mallett's reports from Melbourne discouraged this idea. Low prices plagued the lumber market in the island continent; it would be better, he suggested, to send lumber cargoes to Australia only under charter. Cargoes sent on owner's account were apt to prove costly. Ainsworth was convinced: when Mallett received an offer to charter his vessel for the following year, Ainsworth instructed him to accept it.[56]

The Australia–Hong Kong coal trade proved even more of a disappointment. Ainsworth had directed his agents to reject any offers of less than twenty shillings per ton. Mallett arrived in Melbourne to find that steamships had carried so much coal to Hong Kong that prices there were depressed and there was little demand for tonnage to haul more to the China coast. The best offers were at thirteen shillings per ton. This, coupled with a fear that the delay attendant upon a charter to haul coal might prevent him from reaching Hong Kong in time to fulfill the terms of his charter from there to Portland, led Mallett to speed his vessel northward in ballast.[57]

Bad luck continued to plague *Kate Davenport*. Once the vessel was in Hong Kong, the charterers found themselves unable to find enough emigrants to fill the space they had contracted. After holding the ship on demurrage for six days, they finally decided to let it go with what passengers they had. The *Davenport* had room for 425; she sailed with 195 aboard.[58]

With all that went wrong, it seems a wonder that *Kate Davenport* turned a profit on her first voyage for Ainsworth. The return was small, however. The ship had cost Ainsworth $35,000; the profit on the first voyage—$3,185.12—was a far cry from the rate the *Coloma* and *Alden Besse* had been earning.[59]

The profits themselves presented a problem. When Captain Mallett set about to return the money on credit to his vessel's account in

55. Mallett to Hawkins, 21 Jan. 1882, Mallett to G. Ainsworth, 23 Jan. 1882, Mallett to J. Ainsworth, 23 Jan. 1882, ibid.
56. Mallett to G. Ainsworth, 23 Jan. 1882, Mallett to Hawkins, 21 Jan. 1882, ibid.
57. Mallett to G. Ainsworth, 23 Jan. 1882, Mallett to J. Ainsworth, 23 Jan. 1882, ibid.
58. Mallett to G. Ainsworth, 7 June 1882, ibid. The vessel also carried 975 tons of cargo.
59. Vessels' journal and day book, APSA, statement showing earnings, in front.

Melbourne, he found that no drafts on San Francisco were available. To remit the funds he found it necessary to purchase a bill of exchange drawn on London. The only value such a draft would have to Ainsworth, who had no trade with Britain, would be what he could gain from passing it on to someone else. To do so would mean that the bill would be discounted below its face value.[60]

On its next voyage *Kate Davenport* was to carry lumber from Burrard Inlet to Melbourne. Captain Mallett took steps to insure that this second voyage would not be a repetition of the first. He had large bow ports cut to facilitate proper stowing of the cargo. He personally supervised both the trimming of the ballast and loading of the cargo in order to insure that this time the vessel would not prove to be crank.[61] Payment was to be on a piece basis, and the captain managed to get roughly one million feet of lumber aboard in addition to some two thousand cases of canned salmon, which the charterers wished to send to Australia as a trial shipment.[62]

The captain's efforts were rewarded: upon completion of the second voyage the accounts of the *Davenport* showed profits on the trip of $6,776.99. The improvement came on the first leg of the voyage; the Australia–Hong Kong and Hong Kong–Portland legs were no more successful than they had been before.[63]

The usefulness of *Kate Davenport* to John Ainsworth was short-lived. Captain Mallett returned from his second charter to find no business available for his ship. The vessel anchored at Tacoma, where she lay for several months running up bills and earning nothing. Once, the ship drifted loose and the owners had to pay $2,000 to the tug that returned her to her berth.[64] Finally, when the Hong Kong and Whampoa Dock Company[65] and F. Blackhead and Company, Hong Kong commission merchants, offered charters to the *Coloma* and *Alden Besse*, Ainsworth insisted that the orders be carried on the *Davenport*.[66]

---

60. Mallett to J. Ainsworth, 9 Feb. 1882, Mallett to G. Ainsworth, 9 Feb. 1882, APSA, 3:*Davenport* 3.
61. Used in a nautical context, "crank" means "lacking stability." This usually results from a vessel's being too lightly laden in the lower hold or insufficiently ballasted.
62. Mallett to G. Ainsworth, 30 Apr., 21, 25 Aug. 1882, Mallett to Hawkins, 3, 13, 17 Aug. 1882, APSA, 3:*Davenport* 3.
63. Vessels' journal and day book, APSA, statement of earnings, in front.
64. Statement from steamer *Lucy*, dated 29 Apr. 1884, APSA, 4:accounts.
65. Successor to the Whampoa Dock Co. and the Spratt dock.
66. Copies of telegrams, various dates, APSA, 4: *Davenport* 5. See also C. Walker to W. H. Talbot, 13 and 16 July 1885, PTP, Walker corres.

Business for the smaller vessels could be obtained elsewhere. Following this trip, her only direct voyage to Hong Kong under Ainsworth's colors, the *Davenport* was sold to William Renton, who put her to work hauling lumber for the Port Blakely Mill Company.

The mid-eighties were a difficult time for other vessels besides *Kate Davenport*. Pacific lumber markets were depressed, charter rates down. Cyrus Noyes wrote from Hong Kong: "Business is very dull here for ships. Some ships been here since last June, wont accept the rates offered. I think the 'Besse' and 'Coloma' Are fortunate to have a steady trade." They were fortunate, indeed; while others were tying up for lack of business, the younger Noyes was able to write: "I think we shall do well on Lumber and Spars next year as evry one knows we will come back againe. We have everything fixed for orders from here." [67]

However, even the *Coloma* and *Alden Besse* felt the impact of changing conditions. Railroad construction shifted to the north after Portland's transcontinental connections were completed, and much of the trade in Chinese passengers and provisions shifted with it. In 1883 on his fourth voyage for Ainsworth, Allen Noyes ran his bark to Victoria, rather than Portland, on the return voyage. Though expenses proved high on this initial trip to Vancouver Island, Noyes thought it offered good opportunities for traffic in Chinese passengers. "Lots want to go [back to China] if they can get the money to pay their passage," he wrote. [68] In view of the northward shift of business, the vessels now began to take on lumber at mills on Puget Sound rather than the Columbia. [69]

Gradually other changes came to Ainsworth's trade. In 1884 Allen Noyes left the *Alden Besse* to become marine surveyor in Portland, though he continued as a one-eighth owner. John O'Brien, who had long been serving as first mate, took over as master. [70] Later in the same year, the *Coloma* and *Alden Besse* began to stop off at Honolulu on their

67. C. Noyes to J. Ainsworth, 4 Apr. 1884, APSA, 4:*Coloma* 5.
68. A. Noyes to Hawkins, 13 July 1883, APSA, 4:*Besse-Coloma* 4.
69. C. Walker to W. H. Talbot, 13 and 16 July 1885, PTP, Walker corres.
70. O'Brien was to become one of the most famous shipmasters on the Pacific Coast. Born in Ireland in 1851, he went to sea at age sixteen. *Alden Besse* was his first command. Later he was master of the bark *Alice B. Dickerman* and steamers *Premier, Eureka, Edith, Victoria, Buford,* and *Rosalie.* He appeared with Buster Keaton in *The Mariner.* John O'Brien, *A Captain Unafraid: The Strange Adventures of Dynamite Johnny O'Brien* . . . (New York: Harper & Bros., 1912); Gordon Newell, ed., *The H. W. McCurdy Marine History of the Pacific Northwest* (Seattle: Superior Publishing Co., 1966), pp. 16, 96, 145, 334–35, 413–14.

westward crossings in hope of picking up Chinese passengers to supplement the dwindling numbers from the Pacific Northwest. Results were not altogether satisfactory, for numerous other vessels, eager for business, were doing the same. Ainsworth's agents in Honolulu managed to obtain only fifteen passengers for *Alden Besse* on its first attempt at entering the Hawaii–Hong Kong passenger business, too few to cover the cost of the stop. Later attempts were slightly more successful, but, at best, calls at Honolulu provided only a slight supplement to the income of the vessels and a chance to take on fresh water and provisions in mid-crossing.[71]

As railroad construction passed its peak, the number of Chinese returning to their homeland from the Northwest grew. This emigration, coupled with the restrictions placed on the immigration of Chinese into the United States that were adopted in 1882, led to a net decline in the number of Chinese resident in the Northwest. Chinese merchants in Hong Kong found themselves with less and less freight to ship to Portland as the demand for rice and Chinese provisions grew smaller in Oregon and areas to the north. By 1885 Ainsworth's new agents in Hong Kong, Melchers and Company, were writing: "The demand for tonnage to load here in Hong Kong is very slack. . . . For Portland it seems there is very little cargo offering this year and Chinese talk that there was scarcely sufficient cargo to fill another small vessel." [72]

In an attempt to fill the space that goods destined for purchase by Chinese residents in the United States had once taken, the owners of the *Coloma* and *Alden Besse* began to ship quantities of rattan furniture, camphor-wood chests, and curios, which they hoped to sell to the general public in the Northwest. The change created problems. Profits tended to be small, and the markets in Portland and Victoria were easily flooded if the captains did not keep a close eye on the supplies on

71. Laine & Co. to Hawkins, 15 Nov. 1884, APSA, 4:general accounts 6; 15 Dec. 1884, APSA, 6:general accounts 6; 14 Nov. 1885, APSA, 23:*Besse* 7; 21 Nov. 1886, APSA, 24:*Coloma* 8; J. O'Brien to Hawkins, 8 Nov. 1884, APSA, 4:general accounts 6.

72. Melchers & Co. to Ainsworth & Co., 17 Apr. 1885, APSA, 5:*Coloma* 6. For information on Melchers & Co., see Arnold Wright, ed., *Twentieth Century Impressions of Hong Kong, Shanghai, and Other Treaty Ports of China: Their History, People, Commerce, Industries, and Resources* (London: Lloyds Greater Britain Publishing Co., 1908), pp. 207–10, 618–20, 742. As Wright put it, "There are few places of any commercial importance in the Far East where branches of the well known firm of Messrs. Melchers & Co. are not to be found." Wright also briefly discusses the Hong Kong and Whampoa Dock Co., Rozario & Co., and F. Blackhead & Co. See ibid., pp. 196–98, 218–20, 223.

hand and choose a diversity of goods corresponding to demand.[73]

Even the lumber trade was becoming more difficult. In 1883 a market for squared timbers developed in Hong Kong as the government of the colony planned to construct a new dock. At the charter rates offered, however, Allen Noyes thought the trade would not pay. He and his associates let the business go to others.[74]

In the face of declining charter rates and dwindling traffic in passengers and eastbound freight, John Ainsworth began to look for other uses to which at least one of his barks might be put. He received offers of charters to haul coal and to carry lumber to San Pedro, but at the rates offered the profits would have been small.[75] Ainsworth decided to sell *Alden Besse*, instead. The two-way trade that had attracted him into deep-water navigation and had brought him returns in excess of what conventional lumber droghers might expect had nearly vanished. Old and racked with illness, Ainsworth apparently had no desire to enter new, less certain fields. Ainsworth's son George arranged to sell *Besse* to John D. Spreckels for $18,000. The elder Ainsworth was pleased. "This is a big price under the circumstances," he wrote. "Capt. A. Noyes will be pleased, as he named 14,000 as the price that would be satisfactory to him, when I last saw him." [76]

While *Alden Besse* entered the sugar trade, the *Coloma* continued on the Hong Kong run with Cyrus Noyes as her master. Just how long John Ainsworth remained the principal owner of the latter vessel is not clear, for new accounts, which have not survived, were started for the *Coloma* at the time the other bark was sold. However, at some time prior to 1895 J. McClacken of Portland became the principal owner.[77] McClacken continued to run her in the China trade, but the trade itself was a pale shadow of what it once had been. As the American consul in

73. Manifests, bills of lading, and receipts, various dates, APSA, *passim*; A. Noyes to J. Ainsworth, 20 Apr. 1883, APSA, 4:*Coloma-Besse* vouchers. Assorted documents in box 4 also reveal that in order to move goods the owners sometimes had to resort to public auction once the market had been saturated.

74. A. Noyes to J. Ainsworth, 20 Apr. 1883, APSA, 4:*Besse-Coloma* 4. Moreover, the dockyards to which Ainsworth and his associates had been selling appear to have been stagnating by this time. Whether this was a result of poor management or increasing competition from newer docks is not clear. See Tientsin *Chinese Times*, 16 July 1887; *Peking and Tientsin Times*, 21 Apr. 1894.

75. O'Brien to Hawkins, 5 June 1886, APSA, 23:*Besse* 7; 8 June 1887, APSA, 24:*Besse* 8.

76. J. Ainsworth to Hawkins, 7 Apr. 1888, APSA, 25:general accounts; Portland *Oregonian*, 24 Jan. 1903. For information on George Ainsworth and on the later career of *Alden Besse*, see Newell, ed., *McCurdy Marine History*, pp. 200, 214, 314.

77. U.S. Bureau of Navigation, Treasury Department, *Merchant Vessels of the United States, Annual List for 1895* (Washington, D.C., 1896).

Hong Kong described it: "A small quantity of Oregon pine is brought from the United States. . . . The . . . bulk of it being brought by one vessel, a bark of about 800 tons, which makes one voyage a year."[78] The *Oregonian* was more specific: "The Coloma would remain here during the Summer and in the Fall would sail for Hong Kong, often carrying a passenger or two, and in May she would return loaded with rice, Chinese passengers, Chinaware, rattan chairs, camphor-wood chests and curios of all kinds. Her sailing away and arrival here were the events of the year."[79]

In time there emerged a new pattern of trade between the West Coast and the Orient, in which steamships played the dominant role.[80] Even a "timber famine" in Hong Kong in 1898 failed to rejuvenate the old trade. Unable to compete, the *Coloma* was sold to San Francisco interests and entered the coastal lumber trade. Cyrus Noyes, his family now living in Hong Kong, left the bark on which he had served so long and took a position as an officer on a steamer running between San Francisco and Hong Kong. Both ship and master had come to terms with steam.[81]

From first to last John Ainsworth's participation in the Pacific lumber trade stood outside the mainstream of the business. He was an entrepreneur seeking return on an investment in merchant vessels. As such, he sought to dispatch the most profitable cargoes he could find. So long as he could maintain a two-way traffic in goods and find charterers who would pay well, he continued in the trade. But ships, not mills, were his fixed investments; and if something other than lumber and spars had provided his most profitable westbound cargoes, he would not have hesitated to load it in place of forest products. For the mill companies the problem was different: they had to ship lumber if it was at all possible, for their fixed investments ashore had to be kept busy if they were to turn a profit. Their vessels might show a loss, but so long as their overall operations showed a gain, the giants of the lumber industry continued shipping lumber to transpacific markets.

78. USBFC, *Special Consular Reports*, vol. 11: *American Lumber in Foreign Markets* (Washington, D.C., 1894), pp. 87–88.
79. Portland *Oregonian*, 24 Jan. 1903.
80. For example, see Frazar & Co. (China) to J. L. Howard, 1 July 1885, Frazar & Co. to E. Smith, 18 Aug. 1885 and 12 Oct. 1887, Frazar & Co. to J. Muir, 18 Nov. 1885, Frazar & Co., memos of cables, 12, 16 Sept., 2 Oct. 1885, OICP. See also Newell, ed., *McCurdy Marine History, passim*, and below, chap. 12.
81. Hong Kong *China Mail*, 15 June 1898; Portland *Oregonian*, 24 Jan. 1903. On the fate of the *Coloma*, see Newell, ed., *McCurdy Marine History*, p. 127.

Ainsworth's business was also less restricted geographically than was that of the millowners. He bought where he could obtain the best rates. On later voyages his vessels carried lumber from mills other than Weidler's, including the Tacoma Mill Company and the Moodyville mill on Burrard Inlet.

At the same time, Ainsworth's trade was typical. Like the profits of the mill companies, the returns from Ainsworth's activities brought further wealth into the Northwest, and his sales in distant ports created additional demand on the mills of the West Coast. Moreover, the problems he and his captains faced in marketing lumber in Hong Kong and Melbourne were largely the same as those faced by all engaged in marketing Douglas fir or redwood around the periphery of the Pacific. The problems of finding and satisfying customers, of meeting competition, of finding ways to return profits to the West Coast, of predicting prices and estimating returns on the various items available for shipping, of minimizing overhead and maximizing profits were problems Ainsworth shared with all who were engaged in the lumber trade of the Pacific, with the giant operations that owned their own fleets, such as those of Pope and Talbot and Asa Mead Simpson, as much as with the captain of the *W. H. Besse*, seeking to send his little bark where it might earn him a living.

Taken by themselves, the cargoes of lumber carried out across the bar of the Columbia by Ainsworth's vessels could have had but little effect on the lumber mills along the river. But Ainsworth's were not the only vessels carrying lumber from the mills of the area. As railroad construction and Oregon's economic growth increased local demand, the sawmills servicing this demand grew. With an increasing number of vessels calling at the river, with the mills growing, and with vessels hauling cargoes of lumber to Hong Kong, it was only a matter of time before others began to carry similar cargoes to markets elsewhere around the Pacific Basin. Though they were smaller than the mills that dominated around Puget Sound and along the redwood coast, and though they were more dependent than the giants on local capital and local markets, by the time the transcontinental railroads arrived at Portland in 1883 the major mills along the lower Columbia already had acquired a relatively broad market base. They had become significant, if not dominant, in the Pacific lumber trade. The Willamette Steam Mills proudly stated on its letterhead that it had lumberyards in

Shanghai and Melbourne.[82] Ocean-going vessels were loading regularly at this and other mills along the river.[83] By making clear the potential that such trade held, John Ainsworth had helped to bring this about.

82. The records of the mill apparently burned with the plant in the 1890s; however, a few items have survived in other collections. See APSA, 1:*Besse* 1; OICP, *passim*.
83. Portland *Oregonian*, 20 Nov. 1877, 1 Jan. 1882. For a more complete discussion, see Thomas R. Cox, "The Lower Columbia Lumber Industry, 1880–1893," *OHQ* 67 (1966):160–78.

CHAPTER VIII

Bar Harbors
and Outports

WHILE large firms that were to dominate the lumber industry of the West Coast were growing around the quiet waters of Puget Sound and while the area of Portland was moving in directions of its own, sawmills were also springing up at many other spots along the long, forbidding coastline between the sound and San Francisco Bay. Development came to some of the tiny bays and inlets scattered along this coast sooner than it did to others, but in time it came to nearly all of them.

It was not easy to succeed on such a coast. From the Strait of Juan de Fuca in the north to Humboldt Bay in the south, anchorages were, almost without exception, bar harbors, the entrances of which had been clogged with sand by ocean currents moving along the coast. Ingress and egress were often impossible and always dangerous. Many a vessel was lost on the sand-choked entrances even after steam tugs were put into operation to aid sailing vessels over the bars.[1] A letter reprinted in the *Oregon Statesman* indicates the conditions at the mouth of the Umpqua: "The *Ortolan* arrived a few days since, other vessels are now due. The *Ortolan* came up in four days from San Francisco. Various may be the opinions as to her entering the river, hence, a remark in point. She came in safely, but how it happened I cannot tell; the better opinion, however, is that she came in on wheels. . . ."[2] Many other bars were worse.

1. James A. Gibbs, Jr., *Shipwrecks of the Pacific Coast* (Portland: Binfords & Mort, 1957), pp. 272–84 and *passim*.
2. Salem *Oregon Statesman*, 25 Apr. 1853. Not all vessels calling at Winchester Bay were as fortunate as *Ortolan*. In December of the following year Lafayette Balch's schooner *Demaris Cove*

Along California's redwood coast the problem took on a different, but not less dangerous, aspect. Here so-called outports dominated. Though these anchorages presented no dangerous bars to cross, they provided scant protection from storms and surf; if vessels were to receive their cargoes, they had to put in close to the rocky shoreline, where a slight miscalculation on the part of the captain, an unanticipated swell, or a sudden lull or shift in the wind might spell disaster.[3] At most of the anchorages on the redwood coast it was impossible to load from a dock or even from lighters. Long apron chutes were extended from rocky headlands and lumber plummeted down them to waiting vessels. Loading from these chutes, like calling at these harbors, was fraught with risks, as an account left by Captain Carl Rydell of the steam schooner *Navarro* makes clear:

> The lumber is sent down the chute, near the end of which a man operates a brake to check the force with which the lumber descends. The seamen stand ready to catch the lumber as it leaves the chutes. As each man gets a piece of timber he runs with it, lays it down exactly where it belongs, and returns to the chute. [It is] difficult for a man below to catch a timber at the right instant and to get the right hold. If he makes a single slip, or if the man at the brake does not apply it in time, he may be injured or killed. . . . Loading ties, which we called "sinkers," is particularly hard and dangerous work. If watersoaked, as they usually are, one of them is as much as a strong man can carry.[4]

Residents of the coast repeatedly sought funds for harbor improvements from the federal government; but, even when these were obtained, conditions were only slightly better. With or without the improvements that were possible at that time, the harbors remained dangerous to vessels large enough to reap a profit when put to hauling lumber from these ports. Storms and fog, against which the govern-

---

wrecked trying to enter the harbor. A local store, which had a large shipment of goods aboard the schooner, inserted the following announcement in the local paper: "Brown, Drum & Co. have the pleasure to inform their friends that they have made a second deposit in the Sand Bank at the mouth of the Umpqua; in consequence of which they are compelled to request immediate payment of all demands due them." Scottsburg (Ore.) *Umpqua Weekly Gazette*, 23 Dec. 1854. For a similar disaster, see Roseburg (Ore.) *Plaindealer*, 28 Sept. 1878.

3. Karl Kortum and Roger Olmsted, " '. . . it is a Dangerous-Looking Place': Sailing Days on the Redwood Coast," *California Historical Quarterly* 50 (1971):43–58. Humboldt Bay was an exception: like the Northwestern ports, it was a bar harbor.

4. Quoted in Kortum and Olmsted, "A Dangerous-Looking Place," p. 45. See also *Northwestern Lumberman*, reprinted in *Canada Lumberman* 4 (1884):356; Silas B. Carr, transcript of interview, March 1953, Bancroft Library, Berkeley, Calif., p. 4; Richard H. Tooker, "A Revised Preliminary Report on the Operation of Slide Chutes on the Mendocino Coast" (mimeographed; n.p., n.d., San Francisco Maritime Museum, San Francisco).

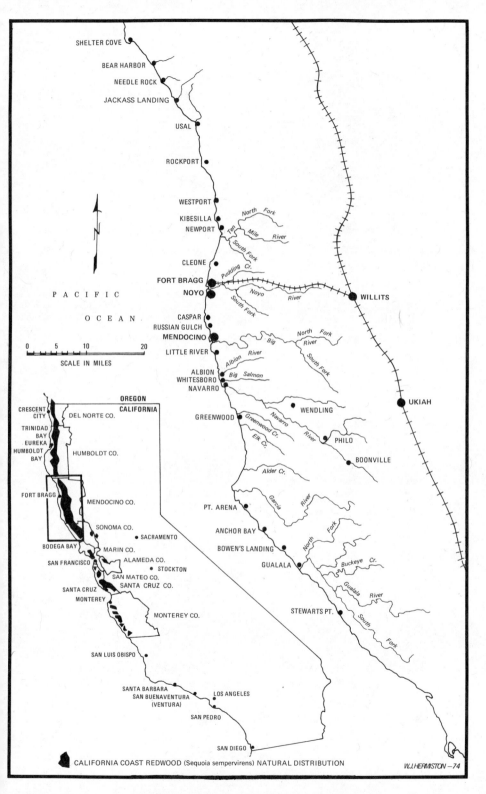

CALIFORNIA COAST REDWOOD (Sequoia sempervirens) NATURAL DISTRIBUTION

W.J.HERMISTON –74

Map 3. California and the Redwood Coast

ment was helpless, continued to plague the area. One sea captain, caught in an unseasonable gale with insufficient ballast, took twenty-three days and 2,017 miles to make it from San Francisco to Humboldt Bay.[5]

Yet along the coast there was timber, and, in spite of the difficulties posed by sand bars and rocks, entrepreneurs were drawn to these areas by the promises of profits that the forests held out. Describing the area around Grays Harbor, one commentator wrote: "All along the Chehalis river and the various streams tributary to it grow the most magnificent forests of fir and cedar." [6] Another described the timber stands along the coast of southern Oregon in even more glowing terms:

I have been nearly thirty-one years in Oregon and Washington Territory, and have a very fair acquaintance with the timber regions of the mouth of the Columbia from the Washougal to the sea, and have seen a good deal of the timber land of Puget Sound and Shoalwater Bay [later Willapa Bay], and in all places have never seen the Coos County forests excelled for density or quality of timber; and, indeed, the white, or Port Orford, cedar of Coos is for fineness and excellence for finishing timber, the best we have in Oregon. Tributary to the Coquille River alone are millions of feet of that variety of timber waiting the lumberman's axes. . . .[7]

More than difficulties of transportation would have had to stand in the way to prevent the tapping of such stands.

Millmen who had too little capital to compete in the areas of the richer stands or who arrived after they had been claimed set up sawmills in places where the stands were smaller and less dense. However small a coastal tract, sooner or later someone always seemed to appear to cut it. Thus, in addition to the large mills that emerged in the more favored locations, a host of small sawmills came into existence along the coast. The topography of the land between the Coast Range and the sea encouraged the trend. Rivers tended to be short and turbulent, providing access only to limited areas. Rocky headlands separated one harbor and its tiny hinterland from the next. The operators of these outposts of industry followed the procedure of the larger mills in shipping their cut to San Francisco. Along this forbidding coastal strip, the local market remained small.

Typical of the larger and more successful of these coastal operations

5. W. Woodley to J. Kentfield & Co., 19 June 1878, JKCP, 3:3.
6. Portland *Oregonian*, 1 Jan. 1882. See also Seattle *Daily Intelligencer*, 17 Oct. 1877.
7. *West Shore* 11 (1885):114–15. See also 2 (1877):150.

was the Gardiner Mill Company, a firm that grew up beside Winchester Bay at the mouth of Oregon's Umpqua River. When the first plant of what was to become known as the Gardiner Mill Company was erected in 1864, the area had already been settled for a number of years. In 1850, filled with hopes of profiting from trade with gold mining areas in the interior and from land speculation, a group of Californians formed the Umpqua Townsite and Colonization Company. In September of that year they dispatched the brig *Kate Heath* to the area with one hundred would-be settlers, merchandise, and machinery for a sawmill. By 1853 Winchester Bay had become a port of call for the steamer *Fremont*, and Captain Josiah B. Leeds had put his schooner *Francis Ellen* into regular service between the Umpqua and San Francisco. The steamer provided mail, passenger, and fast freight service; the schooner furnished an inexpensive way to get the area's lumber to market. In spite of this promising beginning, the area around the mouth of the Umpqua soon went into decline. New land laws passed by the legislature of Oregon killed the land speculation schemes of the Umpqua Townsite and Colonization Company's founders. Trade routes to the interior proved long and arduous. The bar at the entrance to Winchester Bay added a further difficulty, and the gold rush to the interior dwindled. In the face of it all, the company failed and settlers drifted away.[8]

Then, in 1864, entrepreneurs erected a second sawmill on Winchester Bay. Of the new mill, one observer reported: "No better lumber is produced on the Pacific coast. . . . The logs are worked up close [i.e., carefully utilized], all the slabs and refuse lumber being cut into lath. The establishment is not so large as some others, but there is none more complete in its arrangements."[9] With this spur to its economy, the area around Winchester Bay was soon moving ahead once more.

Exportation of lumber from the mill began in September 1864, when 53,907 feet of lumber were sold to the captain of the *Flying Dart* for $11 per thousand at dockside. The captain sailed to San Francisco to market his purchase.[10]

8. Verne Bright, "The Lost County, Umpqua," *OHQ* 51 (1950):211–26; Wright, ed., *Lewis and Dryden's Marine History*, pp. 47, 185; Bancroft, *History of Oregon*, 2:174–83; A. G. Walling, *History of Jackson, Josephine, Douglas, Curry and Coos Counties in Oregon* (San Francisco, 1884), pp. 401–4; Scottsburg *Umpqua Weekly Gazette*, 16 June 1854, 31 Mar. 1855; Harold A. Minter, *Umpqua Valley, Oregon, and Its Pioneers* (Portland: Binfords & Mort, 1967), pp. 75–83.
9. Scott, *History of the Oregon Country*, 3:40–42.
10. Gardiner Mill Company, journals, Gardiner Mill Co. Papers, University of Oregon library, Eugene, Ore. (hereafter GMC journals), 1:20.

The sale to the captain of the *Flying Dart* brought only $592. Apparently the owners of the mill—Gardiner Chisholm, George Bower, Dave Morey, and John Kruse[11]—felt that greater profits could be earned shipping lumber to San Francisco on the mill's account. They chartered the schooners *Mary Cleveland* and *W. F. Bowne* to haul their cut when more was ready. In San Francisco they turned the cargoes over to Charles Hanson and Company to sell on commission. This proved hardly more satisfactory than dockside sales. The cargo aboard *Bowne* sold for $3,273.04, but expenses included $1,321.56 for freight (the charge came to $8 to $9 a thousand board feet and $163.75 commission. Expenses on the shipment aboard *Mary Cleveland* were so high that only $2.40 a thousand was left to cover the cost of manufacture.[12]

In the face of such problems the owners of the mill decided to sell. Charles Hanson purchased the plant in 1868, perhaps because its owners lacked the capital to rebuild after a fire had leveled the plant the year before and Hanson saw no other way to recover what the proprietors of the mill already owed him. Hanson kept the mill only three years, selling it in 1871 to Asa Mead Simpson, a rising young lumberman from San Francisco.[13]

The mill prospered under Simpson. He had ample money available from other enterprises to underwrite the expenses of operation. Moreover, Simpson owned his own fleet of sailing vessels to haul the cut to San Francisco, his own tugs to insure safe crossings on the bar of the Umpqua, and his own lumberyards in California through which to market the lumber. By eliminating middlemen whose exactions had plagued the founders of the mill at Gardiner, Simpson had insured that this, and his other mills, would prosper. His fleet of lumber vessels was already one of the most extensive maritime operations on the Pacific Coast. Both it and the chain of sawmills it serviced were to become even larger.[14]

Profits from the operations at Gardiner grew not just because

11. Roseburg *Plaindealer*, 11 June 1899.
12. GMC journals, 1:17, 20, 23, 36, 93–94, 116.
13. Roseburg *Plaindealer*, 11 June 1899.
14. See Stephen Dow Beckham, "Asa Mead Simpson, Lumberman and Shipbuilder," *OHQ* 68 (1967):259–73; Portland *Oregonian*, 13 Jan. 1915, 7 July 1947; A. M. Simpson, "Lumber History of the Pacific Coast," North Bend *Coos Bay Harbor*, 25 Sept. 1913; Bancroft, *Chronicles of the Builders*, 4:429–49; *Portrait and Biographical Record of Western Oregon* (Chicago: Chapman Publishing Co., 1904), pp. 231–35; Walling, *History*, pp. 436–37; Stephen Dow Beckham, *The Simpsons of Shore Acres* (Coos Bay, Ore.: Arago Books, 1971), pp. 1–19.

middlemen were eliminated but also because production was increased. From its opening in September to the end of 1864, some 240,000 board feet of lumber were shipped. In 1865, the figure rose to 1.8 million feet and in 1866 to nearly 3 million feet. By 1866 the schooners *Pacific*, *Noyo*, and *W. F. Bowne* were kept steadily at work hauling away the cut of the mill.[15] The mill suffered from fire in 1867, but when it reopened, production again began to rise. It climbed sharply when Simpson added a new mill—valued at $35,000—in 1876.[16] The following year, he dispatched fifty-four cargoes of lumber from the two mills. Simpson's old plant shipped 6 million feet, the new mill 6.5 million. The smallest cargo, 102,000 feet, went to Santa Monica on the schooner (?) *Ingalls*; the largest cargoes, 410,000 and 450,000 feet, went to San Francisco on Simpson's schooner *Webfoot*. The total shipped in 1877 was smaller, reflecting a general decline in the lumber market. The old mill shipped 5.7 million feet and the new mill 4.8 million. Together the two produced forty-four cargoes of lumber. The greatest portion went to San Francisco, though occasionally cargoes would be sent to minor ports in the area around San Francisco Bay, and in 1877 Simpson dispatched seven to ports in southern California. One cargo was also shipped to San Luis Obispo on the coast of central California that year and another the next. Sales in the area around Winchester Bay were small, constituting only about one-tenth of the total production even after the commencement of salmon canning operations increased local demand. Earlier local sales amounted to even less.[17]

15. GMC journals, vols. 1 and 2 *passim*.
16. Ibid., 36:fly leaf; G. S. Hinsdale to G. S. Hinsdale & Co., 7 Sept. 1878, GMC journals, 36:110–11; Wright, ed., *Lewis and Dryden's Marine History*, p. 185; Walling, *History*, p. 532; *West Shore* 1 (1876):19; Courtland Matthews, "The White City," *Four L Lumber News* 12 (1931):9.
17. GMC journals, 1, 16–17:*passim*; 25:196; 26:326–29, 332–33. Walling, *History*, p. 437. Most voyages from the Umpqua to San Francisco took from ten to fifteen days, though the schooner *Sparrow* once made the run in three. Often more time was spent waiting for favorable conditions than sailing. The barkentine *Webfoot* once waited off the bar of the Umpqua thirty-six days before entering; *Sparrow*, loaded and ready for sea, waited at dockside for three weeks for storms to abate. Roseburg *Plaindealer*, 16 Feb., 2 Mar. 1887.
    Apparently the bulk of the lumber marketed along the central and southern California coasts came from Simpson's mill in Santa Cruz County rather than his northern mills. Trip records of the schooner *California* (GMC journals, 13:2–30) provide good insights into this trade. During 1869–73 the *California* was kept busy hauling lumber, shingles, posts, and other forest products on her southbound passages, and wheat, beans, wool, and other agricultural products to San Francisco on the return. San Luis Obispo, Cayucos, and San Buenaventura were the usual destinations of southbound shipments. By 1873 timber stands accessible from Simpson's Santa Cruz mill were running out, if one can judge from the products being shipped. Shortly thereafter the *California* was transferred to the Umpqua River–San Francisco run.

Costs of labor were always a major part of the expense of production. In December 1877 the payroll at the mill totaled $1,128.62. Wages ranged from $1.15 to $3.27 per day. The number of man-days worked fluctuated with the seasons, but not widely. Payroll records show the mill employing around 570 man-days of labor during summer months and approximately 500 to 530 in winter months.[18] Thirty-two individual employees appear on the mill records for a typical month in 1877. For the year, the company paid out $13,081.20 for 6,489 man-days of labor. Wages had been higher the year before when workers earned $21,122.23 for 6,462 man-days of labor. Wage cuts apparently were one of the easiest ways for millowners to adjust to declines in the price of lumber such as occurred in 1877.[19]

If names are any indication, the labor force was primarily Caucasian. A few Chinese names appear in the records, and the number increased noticeably in the early 1880s. Occasionally Indians worked at the mill; Indians also sold rafts of logs to the mill from time to time.[20]

The lumbermen appear to have been a rough lot. In 1864 Royal A. Bensell was one of a detachment of soldiers sent to the nearby area of Coos Bay to return runaway Indians to the reservation at Alsea. In his diary Bensell recorded: " 'Tyee' Jim and two men go up to Capt Hamiltons. Find some lumbermen on a spree. Jim narrowly escapes being shot. The boys tho't they were short handed and come back." A few days later, having seen even more of the area's inhabitants, he wrote: "The lumbermen up these bayous and Sloughs are the roughest of men. Nearly all are married to Squaws or else have a written obligation that [they] will marry rather than allow the Ind[ian] Agt to deprive them of their concubines. They conceal the Indians, warn them, and otherwise enhance the difficulties of catching the red devils." Disenchanted with the whole business, Bensell added, "The fact of the business is, this rowing after Siwash is no [proper] part of a soldier's duty."[21] There are numerous indications in the records of the

18. GMC journals, 18:*passim*; 26:7–8, 330, 331; 43:49; 44:219. Unlike sawmill operations, logging camps ceased work during the wet winter months, for mud made it impossible to get logs out of the woods. The mill's winter operations were made possible by logs stockpiled during the dry season.

19. Ibid., 26:7–8, 330.

20. Ibid., 1:48 and *passim*; 14:42, 45, 76 and *passim*; 15:4, 6, 63, 399; 17:332; 35:399–426, 580–85.

21. Royal A. Bensell, *All Quiet on the Yamhill: The Journal of Royal A. Bensell . . .* , ed. Gunter Barth (Eugene: University of Oregon Books, 1959), pp. 145, 146. Cf. *Canada Lumberman* 5 (1885):235. Bensell later had the first mill on Yaquina Bay. By 1868 he was shipping lumber to San Francisco. See Bensell, *All Quiet on the Yamhill*, pp. 194–95.

Gardiner Mill Company that its labor force was little, if any, more polished.

Yet, the type of men described by Bensell did not make up the entire population of the coastal logging communities. C. B. Watson, collector of customs at Coos Bay in the 1880s, remembered the area far more fondly than did Bensell:

This little bay affords an interesting and lively appearance. Its people are largely from Maine, Massachusetts and other northeastern states. Many of them have been reared to a seafaring life and many more come from the Maine woods. The manners are essentially of the New England type and the people are noted for their generosity and hospitality so characteristic of the country from whence they hail. Many are from Boston and never allow you to depart without impressing that fact on you. But they are good people and no more enjoyable time could be had than at a New England clam-bake and celebration at Coos Bay.[22]

Even when allowance is made for the increased degree of culture that the passage of time undoubtedly brought, it seems that Bensell and Watson were describing different elements of the population. Perhaps time had romanticized what Watson remembered as well. The gulf between common laborers and the upper levels of society was undoubtedly great. This was probably as true at Gardiner and the other lumber ports as at Coos Bay.

Logs were an even larger item of expense than labor. Log prices appear to have been somewhat more stable than wages, though they did fluctuate. Prices varied with length, quality, and species. In 1864, when the old mill began operation, prices paid for logs ranged from $1.50 to $3.90 per thousand feet. Cedar logs were at a premium: at least one purchase of cedar logs was made at $5.00 per thousand. In 1868 and 1869 log prices rose temporarily. Most purchases were for $4.50 to $5.00 per thousand for good-quality fir logs, which would have brought $3.00 to $3.50 in 1864. By 1877 the price had dropped to $3.50 or $4.00. In 1876—when the cost of labor at the two mills was $21,122.23—the mills purchased logs containing 6,636,098 feet for $29,993.36. That is to say, logs cost about 17 percent more than the labor used in the mills. In 1877 the differential was even greater. With $24,549.73 paid for logs containing 5,448,539 feet,[23] and wages

22. C. B. Watson, *Prehistoric Siskiyou Island and Marble Halls of Oregon* (n.p., 1909), p. 103.
23. GMC journals, 1:11, 19–20; 3:34; 4:27; 18:36; 19:213; 23:*passim;* 25:201; 26:5, 20–21, 340–43. See also Portland *Oregonian,* 1 Jan. 1882. These charges were for logs delivered at the mill, that is, they included costs of boomage, towing, and the like.

declining to $13,081.20, the log costs were about 30 percent higher than the cost of labor in the mills.

Fixed expenses helped to make log costs less flexible than wages. Perhaps, too, mill operators were inclined to make millworkers absorb more of the decline in lumber prices than the loggers because they knew that laborers would be easier to replace than loggers, who had the equipment and "know-how" to get enough sawlogs out of the woods to keep the mills running. It probably seemed sensible to try to keep the loggers who supplied the mills contented. Just how many loggers there were is not clear, but records for the early 1880s show that there were at least seven logging camps supplying the Gardiner Mill Company at that time.[24]

The above statistics indicate that the Gardiner Mill Company was exporting from the area of Winchester Bay items with a relatively high resource content and low labor content. In view of the fine resource base of the area and the small population, this is hardly surprising. Regions tend to export those things that contain high percentages of the factors of production of which they have an excess. Rough, green lumber with its high resource content and relatively low labor and capital content furnished Winchester Bay with such an item.

Unfortunately, analysis of the operations of the Gardiner Mill Company must be left incomplete. The accounts of Simpson's fleet of lumber droghers are not in the company's records, nor are the records of the sales of lumber through Simpson's yard in San Francisco. Presumably both sets of records were kept in San Francisco, where Simpson had his central office, and were destroyed by the earthquake and fire of 1906. Perhaps they were kept instead in the Simpson mansion near Coos Bay, which burned in 1921. Either way, they are no longer extant.

Whatever the actual cost of transportation and whatever may have been the price at which the lumber sold in San Francisco, it is clear that the mill was making money. The Gardiner Mill Company, like Simpson's business as a whole, grew larger. Not including labor, the old mill cost $5,167.98. The new mill added in 1876 cost $35,000. Yet, in spite of a fire that destroyed the mill and thirty-nine buildings in Gardiner in 1881, by 1885 the two, together with landholdings and merchandise in the company store, were valued by the county assessor

24. GMC journals, 35:536.

at $46,150. In 1890 they were assessed at $68,695. The operation clearly was not languishing.[25]

The empire in lumber and ships built by Asa Simpson did not begin with the Gardiner Mill Company. In the fall of 1850 Simpson and an old friend from Simpson's home state of Maine, S. R. Jackson, joined forces to open a lumberyard in Stockton. Simpson was soon looking for sources of lumber from which to stock their yard. He sailed northward along the California and Oregon coasts to find them and in Astoria purchased an unfinished mill. In 1852 he opened another lumberyard, this one in Sacramento, and again sailed along the coast of Oregon searching for sources of lumber. During the trip he called at Winchester Bay, where his crew cut pilings to haul back to San Francisco.

The mill at Astoria was a short-lived operation, but during his travels Simpson had become convinced that a lasting sawmill operation could be built at Coos Bay. Vessels could be kept busy hauling not only lumber but also coal from Coos Bay, an advantage which set it apart from other harbors along the coast of Oregon.[26] In 1856 Louis P. Simpson, a brother, sailed the brig *Quadratus* to Coos Bay with machinery and supplies for founding a mill. After a delay caused by grounding on the bar and the drowning of Louis Simpson, the mill was built. Asa Simpson's sawmill empire had been started. Like those of Pope and Talbot and other leaders on Puget Sound, it emerged from the vertical integration backward of a California lumber dealer's operations.[27]

Simpson's holdings expanded rapidly. By 1868 he and his associates had the largest mill on Coos Bay, a mill on the Umpqua, and another

25. Ibid., 1:7; 36:fly leaf. Roseburg *Plaindealer*, 31 July 1880. Douglas County Assessment Roll, Douglas County Court House, Roseburg, Ore., 1885 volume, 1890 volume. Walling, *History*, pp. 436–37. Matthews, "The White City," p. 36.
26. Beckham, "Asa Mead Simpson," pp. 259–63; Bancroft, *Chronicles of the Builders*, 4:432–33; *Portrait and Biographical Record*, pp. 232–35; North Bend *Coos Bay Harbor*, 25 Sept. 1913. The harbor and coal fields at Coos Bay were surveyed and given publicity by Nathaniel Crosby, Jr., either in 1853 or 1854. Crosby had maps of the harbor printed and made available to the public. Coal was an important item of commerce during the first two decades of commercial activity at Coos Bay. See Crosby to Joseph Lane, 22 July, 13 Aug. 1854; Crosby to Waterman, 9 Apr. 1854, reprinted in Scottsburg *Umpqua Weekly Gazette*, 12 May 1854; Portland *Oregonian*, 29 Mar., 6 Apr. 1854; *West Shore* 8 (1882):26, 35.
27. *Portrait and Biographical Record*, p. 232; Salem *Oregon Statesman*, 13 May 1856; James F. Imray, *Sailing Directions for the West Coast of North America between Panama and Queen Charlotte Island* (London, 1868), p. 253; Beckham, "Asa Mead Simpson," pp. 263–65; Bancroft, *Chronicles of the Builders*, 4:435.

being built near Santa Cruz. They also owned a shipyard at North
Bend and twenty-four vessels, sixteen of which traded regularly to
Coos Bay and the Umpqua, five to the Columbia, and three to Puget
Sound.[28] Though Simpson had rid himself of the plant at Santa Cruz,
by 1882 his holdings had increased to seven sawmills: one each at
Shoalwater Bay, Knappton, Coos Bay, Crescent City, and Boca, and
the two plants of the Gardiner Mill Company on the Umpqua. Later
he erected the Northwestern Lumber Company mill at Grays Harbor.
The sawmills provided Simpson's yards with all the major commercial
varieties of lumber: Douglas fir and spruce from the northern mills,
Port Orford cedar from Coos Bay, redwood from Crescent City, and
sugar pine from Boca in the Sierra. The mills reportedly employed five
hundred workers. In addition to his yards and mills, Simpson had at
least fifteen vessels, two shipyards, and a central office located on
Market Street in San Francisco.[29] If anyone could lay claim to the title,
"King of the Lumber Coast," it was Asa Mead Simpson.

More than a quest for increased sales lay behind the growth of the
house of Simpson. Equally important was the captain's desire to free his
enterprises of dependence on others, a dependence which, if nothing
else, would allow outsiders to profit from his activities. This was money
Simpson preferred to keep for himself. He acquired sawmills to supply
his lumberyards with all the types of lumber for which there was apt to
be a call. Then he acquired a fleet of vessels to link the two.[30] Knowing
that other vessels would have to be added if the fleet were to expand
and losses be replaced, he established shipyards of his own. Realizing
that steam tugs were necessary if entry into the harbors of the coast
were to be made safe, and that others would supply them if he did not,
Simpson purchased tugs and put them to work along the coast. He
would charge others for towing their vessels across the bars, not let
them charge him. In the terminology of a later generation, Simpson
sought to integrate his holdings both vertically and horizontally.[31]

28. Portland *Oregonian*, 1 Aug. 1868.
29. Ibid., 1 Jan. 1882. *West Shore* 1 (1876):1; 8 (1882):26, 35. Hittell, *Commerce and Industries*, p.
594. Wright, ed., *Lewis and Dryden's Marine History*, p. 179. See also Bancroft, *Chronicles of the
Builders*, 4:439–41.
30. In reference to his own holdings, Robert Dollar reported that he was able to cut his
transportation costs in half by owning his own vessels. See Robert Dollar, *Memoirs of Robert
Dollar*, 4 vols. (San Francisco: for the author by W. S. Van Cott & Co., 1917–25), 1:321.
31. Robert E. Johnson, "Schooners Out of Coos Bay" (master's thesis, University of Oregon,
1953), pp. 5–6. Bancroft, *Chronicles of the Builders*, 4:436–37. Beckham, "Asa Mead Simpson,"
pp. 265–67. *Portrait and Biographical Record*, p. 232. Wright, ed., *Lewis and Dryden's Marine*

The tug *Triumph* bringing in the schooner *Onward* across the Coquille River bar. The outbound crossing, deep-laden with lumber, could be even more hazardous and was sometimes impossible for weeks on end both on this and other bar harbors. Courtesy of the Victor C. West Collection

Ox teams could pull unbelievable loads over skidroads, as this picture taken at Robert Dollar's woods operations near Usal, California, makes clear. Eighteen logs, totaling 12,000 board feet, are here en route to the mill. Courtesy of the Bancroft Library, Berkeley, California

Railroads transformed woods operations in Washington just as they did elsewhere. This ten-ton Porter rod engine is said to be the first acquired by the Polson Logging Company, a firm that also pioneered in the use of splash dams and steam donkeys. Courtesy of the Photography Collection, Suzzallo Library, University of Washington

The West Coast Lumber Company (formerly the West Coast Redwood Company) in San Diego. Its stock of lumber can be glimpsed behind its office, shown here in the 1890s. Courtesy of the Historical Collection, Title Insurance and Trust Company, San Diego

Simpson's attitude toward insurance is illustrative of his overall outlook. From first to last he refused to insure either his vessels or their cargoes. As he once explained:

I have never insured and am now interested in twenty five craft steam and sail all actively employed with no insurance. My experience has been one of bad luck having been interested in more than thirty vessels since "forty nine" that have been wrecked and still I am largely ahead in my insurance account. Counting legitimate wrecks only, two per cent would have left a handsome profit on all my shipping business but the largest of my losses have been caused by a lack of caution so outrageous that it amounted to Vandalism almost if not quite. No less than eight vessels of the number were run ashore on the broad side of America without any sort of reason whatever and still three pr cent will a good deal more than cover all losses. . . .[32]

In other words, Simpson believed it less costly to cover his own losses than to pay insurance companies to do so.

This does not mean that Simpson was a gambler. Indeed, he was quite the opposite. When asked for advice on how to handle the vessels of Henry Villard's Oregon Improvement Company, Simpson wrote that the company should "stop plunging. . . ." It is better to be slow and cautious rather than precipitous, he added, for "the world was not made in a day and time is not always money. . . . Avoid all rush and the risk is less than $1\frac{1}{2}$ pr cent on this Pacific Ocean to safe ports." [33] In a second letter, he made his point even more strongly. He detailed how a vessel in which he held a small interest had been insured at the insistence of Simpson's partners. The vessel, the barkentine *Portland*, was struck by a steamer. Simpson wished to settle out of court with the owners of the steamer, but the insurance company demurred. The case was taken to court where it dragged on for three years before being settled in favor of the steamer even though, according to Simpson, any seaman could see that all the evidence was against her. Simpson summed up his feelings: "*Moral* I don't want any law suits; I don't want any insurance, one is uncertain, the other is N. G." [34]

Simpson was no armchair administrator. A sea captain by training,

---

History, pp. 75, 176–77. William Tichenor Papers, Oregon Historical Society, Portland, 1:40; 2:22–26. Integration was common in the lumber industry. See Wilson Compton, *Organization of the Lumber Industry* (Chicago: University of Chicago Press, 1916), pp. 49, 51–52; Willard L. Thorp, *The Integration of Industrial Operation*, Census Monograph 3 (Washington, D.C.: U.S. Government Printing Office, 1924), pp. 61–62, 246–48, and *passim*.

32. Simpson to Elijah Smith, 5 Nov. 1895, OICP. Cf. A. Pope to Wm. Pope & Sons, 4 June 1857, PTP, Pope corres.

33. Simpson to Smith, 5 Nov. 1895, OICP.

34. Simpson to Smith, 20 Nov. 1895, OICP.

he long continued to serve as master of vessels in his growing fleet.[35] He knew both ships and shipbuilding and did not hesitate to give detailed advice to the master shipbuilders who ran his yards. The "Shipyard Lumber Book" of Simpson's mill at Hoquiam is full of copies of instructions from the captain regarding vessels under construction there. They included advice on how to keep costs down, what sort of wood to use, where to put various fittings, the method to use in putting in ship's knees, and what the general lines of vessels should be. Practical considerations were behind his directions. For example, he wanted a vessel that was being built for use on runs to bar harbors to "have a long straight part on her bottom to rest upon in case of grounding with a full load on." Simpson had seen vessels with overly curved keels break their backs when they grounded while heavily laden. He also ordered that his vessels be designed with sufficient dead rise to allow them to sail without ballast when empty and, to add versatility, so that they could haul coal or grain if he should choose to use them in that manner.[36]

Simpson had strong opinions about the vessels turned out in his yards, too. He called the *Ranger* "a very bad job" and labeled the work of a man reputed to be one of the best designers on the coast "a bad bungle." Simpson added, "I am afraid he is a humbug." [37]

Simpson's master builders sometimes had objections to the captain's instructions. Robert E. Johnson interviewed the son of one of Simpson's master shipbuilders and described an incident related to him in these words:

A. M. Simpson, a very dignified man who was always formally attired, came aboard in his usual dress and told [master builder K. V.] Kruse that they would put the crew's head under the forecastle deck instead of in the forward deckhouse. Kruse protested that the deck was too low, so Simpson undertook to prove him wrong. Squatting down, he backed into the space but was unable to lower his trousers. He crept into the open and stood erect to half-mast his nether garment and again backed under the forecastle.

35. Tichenor Papers, 1:40; 2:27–31.
36. Simpson to [George H. Emerson (?)], 12 July 1887, 28 Mar., 2, 10 Apr., 12 Nov. 1888, 5 Dec. 1890, 16 Apr. 1891, and other dates, "Shipyard Lumber Book," Hoquiam Mill Co. Papers, University of Washington library, Seattle. See also G. Emerson to Simpson, 20 Apr. 1891, EL; Bancroft, *Chronicles of the Builders*, 4:436–37; Johnson, "Schooners Out of Coos Bay," p. 32; Newell, ed., *McCurdy Marine History*, p. 48; G. F. Matthews, "The Shipbuilding History of the Matthews Family" (mimeographed copy, Bancroft Library, Berkeley, Calif.), 1:9–10; Seattle *Daily Intelligencer*, 15 Aug. 1877.
37. Simpson to [G. Emerson (?)], 21 Mar. 1888, "Shipyard Lumber Book," Hoquiam Mill Co. Papers; Simpson to G. Emerson, 16 Apr. 1891, ibid. Cf. Pope & Talbot to Puget Mill Co., 18 July 1890, AC, incoming corres.

This time his head hit a deck beam with a resounding crash. He considered his point proved, but directed Kruse to have a niche cut in the deck beam so that no more heads would be cracked on it. Simpson was leaving for San Francisco the next day, so Kruse decided to wait until he had departed and then install the head in the foreward deckhouse. Simpson must have sensed something of Kruse's reluctance, however, for early the next morning he went aboard the new vessel with an axe and notched the offending beam himself. As soon as he had sailed, the workmen replaced the notched beam and put the head in the place Kruse thought suitable.[38]

Thwarted though he was this time, Simpson's influence must have played a major role in helping his yards to achieve their pre-eminent position among West Coast shipyards. His yard at North Bend turned out the first four-masted schooner on the West Coast and the first five-masted schooner built in the United States for ocean service. The yard also produced the first four-masted barkentine. His vessels had less sheer than most built on the West Coast, probably, as has been indicated, in order to make it less likely that their backs would break if they should go aground crossing a sandbar.[39] By and large the products of Simpson's shipyards proved themselves both sturdy and practical. These qualities were recognized by others. Nearly half the vessels built at the yard in North Bend were sold rather than kept for use in Simpson's own fleet.[40]

The finest product of any of Simpson's yards and the highest tribute to the captain's talent as a marine engineer was the clipper ship *Western Shore*, one of only three full-rigged ships produced on the Pacific Coast. It was the only thorough-going clipper. At the time of her construction, the 1,188-ton *Western Shore* was the largest vessel yet produced on the West Coast. The vessel had a brief but brilliant career, setting, or very nearly setting, records on every run. When *Western Shore* visited Puget Sound in 1877, the Seattle *Daily Intelligencer* welcomed her as "one of the finest that ever entered our harbor, and [one which] enjoys the enviable reputation of being the fastest sailing ship afloat." [41] Yet, when *Western Shore* had first arrived in San Francisco to load grain for England, many were contemptuous of this not-particularly-beautiful ship from Coos Bay that Simpson dared to call a clipper. Simpson was unperturbed by the scoffers. As

38. Johnson, "Schooners Out of Coos Bay," pp. 34–35.
39. Ibid., pp. 25–26, 29–31; Simpson to [Emerson (?)], 12 Nov. 1888, "Shipyard Lumber Book," Hoquiam Mill Co. Papers.
40. Portland *Oregonian*, 13 Jan. 1915.
41. Seattle *Daily Intelligencer*, 5 Feb. 1877.

one contemporary reported, "The quiet and unpretentious Yankee who designed her . . . sat back and silently munched his quid in the solitude of his little office on Market street. He was 'waiting the verdict.' " [42] On the voyage to Liverpool, *Western Shore* decisively beat two ships with established records as fine sailers, the *Three Brothers* and *British King*. As if to prove this was no fluke, Simpson's little clipper also beat them back. Other fine passages followed. *Western Shore* cost $86,000 to build; once she had proven her worth, Simpson could have sold her, according to the *West Shore*, "any day of the week" for $100,000.[43]

In 1878, only four years after she had been launched, *Western Shore* struck Duxbury Reef off San Francisco while doing twelve knots. With much of the ship's bottom torn out, the crew abandoned her. She soon broke up. Simpson never replaced her. He abandoned the grain trade, for participation in which she had been built, and concentrated henceforward on the lumber trade. *Western Shore* has relevance to this study not as a carrier of lumber but as an example of Simpson's knowledge of the sea and ships. It was a knowledge that served him as well in the lumber trade as it did in the grain trade, and for a far longer period of time.[44]

Simpson played as active a role in directing the operation of his sawmills as he did in directing his shipyards. He once removed John Wood as manager of his mill on Shoalwater Bay because Wood had replaced the inefficient Chinese laborers that had been used at the mill with more highly paid white workers without first consulting Simpson. Wood was replaced by R. B. Dyer, who in time also ran afoul of Simpson. By the end of the nineteenth century the mill was old and in need of repairs. Dyer closed the mill, had it torn down, and replaced it with an up-to-date plant. When the news of what was being done reached Simpson, he took the first vessel headed for Washington.

42. *West Shore* 1 (1876):9.
43. Ibid.
44. Ibid.; Johnson, "Schooners Out of Coos Bay," pp. 19–21; North Bend *Coos Bay Harbor*, 13 May 1915; Lubbock, *The Down Easters*, p. 87; Bancroft, *History of Oregon*, 2:728; Wright, ed., *Lewis and Dryden's Marine History*, p. 264; Beckham, "Asa Mead Simpson," pp. 267–68; Marshfield (Ore.) *Sun*, 13 July 1939; Henry Hall, *Report on the Shipbuilding Industry of the United States* (Washington, D.C., 1884), p. 133; Bancroft, *Chronicles of the Builders*, 4:437–38. Hall's claim that the *Western Shore* once made the run from the Columbia River to Liverpool in 97 days is not corroborated by other sources, but the 103- and 110-day passages from San Francisco to Liverpool and 100-day passage from Astoria to Liverpool that others acknowledge were spectacular enough. Perhaps Hall's report refers to time from outside the bar of the Columbia to Liverpool, while the 100-day report refers to time from dock to dock.

However, by the time he arrived the new mill was already under construction. Angered, Simpson transferred Dyer to his mill at Knappton. The captain's ire passed when the new plant Dyer had laid out proved a major improvement over its predecessor, exceeding all previous production records.[45]

The organizational details, and perhaps even the true extent, of the holdings of Asa Simpson will probably never be known because of the destruction of the bulk of his records. It is clear, however, that Simpson had varying degrees of interest in a host of enterprises. There was no all-embracing corporate entity to bring order into the structure. At base, it was an old-fashioned business empire built on a maze of proprietorships and partnerships. It tended to be a many-faceted and dynamic empire because Simpson himself was many-faceted and dynamic. It is hardly surprising that this structure, the length and breadth of which were so very nearly the length and breadth of the shadow of its founder, did not long survive his death.[46] Rugged individualist that he was, it somehow seems appropriate that Asa Mead Simpson made his fortune on that rugged coast of bar harbors and outports where so many others failed.

Simpson was important, but he was hardly typical. There were dozens of sawmill operations scattered along the coast between San Francisco and Puget Sound. A few were large; most were small. Information on the smaller mills is difficult to come by. Most kept scanty records; many were short-lived; numerous mills burned and their records were destroyed with them. The locations, even the names, of a large number have doubtless been forgotten. Fortunately, however, some sources are extant. One of the most important is the papers of William Kyle. Indeed, Kyle's records give insights not only into the problems of the small coastal sawmill operator but also into those of operating bar tugs and lumber schooners, as well as other enterprises not connected with the lumber industry.

William Kyle was a small-time entrepreneur who resided in the village of Florence at the mouth of Oregon's Siuslaw River. His business interests were varied. At one time or another, they included a salmon cannery, a store in Florence and another upstream at Mapleton, the bar tug *Robarts*, a smaller steamer for hauling goods on the river,

45. Unidentified newspaper account, Dubar Scrapbook no. 117, University of Washington library, Seattle, pp. 46–47.
46. Beckham, *The Simpsons*, pp. 23–36.

the three-masted schooner *Bella*, and the Florence Lumber Company. His ambitions were unlimited. He dreamed of gaining a monopoly on the trade and business of the valley of the Siuslaw River and perhaps of the entire central portion of the Oregon coast.[47] As he once avowed, "There is plenty of capital to be got, and I for one am going to handle some of it." [48]

But while Kyle's ambition was great, his supply of capital was not. He gained his start with his salmon cannery, an enterprise financed by one Michael Meyer of Astoria. The firm of Meyer and Kyle, made up of the various enterprises in which Kyle was interested, was an operation to which Meyer contributed the initial capital and Kyle his time and energy as managing partner. In time Kyle bought out Meyer's share of the business and it became known as William Kyle and Sons. Under both names the operations suffered from a shortage of operating capital, much of which had to be supplied by Kyle's distributors in San Francisco. As one businessman put it, "The manufacturer who needs the jobber as a commercial banker is a weak manufacturer." [49] Kyle was, indeed.

Kyle struggled continually to increase the company's assets. As he put it in 1897, profits the preceding year had been $6,473.80, "but all of that is out in new investments." [50] The company's records show that the history of Meyer and Kyle consisted of one financial crisis after another. The firm's capital seemed always to be "out in new investments"; a shortage of operating capital often resulted. Kyle was the epitome of that ever-confident class of boomers, speculators, and would-be entrepreneurs so common on the American frontier. As a whole, the class was long on hope but short on capital and, often, on sound business judgment as well. Kyle could have served as the archetype.

Kyle's store and cannery operations are beyond the scope of this study, but in the early 1890s, feeling the need for improved transportation facilities in order to get his salmon pack to San

47. Meyer and Kyle to C. E. Whitney & Co., 18 Jan. 1897, WKSP, 170. See also F. Rogers to W. Kyle, 25 Dec. 1896, WKSP, 148. At the time of this writing the Kyle papers have not been processed and, because of the large size of the collection, are extremely difficult to use. When processed, this collection should prove an invaluable source for historians. See also Alfred L. Lomax, "Early Shipping and Industry in the Lower Siuslaw Valley," *Lane County Historian* 16 (1971):34.
48. Kyle to G. Miller, 27 Nov. 1899, WKSP, 174.
49. Quoted in Porter and Livesay, *Merchants and Manufacturers*, p. 130.
50. Meyer and Kyle to First National Bank (Eugene), 27 Sept. 1897, WKSP, 170.

Francisco, Kyle purchased the tug *Robarts*. He hoped that with a tug in operation on the bar of the Siuslaw it would be easier to get vessels to come there. Local millowners had foreseen similar advantages in having a tug on the bar and promised to take a large interest in her, but they later reneged. Whether Kyle was upset by the development is unclear, for his reports of the incident varied as the occasion seemed to demand. When seeking a loan, he told a bank in Eugene: "We would sooner hold it all as it is a paying investment." [51] To a person protesting high charges for the tug's towing services, he wrote: "As Far as a Monopoly of this buisiness [sic] is concerned I Can assure you that we neaver would have had anything to do with it but for the mills here who promised to take ⅓ Each of the Boat when we bought it. And on the strength of these promises we have invested all our means in her. She is not today paying interest on the Money invested in her. . . ." [52] Kyle once stated that it cost $20 a day to run the *Robarts*. If that figure is correct, it seems unlikely that the vessel earned any great profit doing what little towing was available in the area. [53]

Millmen found the quality of the tug's service a source of despair. The vessel seemed to run wherever the quickest profits could be turned. One wholesale lumber dealer in San Francisco asked rhetorically: "When she is passing the Alsea River & Schooners are waiting to be towed out & . . . the Tug passes on & leaves them to wait 6 or 8 days, it dont look as though the Tug wanted to accomodate much, does it?" On another occasion, the same complainant pointed out that "accomodations for towing are so uncertain that vessels are loth to going to the Siuslaw river, Vessels lie off the river 10 to 18 days & no accomodation." At least one vessel went aground while waiting offshore for the tug. [54]

Part of the problem was that Kyle used the tug for hauling coastwise freight as well as for towing. The two uses often conflicted. One millowner wrote to Kyle to complain about the service: ". . . when

51. Meyer and Kyle to First National Bank, 16 July 1894, WKSP, 168. For a description of the *Robarts*, see Kyle to Whitney, 23 Aug. 1897, WKSP, 170.
52. Meyer and Kyle to C. W. Farnum, 18 Sept. 1894, WKSP, 168. See also Meyer and Kyle to Harrison Bros., 19 Sept. 1894, ibid.
53. Meyer and Kyle to Benton County Flouring Mills Co., 20 Oct. 1894, WKSP, 169. The tug earned approximately $35 a day for towing services when working; but overhead continued even when the tug was idle or barbound. See Netarts Bay Lumber Co. to Meyer and Kyle, 6 May 1899, WKSP, 154.
54. Farnum to Meyer and Kyle, 7 June 1895, 27 Dec. 1894, WKSP, 146. See also 13 Sept. 1895, ibid.

ever the bar is smooth she is off somewhere with a few tons of freight
and by the time she gets back it is rough again. We think the freight
might wait a day or two when there are two or three vessels laying
under expenses and mill men laying out . . . their money . . . [;] we
think you are injuring your own business as well as ours by not
attending to vessels better when they come here." [55] Indeed, the
Eugene *Guard* went so far as to charge that Kyle purposely kept tug
service poor in hopes outside vessels would stay away so the *Robarts*
could monopolize the port's traffic in freight. The editor of the
Florence paper dissented, arguing that if the river's lumber trade were
cut off for lack of vessels, the business of the entire valley, which
included Kyle's, would be crippled. The editor thus concluded that
Kyle had no possible motives for keeping service poor, yet agreed that
the existing service supplied by the *Robarts* was "a menace to our
shipping industry" and Kyle would "meet with a just condemnation
from all the people" unless he took steps to improve it.[56] There is no
indication that the complaints led Kyle to change the manner in which
the *Robarts* was managed.

In addition to Kyle's shortsighted policies, other problems also
interfered with the service provided by the tug. The vessel's crew
apparently left much to be desired. One customer wrote, "Why the
Devil don't you get a crew, that won't be the laughing stock of the
country, when they are away from home?" [57] The area's harbors
created problems even for steam tugs. The captain of the *Robarts* once
wrote regarding the Siletz River: "You got to get a good Price to go
ther as ther es Rocks ther, and I dont like Rock. I dont mind Sand."

55. Harrison Bros. to Meyer and Kyle, 23 Apr. 1897, ibid. Kyle found the freight business
frustrating: "It seams there is always some way of getting some other boat to do the work in the
summer months and winter falls to us. We have tried hard for a number of years to build up a
trade on Flour and feed for Coast points . . . between Coos Bay and Tillamook but if some tramp
steemer comes along she generally takes the trade from us, in the summer onley, and leaves us the
Winter months to contend with, with there storms and rough Bars." It did not seem to occur to
Kyle that the low quality of service provided by the *Robarts* might have been the reason shippers
turned so readily to others. Meyer and Kyle to Edwin Stone, 21 Sept. 1899, WKSP, 174.
56. Florence (Ore.) *West*, 30 June 1893.
57. D. Devery to Kyle, 19 Aug. 1899, WKSP, 154. Attracting better crews was probably not
possible, however. An agreement dated 1 Feb. 1896 (WKSP, 168) indicates that gross earnings
of the vessel were to be divided so that $\frac{1}{3}$ went to the owners and $\frac{2}{3}$ to the crew. Of the crew's
share, the captain was to get 18/57, the engineer 18/57, the fireman 11/57, and the deckhand
10/57. The crew had to obtain fuel for the tug and furnish all the labor; the owners furnished the
tug itself and supplies. While the gross earnings do not appear to be indicated anywhere in the
Kyle papers, it seems unlikely that the crew of the *Robarts* was clearing much. If better crews had
been available, Kyle might not have been able to get them to work on his tug for such wages.

Later that same month, he reported from Kernville that he had "slid" across the bar of the Siletz on six feet of water and was barbound: "Al We Wont now es your Lumber Trucks and put them under the Tug then We ar Rady for Bissnis. . . . I will get H—— wen I get out and up to Nehalem but I can Stand et I ges. Well this [is] Pryti good Contry insid but tuf to get here; ef I Cant get out to nite then Im Stuk good . . . this es the Worst hol the Tug has bin in." [58] Even with the best of crews, this would not have been an easy coast to serve.

Poor though the service was, mill operators continued to employ the *Robarts*. Lacking the capital to buy their own tugs, the operators of these tiny coastal mills had no practical alternatives. Both they and Kyle realized the absolute necessity of towing service. As Kyle said, the tug was what made everything else possible.[59] Under the circumstances, he was in a position to drive a hard bargain with those desiring the services of his vessel. When owners of a mill on the Alsea River sounded out Kyle regarding the possibilities of stationing the tug there, he replied by pointing out that the four small mills on the Siuslaw cut twice as much lumber as the plant on the Alsea and that the operators on the Siuslaw had signed a contract giving Kyle the exclusive right to their towing business for five years. Kyle made it clear that he would provide better service to operators on the Alsea only if they were prepared to outbid those on the Siuslaw for first call on the *Robarts*.[60] Kyle sent the tug as far south as Coos Bay and as far north as Tillamook Bay and the Nehalem River in order to obtain work and then complained that it was "running around too much on long trips." [61] The fact of the matter was that there simply was not enough business at any of the harbors on the central coast to keep the *Robarts* busy, yet she could neither make money nor provide good service when she tried to cover them all.[62]

The schooner *Bella* was an even greater source of difficulties for Kyle than was the *Robarts*. Like those of the tug, the problems attendant upon the operation of the schooner illustrate the difficulties faced by the entire lumber industry of this rugged coastal region.

58. F. Johnson to Kyle, 2 and 17 Aug. 1899, WKSP, 154.
59. William Kyle & Sons to A. W. Beadle & Co., 24 June 1904, WKSP, 163.
60. Meyer and Kyle to Harrison Bros., 5 Feb. 1895, WKSP, 169.
61. Ibid., 13 Feb. 1895. See also L. Jones to Meyer and Kyle, 4 Apr., 8 May 1899, WKSP, 153, 154; Meyer and Kyle to F. Johnson, 3 July 1899, WKSP, 172; Meyer and Kyle to Hemple and Wheeler, 23, 31 Aug. 1899, WKSP, 174; Meyer and Kyle to G. Cohn, 28 Sept. 1899, ibid.; Meyer and Kyle to E. Stone, 21 Sept. 1899, ibid.; F. Johnson to Kyle, 6 May 1899, ibid.
62. Kyle himself recognized this dilemma. See Meyer and Kyle to Truckee Mill Co., 31 Oct. 1899, WKSP, 154.

Kyle had found it difficult to get vessels to come to the Siuslaw to haul his salmon pack to San Francisco. Millowners on the river had experienced the same difficulty. The way to solve the problem, Kyle felt, was to build a vessel of his own: "We can do nothing here without one, to have a successful business here we must own our own vessles [*sic*]." [63]

As did so many other of the enterprises into which Kyle entered, the building of the *Bella* left him financially overextended. Kyle had planned to build the schooner in partnership with I. B. Cushman, owner of the Siuslaw River Lumber Company, located at Acme, a few miles upriver from Florence. Cushman agreed to furnish lumber for building the vessel; Kyle was to furnish the necessary funds. The plan broke down when Cushman got deeply into debt at Kyle's store and, in order to protect himself against the possibility of Cushman's defaulting, Kyle took over Cushman's interest in the vessel. [64]

Apparently, Cushman had decided that Kyle had so much tied up in the *Bella* that he would have to complete her, so, rather than put his own limited capital into the vessel, he put it into improvements in his mill instead. With the *Bella* running regularly to the Siuslaw, it would be possible to get the area's lumber to market and improvements in the mill would, in all probability, prove profitable. Kyle suspected that this was what was taking place, but there was nothing he could do about it. "We have been Humbuged with one of these saw mills here[;] they have been promising to pay us there account for some time past but instead of paying us they have put a lot of money into new machinery for the mill." [65] Unable to finish the schooner with the funds on hand, Kyle sought a master for the vessel who would take a one-fourth interest in her. [66] He failed to find one, and the vessel sat for a year

63. Meyer and Kyle to First National Bank, 20 July 1896, WKSP, 170. See also Meyer and Kyle to J. O. Schoulen and Co., 2 Nov. 1895, WKSP, 169; Farnum to Meyer and Kyle, 3 May 1895, WKSP, 146; Beadle to Meyer and Kyle, 18 Jan., 5 Feb. 1898, WKSP, 145; Florence *West*, 30 June 1893, 29 Nov. 1895, 12 Feb. 1897.
64. Meyer and Kyle, statement of assets and liabilities, dated 15 Jan. 1897, WKSP, 170; Florence *West*, 29 Nov. 1895, 14 Feb., 13 Nov. 1896.
65. Meyer and Kyle to George H. Tyson, 30 June 1896, WKSP, 170. As Kyle explained on another occasion, others were supposed to take a part interest in *Bella*, "but when it Came to the Scratch to put up they were not in it with anything substantial, but plenty of wind." Meyer and Kyle to J. Haviside, 11 June 1897, ibid. See also I. Cushman to Kyle, 9 Jan., 22 May 1896, WKSP, 148.
66. Meyer and Kyle, statement of assets and liabilities, dated 15 Jan. 1897, WKSP, 170. It was common practice for a captain to be part owner of the vessel of which he was master. As one who would share in whatever profits the vessel earned, he had an added incentive to make fast passages

while Kyle searched for funds to rig her and put on finishing touches.[67]

Even the vessel's launching was plagued by difficulties. When the builders first sought to launch her in November 1896, the ways collapsed and she stuck fast on the riverside mud. There she sat until the following February when finally freed. Nine months and many difficulties later, the *Bella* finally went to sea. She was to be the last ocean-going sailing craft built on the Siuslaw.[68] With the completion of the *Bella*, Kyle's dream of controlling trade on the central coast of Oregon was brought one step closer to realization.

But Kyle's problems were far from over. As launching neared, Kyle expressed the belief that the worst was past: if the mills "prosper so do we."[69] But the mills did not prosper. Cushman's plant soon closed down, and the others were unable to furnish enough lumber to fill a schooner the size of the *Bella*.[70] Kyle had his vessel, but he could not run her to the Siuslaw profitably. He would have to charter her for use elsewhere until such time as sawmill capacity on the river was enlarged. He consigned the schooner to A. W. Beadle and Company, a commission house in San Francisco, for chartering.[71]

A. W. Beadle, managing partner of the firm bearing his name, kept the *Bella* busy. He found work for her at Tillamook Bay in Oregon; Grays Harbor and Olympia in Washington; and Usal, Albion, Needle Rock, Crescent City, and Eureka on the redwood coast. Letters written to Kyle by James T. Smith, a semiliterate sea captain who took a one-sixteenth interest in the *Bella* and served as her master, graphically illustrate the difficulties faced by those serving the bar harbors and outports of the lumber coast.[72] Tillamook, he reported, is one of the "varst Ports on [the] Cost." The *Bella*, a flat-bottomed

---

and otherwise run the vessel efficiently. A master's share generally was less than one-fourth, however.

67. Meyer and Kyle to Beadle, 1 June 1897, 13 Jan. 1898, ibid. See also Meyer and Kyle to Whitney, 21 Oct. 1897, ibid.; Meyer and Kyle to E. Wirschuleit, 16 Sept. 1896, ibid.; Meyer and Kyle to Haviside, 11, 29 June 1897, ibid.; Kyle to H. Jacobson, 29 July 1896, ibid.

68. Florence *West*, 13 and 27 Nov. 1896, 12 Feb., 6 and 26 Nov. 1897. Gordon Newell says the *Bella* was completed in 1896. He also has Kyle residing at Yaquina Bay. Both statements are in error. See Newell, ed., *McCurdy Marine History*, p. 4.

69. Meyer and Kyle to First National Bank, 27 Sept. 1897, WKSP, 170.

70. Cushman's mill, which shipped some 5 million feet of lumber to San Francisco in 1892, was by far the largest on the river. See Florence *West*, 5 May 1893, 3 Feb. 1899.

71. For another lumberman's view of this method of managing ships, see A. Pope to Wm. Pope & Sons, 4 May 1857, PTP, Pope corres.

72. Kyle reported to the local paper that Smith took a one-eighth interest in the vessel, but his records show otherwise. See Florence *West*, 1 Oct. 1897.

vessel with a retractable centerboard, had been specially designed for the shallow bars of the Siuslaw and its neighboring harbors; even so, it had to wait off the Tillamook bar eleven days before conditions were favorable for crossing. Smith told of one schooner, standing offshore when he arrived, that waited thirty-five days to cross the bar, only to give up in the end and sail to Portland. He would not send a vessel of his to Tillamook Bay, Smith added.[73] From Crescent City he wrote: "Here are very bade Place. I was werry Freighth of the wessel. The Foremand on the Warf told my that Sailing wessels have no bisinus here this time of yare. and I think myself that are to big resck to teck for nothing." A few days later he repeated his assessment of conditions in the port: "[We have] had hard time of it her. the Sea have ben Breacket over her for Days and she dregged the morings and Ankers and brock the vindles [windlass]. . . . Tel Beadle that yow dont vont the wessel to go Outside Port no more this Year." From Needle Rock, he wrote that the *Bella* was too big for use in the outside ports; "she might make a few dollars this way and wi might loose it al." [74]

The possibility of a wreck was an important consideration, for underwriters had refused to insure the *Bella*. Kyle heeded Smith's advice and instructed Beadle to cease sending the schooner to such ports.[75] When sent to Olympia to take on a cargo of lumber for San Francisco, Smith found it safer but otherwise no more satisfactory: "I Don't tink this trip are good for Bella. That are ale Reight for a big vessel." [76] Smith knew whereof he spoke. Small vessels such as the *Bella* could not compete with carriers of large capacity on long runs. Their proper area of activity consisted of ports too small for large vessels to enter and close enough to markets so that the expenses of transportation would not exceed the value of the cargo carried. The ports of Puget Sound clearly did not fall in this category.

Beadle kept the *Bella* at work, but the charters he arranged proved unprofitable. When the schooner failed to make money, Beadle blamed the captain, Smith blamed Beadle, and Kyle blamed first one and then

73. J. Smith to Meyer and Kyle, 14 Feb. 1898, WKSP, 148. See also Smith to Meyer and Kyle, 15 Dec. 1897, 19 Feb. 1898, ibid.

74. Smith to Meyer and Kyle, 22, 30 Sept., 4 Aug. 1898, ibid.

75. Smith to Meyer and Kyle, 9 Aug., 30 Sept. 1898, ibid.; Meyer and Kyle to Beadle, 30 Dec. 1897, 2 Sept. 1898, WKSP, 170, 172.

76. Smith to Meyer and Kyle, 1 Apr. 1899, WKSP, 153. See also Beadle to Meyer and Kyle, 19 Apr. 1899, ibid.; Smith to Meyer and Kyle, 22 Apr. 1899, ibid.; Olympia *Washington Standard*, 28 Apr. 1899.

the other. The voluminous correspondence that resulted leaves one with the suspicion that Smith did, indeed, contribute to the vessel's failure to earn profits. He was apparently honest, sober, and reasonably competent, but not the hard-driving type of master necessary to push the vessel along at speeds sufficient to make it possible to turn a profit.[77]

However, Beadle clearly contributed to the failure, too. Anxious to gain commissions on charters, he agreed to send the *Bella* where she had no business going. Moreover, there are indications Beadle managed the schooner in such a way as to further his own enterprises. Both Smith and Kyle suspected as much, but Kyle was so financially overextended that he was in no position to clear himself of the debts he had run up with Beadle so that he could seek a new agent. He tried it briefly, but was virtually blackmailed into accepting Beadle's services once more.[78] As Captain Smith put it, "I see now Beadle work for Beadle as you told me once." [79]

Yet the fault did not lie entirely with Beadle and the captain. The *Bella* was a sorry vessel. The millstuff Cushman had supplied was in short lengths. As a result, the vessel lacked the rigidity and strength that it ought to have had, thus increasing the cost of keeping it seaworthy. In addition, the lack of rigidity meant that the seams worked more than they should have, so that the schooner was often slowed, and sometimes endangered, by leaking. More basically, the *Bella* was poorly proportioned, a fault that both slowed her and hampered her maneuverability.[80]

77. C. D. Bunker and Co. to Meyer and Kyle, 2 Nov. 1897, WKSP, 145; Beadle to Meyer and Kyle, 21 Jan., 5, 9 Feb. 1897, ibid.; Beadle to Meyer and Kyle, 26 Aug., 1 Oct. 1898, WKSP, 150; Beadle to Meyer and Kyle, 9 Nov. 1898, 19 Jan. 1899, WKSP, 153; Meyer and Kyle to Beadle, 19 Nov. 1898, WKSP, 172.
78. Smith to Meyer and Kyle, 9 Aug., 10 Sept. 1898, WKSP, 150; Smith to Kyle, 19 July 1899, WKSP, 154; Beadle to Meyer and Kyle, 6, 17 Sept. 1898, 22 May 1899, WKSP, 150, 154; Meyer and Kyle to Smith, 2 Sept. 1898, WKSP, 172; Meyer and Kyle to Beadle, 20 Sept. 1898, 21 Mar. 1899, ibid.; Meyer and Kyle to S. H. Harmon Lumber Co., 22 Mar., 3 Sept., 13 Dec. 1899, ibid. In the early nineteenth century manufacturers had been men of limited experience and few contacts who needed agents to bring them into contact with customers outside their immediate area. Though most manufacturers had long since developed more efficient means of marketing, Kyle and the other operators of small mills on the Pacific Coast were still utilizing the old system.
79. Smith to Meyer and Kyle, 9 Aug. 1898, WKSP, 150. I. B. Cushman also felt Beadle was profiting unduly at his expense. He once wrote, "I have not heard from Beadle for a week (Expect he is moving in an other safe)." Cushman to Meyer and Kyle, 10 Jan. 1898, WKSP, 145.
80. H. Hansen to Kyle, 9 Mar., 4 Apr. 1898, WKSP, 150; Smith to Meyer and Kyle, 20 Dec. 1897, 9, 20 Jan., 14 Feb., 13 Mar., 14 Nov. 1898, WKSP, 148, 150, 153; Beadle to Kyle, 17, 20, 21 Jan. 1898, WKSP, 145; Beadle to Meyer and Kyle, 9 Feb. 1898, ibid.; Meyer and Kyle to

After fourteen charters the *Bella* was still operating in the red. Not until Kyle obtained a sawmill of his own on the Siuslaw and began to run the schooner to it on a regular basis did the little vessel make money.[81] The operation of lumber droghers on a charter basis would appear to have been a marginal business at best, if Kyle's experience is any indication.

Once Kyle began to haul his own lumber from his own mill, the situation changed rapidly. The *Bella* made twenty-six voyages during the four-year period that commenced when Kyle obtained his mill. She failed to turn a profit on only five trips. Total gain on the twenty-six voyages came to $4,706.40, an average of over $180 per trip. In addition, Kyle's sawmill, stores, and cannery profited from the dependable transportation that became available to them when the *Bella* commenced running to the Siuslaw on a regular basis. Moreover, with increased trade on the river, it became possible to station the *Robarts* at Florence permanently, thus eliminating the costly runs along the coast that had run up expenses while undermining the quality of the service.[82] The integration of holdings, a course which had proven so profitable to Asa Simpson, was beginning to pay dividends to William Kyle as well.

The manner in which Kyle acquired his sawmill is revealing of the code of business ethics on this seacoast frontier. Always operating with a small capital base, Kyle had been pushed to near-bankruptcy when Cushman put his own money into mill improvements and left Kyle to finance the *Bella* as best he could. Kyle's financial problems continued unabated until he obtained a mill of his own. He obtained it by turning on A. W. Beadle the techniques Cushman had used on him. Ironically, Beadle apparently had by this time entered into a partnership with Cushman.

Beadle, 13 Jan. 1898, WKSP, 170. Captain Smith was aware of the problem: "Yow saed Yow expected my to do my best and I do my dear frend, but ven things are not made in proportion that is hard to get out of Axedent and otter Trobble." Smith to Meyer and Kyle, 14 Feb. 1898, WKSP, 148. Local boosters painted a very different picture of the *Bella*. See Florence *West*, 14 Feb., 13 Nov. 1896, 6 Nov. 1897; Lomax, "Early Shipping and Industry," p. 35.

81. Profitable operation out of the Siuslaw was aided by construction of a jetty at the mouth of the river in 1897 and revival of lumbering activity in the area early the next year. See Florence *West*, 19 Mar. 1897, 14 Jan. 1898, 3 Feb. 1899; Alfred L. Lomax, "Early Port Development on the Lower Siuslaw River," *Lane County Historian* 11 (1966):37–39.

82. Summaries of voyages, WKSP, 238; Meyer and Kyle to Harrison and Diven, 17 Dec. 1899, WKSP, 174. Efficiency was no doubt increased even further in 1904 when Kyle obtained a second tug. See Florence *West*, 6 May 1904. The records of the *Bella* are not complete to 1905, when she went aground and was lost. See *The West*, 1 Dec. 1905.

Kyle had agreed to take a sizable interest in a steam schooner that Beadle was having built on the Siuslaw. However, once construction was well under way, Kyle informed Beadle that because of financial reverses he would be unable to take his agreed share. He would like to do so, Kyle insisted, but he lacked the money. Beadle, financially embarrassed by this turn of events, kept prodding Kyle to pay for his share of the vessel, but to no avail. Kyle then turned around and acquired an old mill at Spruce Point on the Siuslaw, refurbished it, and began sawing lumber. Kyle wrote to Beadle that he had to take over the mill in order to get his money out of debts its owners had run up at his store. Another time he told Beadle that he really was lacking in money, that others had put up the funds for the refurbishing of the mill and that he had merely lent his name to the operation. Kyle insisted that his share of the business was a very minor one, but his records belie this assertion. The tone of Beadle's letters to Kyle indicates that he realized as much, but there was little he could do. In time the steam schooner was finished, save for the installation of machinery, which was to be done in California, and was readied for towing to San Francisco.[83] With perfect aplomb, Kyle loaded it with salmon from his cannery for the southbound passage. Beadle, he knew, needed money and, though chagrined, would gladly accept the freight.[84]

Important as the sawmill was in making Kyle's holdings profitable, it operated under handicaps so basic that it could be nothing more than a marginal operation. The mill was small and ill-equipped. The lumber it sawed was of indifferent quality. The S. H. Harmon Lumber Company, which marketed the cut of Kyle's mill, put it diplomatically: "We are trying to work off your pine which is quite a different quality

83. Florence *West*, 18 Nov. 1898, 24 Feb. 1899. It was not uncommon for steam schooners to be built on minor ports and then towed to San Francisco for the installation of machinery. See Jack McNairn and Jerry MacMullen, *Ships of the Redwood Coast* (Stanford, Calif.: Stanford University Press, 1945), p. 117.
84. Cushman to Kyle, 10 Feb., 3 Mar. 1899, WKSP, 153; Meyer and Kyle to Cushman, 1 Jan. 1900, WKSP, 174; Beadle to Meyer and Kyle, 18 Feb., 1 Mar. 1899, WKSP, 153; Meyer and Kyle to Beadle, 18, 24 May, 6 Sept. 1898, 10 Feb. 1899, WKSP, 172; Meyer and Kyle to Harmon, 3 Sept. 1899, WKSP, 174. However, in the end Kyle's maneuver was less successful than Cushman's. Kyle had put in writing his promise to take a part interest in the steam schooner. To avoid having Beadle take the issue to court, he finally agreed to have Beadle borrow the money for Kyle's share and then charge the interest on the loan against Kyle's share of the earnings of the vessel. Thus, against his will, Kyle became part owner of another vessel, the steam schooner *Luella*. See Meyer and Kyle to Beadle, 21 Mar. 1899, WKSP, 172. The difference between Kyle and the other owners of *Luella* was not made public. Indeed, Kyle's daughter Isabel was chosen to christen her. See Florence *West*, 18 Nov. 1898.

as compared with Puget Sound stock and do not find it easy to sell." [85] The mill piled its cut directly on the wharf; it had no yard for storing large quantities, in part because Kyle was so short on operating capital that he found it necessary to sell the lumber from his mill as quickly as possible to obtain funds to keep the mill running. He could not afford to keep large stocks on hand.[86]

This situation created problems. Even with the *Bella* running to the Siuslaw on a regular basis, Kyle had to charter additional vessels to handle the cut of his mill. Yet, the bar at the mouth of the river was such that many ships were unwilling to come there and those that did were sometimes kept waiting offshore for several days until conditions on the bar changed.[87] Once the bar did change, the waiting vessels would all arrive at the mill at once, all the stock-piled lumber would soon be taken aboard, and the vessels would then have to wait while the tiny mill struggled to saw the lumber necessary to finish loading them. By the time the vessels were loaded, the bar would sometimes be closed down again, thus necessitating another wait and perhaps creating conditions whereby several vessels would finally leave the harbor at about the same time and arrive together in San Francisco, glutting the market for lumber of the quality cut by the mill at Spruce Point.

It is hardly surprising that it was difficult to persuade ships to come to the Siuslaw. In order to do so, Kyle had to offer a higher per unit freight rate than that paid by lumbermen in more favored locations. But even then it was not an easy task. When enough vessels could not be obtained, the wharf sometimes became so filled with lumber that the mill had to be shut down until tonnage could be found to remove it.[88]

85. Harmon to Florence Lumber Co., 6 Oct. 1899, WKSP, 154. See also Beadle to Meyer and Kyle, 12 Aug. 1898, WKSP, 150; Harmon to Meyer and Kyle, 27 Mar. 1899, WKSP, 153; Meyer and Kyle to Harmon, 13 May 1899, WKSP, 172; Florence Lumber Co. to Harmon, 22 Dec. 1899, WKSP, 174.

86. Meyer and Kyle to Harmon, 13, 15 May 1899, WKSP, 172; Meyer and Kyle to Beadle, 26 Oct. 1899, WKSP, 174. Kyle could afford neither to keep a large stock on hand nor to wait sixty days or so for payment from sales, even though it was customary for lumber to be sold on sixty-day terms. Kyle wanted the agent for his lumber to assume these notes and give Kyle immediate credit on their books at a discounted rate. Kyle argued: "Our capital is limited and we can't get along without [the] discounting privilege." Meyer and Kyle to Harmon, 13 Dec. 1899, WKSP, 174. See also Florence Lumber Co. to Harmon, 30 Sept. 1899, WKSP, 174. Discounting privileges were important to other manufacturers besides Kyle. They played an important role in the rise of the Jones and Laughlin Steel Corp. See Porter and Livesay, *Merchants and Manufacturers*, pp. 64–69.

87. The wait itself was sometimes dangerous. It was during such a wait that *Bella* was lost. See Florence *West*, 1 Dec. 1905.

88. Meyer and Kyle to Harmon, 13, 15 May 1899, WKSP, 172; Kyle to P. Snodgrass, 22 Aug.

Moreover, Kyle was restricted in his choice of markets for his cut. As he explained, "We have onley one market place S.F. we cannot load large enough vesseles to go south [i.e., to South America or Australia] or to the Islands, small vessels does not carry enough to make long trips, and large ones cant get over the bar. so we have difficulties that we cant overcome no matter how we figure it out." [89]

Only low wages and log costs kept the mill at all competitive. When Kyle started up the mill at Spruce Point, he was able to get logs for $2.50 per thousand. By the end of 1899 they had risen to $4.00, but this was still well below what millmen were paying on Puget Sound, Coos Bay, and the Columbia.[90]

The problems that beset William Kyle beset other millmen whose plants were located away from the prime areas of production. Some, indeed, were far more bedeviled by them than Kyle. Millowners on the Alsea and Tillamook Bay, possessing neither tugs nor schooners of their own, often had to shut down because they could not get lumber carriers to haul away their cut; and when they heard that Kyle had sold the *Robarts* to parties in Seattle during the Alaskan gold rush, they naturally expressed deep concern for the future of their operations. Without the towing service provided by Kyle's tug, however inadequate it might be, they foresaw the possibility of having to close their mills permanently.[91] Like Kyle's these mills were underfinanced, marginal operations that lost money as often as they made it. As a result, they were sometimes forced into bankruptcy when the market for lumber took a prolonged turn downward.[92]

Perry Whiting, a lumber dealer in Los Angeles, has left an account of one such mill. A "very smooth talker" sought to induce Whiting to

1899, WKSP, 174; Meyer and Kyle to Beadle, 1 Dec. 1899, ibid.; Florence *West*, 6 June, 3 Oct. 1902, 1 Dec. 1905.

89. Kyle to Snodgrass, 22 Aug. 1899, WKSP, 174. The editor of the local newspaper was less realistic. In 1906, noting the presence of four mills on the Siuslaw and growing demand in Japan, China, the Philippines, Australia, Hawaii, and South America, he foresaw a day when Florence would be "one of the foremost lumber ports on the coast." See Florence *West*, 19 Jan. 1906.

90. Kyle to Snodgrass, 22 Aug., 12 Sept. 1899, WKSP, 174; Kyle to G. Miller, 27 Nov. 1899, ibid.; Meyer and Kyle to Beadle, 1 Dec. 1899, ibid.

91. Beadle to Kyle, 18 Jan. 1898, WKSP, 145; Harrison Bros. to Meyer and Kyle, 10, 15 Feb. 1898, WKSP, 146; Cushman to Kyle, 4 Sept. 1898, WKSP, 150; Pacific Lumber Co. to Meyer and Kyle, 22 Apr. 1898, WKSP, 153; G. Diven to Meyer and Kyle, 3 Feb., 11 Apr., 26, 30 May 1899, WKSP, 153, 154; Harrison and Diven to Meyer and Kyle, 26 Sept. 1899, WKSP, 154; F. Johnson to Kyle, 5 Aug. 1899, ibid.

92. For examples, see Harrison Bros. to Meyer and Kyle, 18 Sept. 1894, 28 June, 29 Oct. 1895, 8 May 1896, WKSP, 146.

invest in a mill on the Coquille River that had gone bankrupt under its previous management. He told Whiting the mill could be purchased on time and logs bought at $3.50 per thousand. The mill, he said, cut forty thousand feet per day with a crew of fifteen men, and the lumber would run 20 percent clear. In spite of this bright picture, Whiting was cautious. Rather than purchase the mill, he agreed to rent it with an option to buy. He hired his informant to manage the mill at a salary of $150 per month and one-half of the profits. Whiting then sent his new manager to Oregon with instructions to rent the mill, buy five hundred thousand feet of logs, and wire him when the mill was ready to commence cutting.

In due course the telegram arrived, and Whiting departed immediately for the mill. What he found was hardly what had been described. It took twenty-two men to run the mill. At the end of a week, when Whiting computed production and costs, he found that even if the mill ran steadily, it was only capable of cutting twenty thousand feet per day. In fact, it had been shut down for repairs roughly half of the time. Producing lumber at the mill cost as much as, or more than, buying on the open market. To make matters worse, Whiting found that his manager had contracted for several million feet of logs, a donkey engine, and other equipment. Whiting fired the manager, repudiated the unauthorized contracts, and left an associate to oversee the sawing of the logs already purchased. As Whiting summed up the experience, "Only my caution and personal investigation saved me from going broke in that sawmill deal." Others were neither so cautious nor so fortunate in similar enterprises.[93]

I. B. Cushman, in commenting upon a competitor, inadvertently provided a sketch of operations of this sort, including his own: "I have no objection to Mr Benedict running his mill, only caution them not to pay it all to their men, and for logs for they can make nothing out of the mill, anyway, and If they raise the price of Labor and logs, it will work a hardship on me." Some of the newer operators on the Siuslaw seem to think the return from the sale of lumber is all profit, he complained. "If I get out with 50 cts a M clear I am doing more than I think for . . . [and I] will be satisfied with 25 cts." [94]

93. Perry Whiting, *The Autobiography of Perry Whiting, Pioneer Building Material Merchant of Los Angeles* (Los Angeles, Calif.: Smith-Barnes Corp., 1930), pp. 243–45.
94. Cushman to Meyer and Kyle, 22 Jan., 10 Feb. 1898, WKSP, 145. Cushman's operations clearly turned profits, however. Besides becoming a large investor in *Luella*, he also took a major

The Harrison brothers' mill on the Alsea was no better off. In 1895 one of the owners reported that they had lost money on their latest sale: "After paying for the logs we only had 95 cts per M left to cut it." [95] With the sawmills often closed, and paying low wages when running, many millworkers must have reacted as did one F. J. Kobe, who wrote from Tacoma, "I have left florence for I Could not make a living there." [96]

The bar harbors of the central coast of Oregon were not the only ones in the Northwest on which sawmills sprang up. Grays Harbor, Shoalwater Bay, Coos Bay, and many others had their mills. Development came earlier to some than to others. As we have seen, there were mills on Coos Bay by the middle fifties. In 1867 John Pershbaker started another on that harbor and, like Simpson, began to build ships there as well. By 1872 the *Oregonian* reported that on the average sixteen vessels a week were arriving at Coos Bay for lumber or coal. Ten years later four steam sawmills were in operation on the bay, and so many vessels were calling that two tugs were needed to meet the demand for towing service on the bar. One reason for this early and rather extensive tapping of the forests around Coos Bay was the high quality of the forest resource to be found there. It proved an irresistible lure to men such as Asa Simpson.[97]

To the north, Grays Harbor also had extensive stands of the finest timber in its readily accessible hinterland. However, unlike at Coos Bay, development there lagged. Not until 1882 was the first mill erected for the purpose of engaging in the cargo trade.[98] Barriers to the earlier development of the area were apparently not so much natural as

share in *Acme*, a steam schooner built in San Francisco for service to the Siuslaw, and, even though the insuror refused to cover the loss, he was able to rebuild his mill after it burned early in 1899. See Florence *West*, 3 and 10 Feb. 1899, 23 Nov. 1900.

95. Harrison Bros. to Meyer and Kyle, 28 June 1895, WKSP, 146.

96. F. Kobe to McIntosh and Wolfman, 9 Nov. 1897, WKSP, 147.

97. Emil R. Peterson and Alfred Powers, *A Century of Coos and Curry* (Portland: Binfords & Mort, 1952), p. 429. Wright, ed., *Lewis and Dryden's Marine History*, pp. 177, 203, 210. Portland *Oregonian*, 17 Sept. 1872. *West Shore* 8 (1882):35; 11 (1885):115. Unfortunately, few records from operations on Coos Bay are extant. Those items that do exist seem to indicate that the pattern of operations was quite similar to others along the lumber coast. See Smith-Powers Logging Co., ledger for 1881, University of Oregon library, Eugene; John W. Kruse, "Master Shipbuilder's Account Book," microfilm copy, ibid.

98. There had been small mills in the area earlier, but lumber was not exported from the bay until 1881. See Seattle *Daily Intelligencer*, 17 Oct. 1877; Portland *Oregonian*, 1 Jan. 1882; Hoquiam *Grays Harbor Washingtonian*, 25 Jan. 1913, 6 Aug. 1939; William Joseph Bilsland, "An Historical Comparative Analysis of the Economies of Grays Harbor and Pacific Counties, Washington, to 1920" (master's thesis, University of Oregon, 1966), pp. 32–33, 54–55.

man-made. Speculators had recognized the potential of the forests around the bay and bought up millsites and key tracts of timber. Pope and Talbot, and probably others among the giant firms on Puget Sound, looking forward to the day when they would need new sources of sawlogs, also began to acquire extensive holdings in the area. The speculators "demand a hundred times their value" for millsites, the *Oregonian* complained in 1882; "More than one firm has been frightened away." [99]

When development finally did begin around Grays Harbor, it proceeded apace. The bar, incorrectly shown on most charts as quite shallow, turned out to be one of the easiest and safest on the coast. Large vessels could call at the port. In addition, the numerous rivers emptying into the bay made thousands of acres of the finest timber readily accessible. Stands such as these, tributary to millsites that could be reached by large lumber carriers, called for a far different sort of mill from those erected by men such as Kyle. Plants rivaling in size those on Puget Sound were soon under construction. In the forefront of this development, appropriately enough, was Asa Mead Simpson. [100]

The development of large-scale milling operations on Grays Harbor was a source of concern to millmen with plants on Puget Sound. The mills springing up on Grays Harbor could turn out lumber as fine as any produced on the sound and ship it to markets that vessels from the smaller coastal harbors could never reach, markets which were important to the mills on the sound. Moreover, the cost of sawlogs was considerably lower at Grays Harbor than on the sound. All in all, the development of mills on Grays Harbor was a threat that the established firms on Puget Sound dared not ignore. Millowners on the sound began to move more and more of their operations to Grays Harbor. They did so not only to participate in the profits to be earned in this bonanza in timber, but also "in order to be able to control the log price" and get it "up to some decent figure," as W. H. Talbot put it. "At present," he complained, "with Logs at 4⁵⁰ the mills on the Harbor can turn out their lumber on the wharf at a cost of not over 6.50 which

99. Portland *Oregonian*, 1 Jan. 1882.

100. Hoquiam *Grays Harbor Washingtonian*, 4 Jan. 1900, 6 Aug. 1939; Portland *Oregonian*, 1 Jan. 1882; North Bend *Coos Bay Harbor*, 25 Sept. 1913. Though large vessels could usually enter Grays Harbor, its bar did occasionally cause trouble. At one point during the winter of 1894 George Emerson reported that no vessels had been able to cross the bar for two weeks. Thirteen lumber carriers were in the harbor loaded and waiting to get out. See Emerson to A. Simpson, 17 Dec. 1894, EL.

is less than our Logs cost us." If this were allowed to continue, he predicted, it would soon make millowners there rich and "cause us a heap of trouble in the future." [101]

Spurred on the one hand by defensive investments such as those of Pope and Talbot and, on the other hand, by investments resulting from the natural advantages that the area had to offer, Grays Harbor was soon enjoying a first-class boom. Though its greatest days as a lumber-producing center still lay in the future, it quickly emerged in the 1880s and 1890s as one of the most important lumber-shipping ports on the Pacific Coast. [102]

Important though it was, the lumber industry on Grays Harbor was not typical of that on the bar harbors as a whole. The pattern that developed in Hoquiam, Aberdeen, Cosmopolis, and the other lumber-producing centers on Grays Harbor was, in fact, more closely akin to that which had grown up earlier on Puget Sound. Dominated by large mills, dependent on outside capital, and shipping to widely scattered markets, the mills on Grays Harbor had their problems, but they were hardly the same problems as those which plagued operations on the Siuslaw or the Alsea, on Tillamook or Winchester bays.

Small mills were springing up along the coast of northern California, as well as in the Northwest. Though the nature of the outports and the use of redwood sawlogs caused superficial differences, conditions among the small mills of the redwood coast were at base quite similar to those found among the small mills on the bar harbors of the Northwest. As a whole, they too were small, marginal, underfinanced operations constantly beset by uncertain prices and undependable transportation. By 1882 there were approximately twenty mills located along the Mendocino coast and, besides those in the vicinity of Humboldt Bay, others scattered along the coast to the north and south. [103] Farming, fishing, and stock-raising were pursued along this coast, but the lumber industry was the main support of the economy. According to David

101. W. H. Talbot to C. Walker, 1 Nov. 1887, AC, Walker, incoming corres.
102. *West Shore* 11 (1885):33; 15 (1889):442; 16 (1890):44–45, 908. Bilsland, "An Historical Comparative Analysis," pp. 54–56. Stewart Holbrook, *A Narrative of Shafer Bros. Logging Company's Half Century in the Timber* (Seattle: for Simpson Logging Co. by Frank McCaffrey, 1945), pp. 17–18, 21. Much information about the Northwestern Lumber Co.—the firm that owned Simpson's mills at Hoquiam, South Bend, and Knappton—is to be found in the George H. Emerson letter books.
103. Ryder, *Memories of the Mendocino Coast*, p. 16; Menefee, *Historical and Descriptive Sketch Book*, pp. 276, 333; Thompson, *Historical and Descriptive Sketch of Sonoma County*, pp. 30–32; Kortum and Olmsted, "A Dangerous-Looking Place," p. 44.

Ryder, "when timber products were in demand at good prices, the whole coast prospered; when they were not, everyone felt it." [104]

They did not always prosper. In 1877 a local newspaper commented: "We dislike very much to prophecy hard times but in this case circumstances compel us to stare the truth straight in the face. During our late trip to the coast, the mill men informed us that unless the price of lumber comes up, that all the mills on the coast will shut down. The Garcia, Albion, and by this time probably the Little River Mill and parts of the Noyo mills have already closed down." The following year the *Humboldt Times* reported that only six of the coastal mills in Mendocino County were running steadily. In the face of slumping demand and falling prices, millowners slashed wages. Closures and wage cuts prompted many workers to leave the area. The exodus was so sizable that mills still running sometimes had difficulty filling their crews. [105] Only a handful of the operations that sprang up along this difficult coastline were able to survive these and the many other trials they faced.

Like those in the north, sawmill operations on the redwood coast varied widely. However, Tapping Reeves's mill on Little River was perhaps as typical as any. It had a boom for collecting logs floated down the river to the millsite, a chute for loading ships, tracks for moving lumber in the mill yard, a river lighter, a company store, a workshop, storehouses, twenty-five work oxen, logging camps, and—of course— the sawmill itself. [106]

The leading individual timber holder on the Mendocino coast during the late seventies and early eighties was J. S. Kimball. A leading citizen of Westport, Kimball had a mill there from 1879 to 1885, when he sold his timber claims in what was reportedly a most profitable transaction. [107]

However, Robert Dollar, not Kimball, is the best-known lumberman of the Mendocino coast. Dollar started in the lumber business in Michigan, but later transferred his field of activity to California. In 1888 he joined forces with William Westover to purchase a mill in

104. Ryder, *Memories of the Mendocino Coast*, p. 23.
105. Ukiah *Mendocino County Dispatch*, reprinted in Eureka *Humboldt Times*, 21 July 1877; Eureka *Humboldt Times*, 17 May, 18 June 1878, 3 May, 5, 19 July, 2 Aug. 1879; Melendy, "One Hundred Years of the Redwood Industry," pp. 232–35.
106. Melendy, "One Hundred Years of the Redwood Industry," pp. 229–30. See also *Canada Lumberman* 4 (1 Dec. 1884):356; 8 (Sept. 1888):6; Mendocino *West Coast Star*, 10 Apr. 1874.
107. Melendy, "One Hundred Years of the Redwood Industry," p. 233.

Sonoma County. When the timber there was nearly exhausted, Dollar invested in a mill at Usal, on the northern coast of Mendocino County. The Usal mill, unlike most others on the coast, had a long wharf rather than a chute for use in loading vessels. Completed and in operation in the early nineties, the mill did not prove overly profitable at first, apparently because of the poor quality of sawlogs. In 1894 Dollar took over active management of the operation. He soon acquired his first vessel, the steam schooner *Newsboy*, and had it busily at work ferrying lumber to San Francisco. By 1897 Dollar was actively searching out markets for his lumber in the East. He was one of the first to do so.[108]

Sawmills were active elsewhere along the redwood coast. By the late 1860s two mills were operating at Trinidad Bay in northern Humboldt County, thus reactivating a trade in forest products that had lain dormant since the decline of the excitement over the Trinity gold fields. In 1873 the two mills were consolidated by the Hooper brothers who, through the Excelsior Redwood Company, were to become key figures in the redwood lumber industry and, as allies of the Pope and Talbot interests, to play a major role in the development of outlets for lumber in the San Joaquin Valley and southern California. Trinidad Bay had no bar to plague ships, but its open roadstead provided minimal protection against storms. Two vessels were lost there in one month in 1878. The operation was sold to the California Redwood Company in 1883 and burned to the ground in 1886. Uninsured, it was not rebuilt.[109]

Sawmills were also turning out lumber in Del Norte County by the

108. Robert Dollar, *Memoirs*, 1:13–27; 3:23. *Pacific Coast Wood and Iron* 9 (1888):71; 21 (1894): 138; 24 (1895):1; 25 (1896):1; 28 (1897):129. H. B. Melendy, "One Hundred Years of the Redwood Industry," pp. 241–49. McNairn and MacMullen, *Ships of the Redwood Coast*, p. 17. Gregory Charles O'Brien, "The Life of Robert Dollar, 1844–1932" (Ph.D. diss., Claremont Graduate School, 1969). An intrepid individual, both ashore and at sea, Dollar, through his exploits, soon earned the admiration of the hardy men who populated the redwood coast. David Ryder recounts how Dollar, en route to Usal aboard *Newsboy* with money to pay his crews, was caught in a violent storm. Dollar decided to seek shelter in the tiny harbor at Fort Bragg, a difficult anchorage to enter even under good conditions. Those ashore tried to wave Dollar off when they realized what he was attempting. Oblivious of their warnings, Dollar headed in. Halfway in a huge wave caught *Newsboy* and tossed her over the reef and into the harbor. Undaunted, Dollar rented a horse and rode on through the night to Usal to pay off his men. According to Ryder, the north reef at Fort Bragg is still referred to as the Newsboy Channel. The story may be apocryphal; nevertheless, it gives an indication of Dollar's reputation among those who knew him. Ryder, *Memories of the Mendocino Coast*, p. 62. See also Dollar, *Memoirs*, 1:27–28.

109. Melendy, "One Hundred Years of the Redwood Industry," pp. 179–80; Eureka *Humboldt Times*, 3, 17 Feb. 1878, 14 July 1883; Coman and Gibbs, *Time, Tide and Timber*, pp. 205–9; *Pacific Coast Wood and Iron* 17 (1892):120.

late 1860s. There had been mills in this, the northernmost of California's coastal counties, earlier, but not until 1868 was one erected to engage in the export trade. Two firms dominated lumbering in the county during the last three decades of the nineteenth century: Hobbs, Wall and Company and the Crescent City Mill and Transportation Company. Others appeared from time to time, but the story of these two firms was, by and large, the story of lumbering in Del Norte County. During its first full year of operation, what was to become known as the Crescent City Mill and Transportation Company shipped over two million feet of lumber. In 1871 the second of the major sawmill operations was started. Within five years the two were exporting a combined total of ten million board feet of lumber a year. Both continued active into the twentieth century.[110]

Not every firm operating in Del Norte County fared as well as the two manufactories in Crescent City. R. D. Hume, king of the salmon industry on the southern Oregon coast, sought to broaden his interests during the 1880s. He acquired timberlands and built a sawmill on Smith River. The operation was plagued with troubles from the start, not the least of which was the harbor at the mouth of the river, an even more hazardous anchorage than most others along the redwood coast. The project lost money every year from its founding in 1882 to its abandonment in 1894.[111]

Throughout the last half of the nineteenth century millmen were at work along the rugged coast that stretches from the Strait of Juan de Fuca to San Francisco Bay. A few succeeded; more went the way of Hume's mill on Smith River. But when one plant closed, as often as not another was opening somewhere else. In time, however, the number of these mills did dwindle, as the depletion of timber caused many mills to close and the use of railroads to tap more extensive areas of timberland made possible a degree of consolidation. In few places did this consolidation ever lead to the creation of anything of a size reminiscent

110. Melendy, "One Hundred Years of the Redwood Industry," pp. 118–32; Wright, ed., *Lewis and Dryden's Marine History*, p. 42; Anthony J. Bledsoe, *History of Del Norte County, California . . .* (Eureka, Calif., 1881), pp. 121–31; John M. Childs, *Del Norte County As It Is* (Crescent City, Calif., 1894), pp. 28, 48–58, 99. One reason for the relative success of the mills in Crescent City was, according to Melendy, the high quality of the forest resources on which they drew. Their sawlogs yielded a high percentage of clear and were readily accessible. See Melendy, "One Hundred Years of the Redwood Industry," p. 118.

111. Gordon B. Dodds, *The Salmon King of Oregon: R. D. Hume and the Pacific Fisheries* (Chapel Hill: University of North Carolina Press, 1959), pp. 180–85; Childs, *Del Norte County*, p. 49.

of the giants on Puget Sound. More often what developed was only a larger version of the type of mill that had been present on this coast all along. They were an interesting and in some ways important part of the Pacific lumber trade, but they were never central to it. Even in San Francisco, almost the only market to which these coastal mills shipped, other mills had far more impact than they.[112]

The marginality of these coastal mills was reflected not just in their size but also in their mode of operation. Sedentary seaboard merchants were the primary distributors of manufactured goods in the United States until after the War of 1812. During the years that followed they were increasingly replaced by specialized wholesalers—agents such as A. W. Beadle and Company, which handled Kyle's ships and marketed his lumber, and C. E. Whitney and Company, which marketed his salmon pack. But by the 1850s independent wholesalers were also in decline. Mass production techniques had greatly increased the output of single firms. As a result, manufacturers increasingly created their own marketing agencies to distribute the products of their factories. Moreover, while first general and then specialized wholesalers had been important sources of capital for manufacturers, economic developments during the Civil War and the appearance in the late nineteenth century of significant markets for industrial securities gradually freed producers from dependence on middlemen for their financial needs. The Pacific Coast lumber industry followed this pattern. By the 1870s it was dominated by large firms that had combined manufacturing and marketing functions and met their capital requirements with internally generated funds and the sale of securities in the San Francisco financial community.

But in the face of this, William Kyle and his ilk continued operations that depended upon independent wholesalers. They had little choice. They lacked the wherewithal to erect their own marketing structures and any securities they might have offered would have been unattractive to investors. Under the circumstances they continued to lean upon the weak reed of the commission merchants. Even if A. W. Beadle had managed Kyle's interests in the wisest, most honest manner, there seems to have been no way that he could have made Kyle's mill competitive with those of the larger firms that had developed more satisfactory means of marketing their cut.

112. Some lumber from these mills reached foreign markets after reshipment from San Francisco on larger vessels than could serve the outports, but the additional costs of this approach kept such shipments infrequent. See Seattle *Daily Intelligencer*, 17 Jan. 1878.

The better ports—Grays Harbor, Coos Bay, and Humboldt Bay—developed lasting lumber industries, for there it was practical to construct mills large enough to make possible escape from the old order of commission merchants. In time almost all the other areas along this coast of bar harbors and outports saw their mills close and the lumber droghers depart. Many have hardly been visited since.[113]

113. For the story of these waning years, which is beyond the scope of this study, see McNairn and MacMullen, *Ships of the Redwood Coast*; Melendy, "One Hundred Years of the Redwood Industry," pp. 133–44, 251–74.

CHAPTER IX

Railroads,
the Cargo Mills,
and the Boom
of the Eighties

IN 1883 the last gap in the Northern Pacific Railroad was closed. For the first time the Pacific Northwest had direct rail connections with the East. A result, many commentators would have us believe, was the boom that the region experienced during the eighties. Charles M. Gates has expressed this view most succinctly: "The region had been waiting for a long time for a good boom. The railroads now made it a reality." [1] The lumber industry, as did the region as a whole, prospered through most of the eighties. This, too, has been traced to the coming of the transcontinentals.[2]

Of course, railroads had been present in the Far West for years before transcontinental lines began to reach the Northwest. Railroad construction in California during the sixties had provided a market that played a major role in the rise of firms such as the Washington Mill Company. Short stretches of track had been constructed to aid in logging or getting the cut of mills out to market. Such construction also had its impact, minor and localized though it was. In 1870, when rival lines began building southward down the Willamette Valley, each trying to qualify for the single land grant that was to be awarded on the Oregon-to-California route, the impact was even greater. The

1. Johansen and Gates, *Empire of the Columbia* (1st ed.), p. 322. The portion pertinent to this chapter was written by the late Professor Gates. Treatment of the impact of the transcontinentals is more carefully qualified in the second edition.
2. See, for example, Meany, "History of the Lumber Industry," pp. 127, 137; James N. Tattersall, "The Economic Development of the Pacific Northwest to 1920" (Ph.D. diss., University of Washington, 1960), pp. 137, 183.

construction created a demand for ties and bridge timbers that was without precedent in Oregon. Prior to this time, the only railroads in the state had been short portage lines. George Weidler, an associate of transportation magnate Ben Holladay, soon constructed a sawmill to supply the necessary materials. In time, as the Willamette Steam Mills Lumbering and Manufacturing Company, Weidler's plant was to become the largest, most modern sawmill in Oregon and an important participant in the Pacific lumber trade.[3]

But the situation was different in the 1880s. For the first time transcontinental lines were reaching a major lumber-producing area. As rail connections with the East came nearer to reality, many Northwesterners confidently expected that the transcontinentals would provide access to vast, new markets for lumber in the "treeless plains" beyond the Cascades and Rockies.[4] Railroads were, indeed, to have a major impact on the lumber industry of the Far West, but in ways hardly foreseen at the time and often overlooked since.

The building of the northern transcontinentals had mixed effects. So long as the railroads were under construction, they consumed forest products cut not only at new mills that were built along their routes, but also at existing mills along the coastal rivers and harbors, mills that hitherto had been primarily dependent upon the sea-going cargo trade for their sales. The Port Blakely and Washington mill companies, for instance, sold materials cut on Puget Sound for use in the construction of the transcontinentals.[5] To meet the increased demand, millmen in the Northwest expanded the capacity of their plants. According to some estimates, capacity as much as doubled. Veteran sawmill operators occasionally expressed concern over what would happen to the industry when the lines were completed and "normal" conditions returned.[6]

3. G. Weidler to B. Holladay, 29 Jan., 15, 23 Feb. 1870, Ben Holladay Papers, Oregon Historical Society, Portland, box 3. H. Hampton [secretary-treasurer, Oregon Improvement Co.] to Holladay, 22 June 1876, ibid., box 2. Weidler and W. Starbuck, agreement, 27 Nov. 1880, OICP, agreements. H. Villard to stockholders of Oregon Improvement Co., 9 Apr. 1881, OICP, outgoing corres. *West Shore* 12 (1886):110; 14 (1888):117, 415–16; 16 (1890):107, 321. Meany, "History of the Lumber Industry," pp. 134–35. Ellis Lucia, *Head Rig: Story of the West Coast Lumber Industry* (Portland: Overland West Press, 1965), p. 41.

4. *West Shore* 8 (1882):182; 10 (1884):73. The *Oregonian*, 3 Oct. 1878, compared local lumber supplies to Robinson Crusoe's money, but predicted that with the completion of rail interties they would "become an important part of the rail traffic to the interior."

5. W. Adams to R. Maynard, 25 Dec. 1883; R. Kendrick to Adams, 26 Feb. 1884, WMCP; Berner, "Port Blakely Mill Company," p. 164. See also Meany, "History of the Lumber Industry," pp. 134–36.

6. Adams to Maynard, 25 Dec. 1883, WMCP.

Their concern was justified. The lumber industry of the Northwest prospered during the period from 1879 to 1883; the industry, the Vancouver *Independent* noted, "has not for a long time been as active as it is at present." [7] But the completion of the Northern Pacific brought changes. The laying off of the construction crews of the railroad enlarged the work force available to the industry; however, demand for the products of their mills fell off, forcing those that had been producing ties and bridge timbers during the years of construction to compete in markets others had been supplying. *West Shore*, which only a short time before had been complaining of a shortage of workers for getting logs out of the woods, commented on the excess of unskilled laborers in the Northwest. This long-time booster of the region's development now advised laborers to stay away from the area.[8] Prices began to fall. Though most mills managed to continue operating on a regular basis, profits were down. Some millmen declared they were barely sufficient to warrant running the mills.[9]

No vast, new markets had appeared east of the Cascades to alleviate the situation. Long stretches of the Northern Pacific's tracks passed through or near forest land. Though the *Northwestern Lumberman* predicted that Montana and Idaho should prove to be good markets for the coast's lumber, the area traversed by the Northern Pacific seemed to offer as many competitors as it did buyers.[10] Despite the commonly held view of residents of the area west of the Cascades, the land east of the mountains was no treeless plain. Indeed, boosters in the transmontane area had ideas of their own about expansion of the lumber business. The Montana Improvement Company organized in 1883 and entered into a twenty-year contract to supply the Northern Pacific with all the lumber, ties, and bridge timbers that would be needed between Miles City and The Dalles. In addition, the company was to

7. Vancouver (W.T.) *Independent*, 5 May 1881. *West Shore* (7 [1881]:380) proudly announced that in the woods and mills "good wages are paid and all who wish it can find work." See also *West Shore* 8 (1882):126, 162; 9 (1883):44. *Canada Lumberman* 2 (1 May 1882):129; (15 July 1882):209; 3 (1 Jan. 1883):11.
8. *West Shore* 9 (1883):44, 47; 10 (1884):61.
9. Portland *Oregonian*, 26 Apr. 1886. *West Shore* 9 (1883):203; 10 (1884):222; 11 (1885):193.
10. Reprinted in *West Shore* 11 (1885):291. See also ibid. 6 (1880):35, 57; Portland *Oregonian*, 15 Nov. 1882, 4 Dec. 1883; Palouse City (W.T.) *Boomerang*, 28 Sept. 1888; Elwood Evans, *History of the Pacific Northwest*, 2 vols. (Portland, 1889), 2:216; Robert W. Swanson, "A History of Logging and Lumbering on the Palouse River, 1870–1905" (master's thesis, Washington State University, 1958), pp. 47–54, 59; W. Hudson Kensel, "The Early Spokane Lumber Industry," *Idaho Yesterdays* 12 (1968):25–31.

receive special rates on lumber it shipped on the line.[11] The large quantity of wood being floated down the Clearwater River to Lewiston moved one observer to declare that for over a hundred miles the stream was not so much a river as a "floating woodyard." [12] Though other schemes were less grandiose in conception than that of the Montana Improvement Company and other sights less striking than that of the Clearwater, the rising production east of the Cascades, wherever it occurred, served to lessen the market for lumber from the coastal mills.[13]

However, the greatest barrier to these eastern lumber markets was not transmontane competition but high rail rates. Completion of the Northern Pacific brought a decrease in rates from Portland to Walla Walla from $1.12 per hundred pounds to $.68, and Portland-Spokane freight now ran at $.51 per hundred; but these rates were still too high to allow penetration of the potentially significant markets that lay further east.[14] Until major reductions in rates on eastbound freight were made, lumber from Michigan continued to dominate markets as far west as southern Idaho.[15] In spite of repeated efforts, mill operators in the Portland-Vancouver area—the region that stood to gain the most from reductions—were unable to get the railroad to lower charges to a level that would have made their lumber competitive in territories and states to the east.[16] A few small operators, such as Eugene Semple, who owned the Lucia Mills in Vancouver, found large enough markets in the Inland Empire to help their operations significantly; but to the owners of larger plants the initial results of the arrival of the transcontinentals must have been a disappointment.[17]

Then came the rate wars. The Central Pacific declared a rate of

11. Deer Lodge *New Northwest*, reprinted in *West Shore* 9 (1883):153.
12. *West Shore* 6 (1880):257. See also ibid. 6 (1880):35.
13. Idaho's production rose from 18 million board feet in 1880 to 31 million in 1890. By the end of the eighties, Blue Mountain mills were consuming 30 million feet of logs annually and cutting two hundred thousand board feet of lumber per day. By this same time 4 million feet of logs were being floated down the Palouse River to Colfax and Palouse City each year; the mill in Colfax produced forty thousand board feet per day. See *West Shore* 7 (1881):299; 16 (1890): 73. Meany, "History of the Lumber Industry," p. 114. Swanson, "History of Logging and Lumbering on the Palouse," pp. 47–54, 59. E. Semple to A. Stokes, 28 Apr. 1884, Eugene Semple Papers, University of Washington library, Seattle.
14. Portland *Oregonian*, 14 Oct., 15 Nov. 1882. See also *West Shore* 9 (1883):44.
15. *West Shore* 7 (1881):63; 11 (1885):259; 12 (1886):13.
16. Ibid. 11 (1885):259; Semple to Proebstel Bros., 14 July 1884, Semple Papers.
17. Semple to Proebstel Bros., 23 June 1884, Semple to Reene & Redman, 7 July 1884, Semple Papers.

$ .50 per hundred pounds for freight from California to Denver, $ .60 to Omaha. The Oregon Rail and Navigation Company and its connecting lines countered with a comparable rate of their own.[18] These reductions were crucial to lumbermen seeking markets to the east. The *Northwestern Lumberman* noted that coastal redwood, fir, and cedar could now compete in the Omaha market on even terms with pine from Michigan, Wisconsin, and Minnesota, the area Northwestern lumbermen referred to as the Lake States.[19] Outlets in Utah and Colorado also offered opportunities, though not without competition. Production was high in the mountains of California, and pine shipped east from the Sierra competed with Douglas fir from the Northwest for sales in this new market. Mills in eastern Oregon also sought to gain entry: in 1887 McMurren and Crabill Company of Baker sent thirteen carloads of pine lumber to Ogden.[20]

Undaunted by the competition, George Weidler secured large orders from dealers in Denver and ordered additional logs to fill them. Then, with the first fourteen cars loaded on his siding, the railroad changed its freight rates. Officials of the OR&N explained that the intricacies of the company's pooling arrangement with the Santa Fe made higher charges necessary. The increase from $ .50 to $ .65 per hundred pounds on Portland-Denver freight was sufficient to shut mills in the Portland area out of the markets in Denver and Omaha. Although Weidler was able to persuade railroad officials to allow the loaded cars to be sent at the fifty-cent rate, future sales were out of the question. He found himself with a large supply of logs that he would simply have to hold until other markets for his lumber could be found.[21]

The higher rates did not prove permanent. Within months it was

18. The OR&N line up the Columbia River connected with the Northern Pacific and, via the Oregon Short Line, with the Union Pacific.

19. Portland *Oregonian*, 22 Feb., 26 Apr. 1886; *Northwestern Lumberman*, reprinted in *West Shore* 12 (1886):61; *West Shore* 11 (1885):191, 259. John H. Cox incorrectly identifies this fifty-cent rate as having been adopted in 1893 in response to Great Northern's famed forty-cent rate to St. Paul. Cox, "Organizations of the Lumber Industry of the Pacific Northwest, 1889–1914" (Ph.D. diss., University of California, Berkeley, 1937), pp. 6–7.

20. Portland *Oregonian*, 5, 14 May 1885; Hittell, *Commerce and Industries*, p. 583; *An Illustrated History of Baker, Grant, Malheur and Harney Counties* (Spokane, Wash.: Western Historical Publishing Co., 1902), p. 172.

21. Portland *Oregonian*, 26 Apr. 1886; L. C. Palmer and Eugene Semple to "lumbermen," in Portland *Oregonian*, 28 Apr. 1886. *West Shore* 11 (1885):259; 14 (1888):410, 415–16, 494–99. Meany, "History of the Lumber Industry," p. 137. Meany comments that the competition in the Colorado-Nebraska markets came from southern pine, a somewhat doubtful suggestion for this early date.

again possible to penetrate markets in Colorado and Nebraska. They proved difficult to capture, however. Dealers in these areas, Eugene Semple noted, were wary of tying themselves too closely to a source of supply on the Pacific Coast that might vanish in the face of new rate increases. They preferred to continue to obtain the bulk of their lumber from producers in the Lake States, since rates from there seemed less apt to be changed abruptly.[22]

Moreover, though the OR&N professed to be following policies designed to further trade with the interior, some Northwesterners were convinced, even after the sixty-five-cent rate had been reduced, that the company was in fact charging all it felt the traffic would bear.[23] Such a policy, if it really did exist, certainly would have slowed the rate of increase in lumber shipments to the interior. Impatient with these supposed restraints on the development of their region, *West Shore* urged the development of river transportation facilities on the Columbia so as to break the OR&N's stranglehold on trade by providing an alternate means of reaching the interior.[24]

Portlanders were proud of their achievements in pioneering in these interior markets, but the impact of their shipments on the Northwest as a whole was small. Seattle millman George Stetson attempted to win rates similar to those enjoyed by Weidler, but failed in the effort. In 1890 it was still possible for the Seattle *Times*, in writing of the status of the Northwest's lumber industry, to ignore completely Portland's shipments east by rail. Indeed, the newspaper looked forward to the time when shipments to areas east of the Rockies might begin! [25] Apparently the *Times*'s ignorance of the shipments that had been made was a truer measure of their importance to the economy of the region

22. Eugene Semple to editor, in Portland *Oregonian*, 26 Aug. 1886. The name was misprinted as "Temple," but from contextual evidence it is clear that the letter was written by Semple.
23. *West Shore* 14 (1888):494–99; Portland *Oregonian*, 12 Jan. 1887. David Lavender, in *Land of Giants: The Drive to the Pacific Northwest, 1750–1950* (Garden City, N.Y.: Doubleday & Co., 1958), p. 382, repeats the old charge.
24. *West Shore* 14 (1888):494–99.
25. Portland *Oregonian*, 25 Aug., 3 Sept. 1887. *West Shore* 9 (1884):222; 11 (1885):259; 12 (1886):110. Meany, "History of the Lumber Industry," p. 137. Ames to W. H. Talbot, 10 Nov. 1885, PTP, Ames corres. Alexander Norbert MacDonald, "Seattle's Economic Development, 1880–1910" (Ph.D. diss., University of Washington, 1959), pp. 68–95. Seattle *Times*, reprinted in *West Shore* 16 (1890):376–77. "For some time past," said the *Times*, "heavy freight rates have barred fir timber out of the eastern market or the markets of the middle west." In an editorial note *West Shore* pointed out that mills in Portland had been sending fir eastward for some time. Cf. *Canada Lumberman* 13 (Nov. 1892):10. Cox, "Organizations of the Lumber Industry," pp. 1–8, also seems unaware of these early rail shipments out of Portland.

than were the boasts of Portlanders. In recent studies James N. Tattersall has shown that the total effect of these rail shipments on the Northwest was far less than that of the great increase in exports by sea that occurred during the same period.[26]

Not all the forest products sent east shortly after the completion of the transcontinentals were from Portland. Millmen in Vancouver loaded cars at their plants and had them barged across the Columbia to the OR&N tracks on the south bank of the river.[27] Shingles were shipped from Puget Sound to points as far east as Pennsylvania and New York after the completion of the Cascade Division of the Northern Pacific in 1887. The Atlas Lumber and Shingle Company, established in Seattle in 1889, catered to this trade. The shipments were sometimes quite sizable. One shingle mill near Anacortes received two orders in early 1891 that totaled five million shingles. Bridge timbers were also sent east from the area around Puget Sound. An order sent to Keokuk, Iowa, by the St. Paul and Tacoma Lumber Company included pieces of such length that it was necessary to arrange the railroad cars in tandem to handle them.[28] Similarly, lumbermen in coastal British Columbia began shipping lumber eastward over the Canadian Pacific soon after its completion and by 1889 one observer described the trade as "quite an item." [29]

As was true of the lumber shipments from Weidler's Willamette Steam Mills and from the North Pacific Mill Company in Portland, these first shipments from Washington and British Columbia cannot be shown to have had a major economic impact on the region. The markets opened were too limited, the quantities of goods shipped too small, the number of mills involved in the shipments too few.[30]

Yet, it must be admitted that the lumber industry of the region west

26. Tattersall, "Economic Development of the Pacific Northwest," p. 137 and *passim;* Tattersall, *The Importance of International Trade to Oregon* (Eugene: University of Oregon Press, 1961); Tattersall, "Exports and Economic Growth: The Pacific Northwest 1880 to 1960," *Papers and Proceedings of the Regional Science Association* 9 (1962):215–34.
27. Vancouver *Independent*, 2, 16 Aug., 6 Sept. 1893; Henry V. Poor, *Manual of the Railroads of the U.S., 1890* (New York, 1890), p. 845; *West Shore* 15 (1889):61.
28. Meany, "History of the Lumber Industry," p. 137. Evans, *History of the Pacific Northwest,* 2:124. Coman and Gibbs, *Time, Tide and Timber,* p. 215. Johansen and Gates, *Empire of the Columbia* (1st ed.), p. 391. *West Shore* 13 (1887):42; 14 (1888):117; 17 (1891):171, 255, 281. F. I. Vassault, "Lumbering in Washington," *Overland Monthly,* new series 20 (1892):25–27. Portland *Oregonian,* 1 Jan. 1889.
29. *Canada Lumberman* 9 (Mar. 1889):6. See also ibid. 8 (Aug. 1888):1, 7; 9 (Jan. 1889):1, 5; 9 (May 1889):1.
30. For further development of this argument see Cox, "Lower Columbia Lumber Industry."

of the Cascades benefited almost immediately from the arrival of the transcontinentals. The population of the Northwest grew rapidly following the completion of connections with the East. The railroads, in making it easier and cheaper to reach the Northwest, encouraged immigration into the area, and the publicity attendant upon the building of the lines made more people aware of opportunities present in the Northwest. This growth in population meant not only a building boom in Oregon, Washington, and British Columbia, but also a permanently larger domestic market for products from the area's sawmills. Moreover, the improved transportation facilities and the wealth of publicity the area received helped make investments there more attractive to capitalists in other regions. The capital thus attracted led to more building and a larger demand for lumber.[31]

A significant portion of the wealth that began to flow into the area during the eighties went into the construction of sawmills. As Cyrus Walker put it, "Lots of people [are] getting crazy on the mill question."[32]

Millmen in the Lake States could see that their stands were about gone. Some began to move their operations to the southern pineries or north, into the largely untouched stands of Ontario; some began to look toward the redwood forests of California; many others turned to the forests of the Pacific Northwest. In each area this influx of men with knowledge and capital accumulated in the Lake States brought sweeping changes to the lumber industry already there.[33]

In the Northwest many of these new arrivals turned their backs on the cargo trade that had for so long supported the region's lumber industry and built plants designed to supply the eastern markets, which were reached by rail. The first such plant in the Northwest was the St.

31. *West Shore* 8 (1882):126, 162; 9 (1883):44. *Canada Lumberman* 3 (1 June 1883):166; 3 (15 Sept. 1883):281. Abner Baker, "Economic Growth in Portland in the 1880s," *OHQ* 67 (1966):105–23. Tattersall, "Economic Development of the Pacific Northwest," p. 186. Johansen and Gates, *Empire of the Columbia* (2nd ed.), pp. 316–17, 322, 328–32. Norman H. Clark, *Mill Town: A Social History of Everett, Washington, . . .* (Seattle: University of Washington Press, 1970), pp. 3–29. Cf. Portland *Oregonian*, 24 Mar. 1874. *West Shore* 9 (1883):47; 16 (1889):420.
32. Walker to W. H. Talbot, 10 May 1888, PTP, Talbot corres.
33. Edgar Cherry & Co., *Redwood and Lumbering*, pp. 18–21. Meany, "History of the Lumber Industry," pp. 222–27 and *passim*. James Elliott DeᶜLaugh, *History of the Lumber Industry of America*, 2 vols., 2nd ed. (Chicago: The American Lumberman, 1906–7), 1:452–53 and *passim*. *Canada Lumberman* 8 (Feb. 1888):10; 12 (Feb. 1891):8, 11. Large quantities of this new capital went toward gaining control of timberland as well as into new mills. For examples, see G. Emerson to A. Simpson, 18 Dec. 1896, EL; Ralph W. Hidy, Frank Ernest Hill, and Allan Nevins, *Timber and Men* (New York: Macmillan, 1964), pp. 211–16, 222–25.

Paul and Tacoma mill, erected at Tacoma in 1889. Others soon followed. For years there had been two basic types of mills in the Northwest: the cargo mills, often quite large, which depended upon markets reached by sea; and small mills scattered in the interior and serving local markets.[34] Now there was a third.

In time the presence of these rail mills forced many of the old cargo mills to adjust their mode of operation. Eventually some abandoned the cargo trade altogether and entered upon a new incarnation as rail mills. Others, unable or unwilling to adjust, were forced to close. For better or for worse, the change had been triggered by the arrival of the transcontinentals.[35] This transition began in the 1880s and was to continue in the decades that followed.

As important as the direct and indirect benefits of rail connections with the East was the role railroads played in making forested hinterlands accessible from the ports of the Far West. The transcontinentals played a part, but the many local logging railroads built during the decade were largely responsible.[36]

The development of a network of logging railroads was a prerequisite not just for expansion of the lumber industry but for the very survival of its larger components. *West Shore* noted that "Timber lands of Oregon and Washington Territory that are situate [*sic*] on the banks of navigable streams, bays or inlets are fast being cleared, as the present home and foreign markets take all that is for sale. Logs that are easily shot into the large streams, and the once fine bodies of timber along Puget Sound and the Columbia river, are even now, almost a thing of the past." [37] Captain William Renton of the Port Blakely Mill Company put it more succinctly: timber "contiguous to the Sound is nearly exhausted." To a lesser degree this was also true of the other established areas of production. A few areas remained untapped, but

34. Meany, "History of the Lumber Industry," pp. 128–30. The mills of the interior might be further subdivided, for there were differences among those found east of the Cascades, in the area of Portland and Vancouver, and in the interior valleys west of the Cascades.
35. Coman and Gibbs, *Time, Tide and Timber*, pp. 210–20; Meany, "History of the Lumber Industry," pp. 135–38, 145; Clark, "Analysis of Forest Utilization," pp. 74–87, 91–100; Cox, "Organizations of the Lumber Industry," pp. 1, 3, 8; Hidy, Hill, and Nevins, *Timber and Men*, pp. 219–22. For a description of the Pacific Mill Co. of Tacoma, one of the new rail mills, see *West Shore* 16 (1888):552–55.
36. Tattersall, "Economic Development of the Pacific Northwest," pp. 131, 138, 186; Victor S. Clark, *History of Manufactures in the United States*, 3 vols. (New York: for the Carnegie Institution of Washington by McGraw-Hill Book Co., 1929), 2:482.
37. *West Shore* 8 (1882):63. Cf. Melendy, "One Hundred Years of the Redwood Industry," p. 63; Johansen and Gates, *Empire of the Columbia* (1st ed.), p. 243.

these were exceptions rather than the rule.[38] They were not adequate to keep the large mills long supplied. In the face of growing demand and the disappearance of the type of stands on which the industry had long depended, future supplies, the *West Shore* predicted, would have to come "from the vast untouched forests of the mountain watersheds. These can only be reached by logging railroads. . . ."[39]

Lumbermen were not long in moving to build the necessary railroads. Apparently the first logging railroad in the Pacific Northwest was a narrow-gauge line built in 1881 between Tenino and Olympia. In the first six months of its operation "several million feet" of logs were hauled out over the line, returning a good profit on the builder's investment and demonstrating the feasibility of railroads in Northwestern logging operations. It proved far more satisfactory than the log cars and mules that were being used on tramways in other areas around the sound. By the following year five separate logging lines were in existence around Puget Sound.[40]

Others soon followed. In 1883, for example, the Port Blakely Mill Company began to construct a logging railroad of its own into timberland west of Puget Sound. The company hoped eventually to have a line all the way to Grays Harbor. The object was not only to make available a larger and higher quality supply of logs but also to gain a better bargaining position with the independents from whom the mill had been obtaining its sawlogs. Loggers had been seeking higher and higher prices for their rafts of logs; the owners of the Port Blakely plant hoped, by developing alternate sources of supply, to create competition that would help hold prices down. By and large, the plan was successful. By 1888 seventy-five to eighty cars, carrying some 225,000 feet of logs, were rolling over the company's tracks each day.[41] Renton described the railroad as "the best piece of Property we have.

38. Port Blakely Mill Co. to Renton, Holmes & Co., 17 Jan. 1884, PBMCP. See also *Canada Lumberman* 3 (1 Jan. 1883):11. For discussions of remaining waterside stands, see Seattle *Daily Intelligencer*, 30 Mar. 1877. *West Shore* 2 (Jan. 1877):68; 7 (1881):18; 13 (1887):407–8; 16 (1890):44–45. Vancouver *Independent*, 30 Apr. 1890. Hoquiam *Washingtonian*, 19 Oct. 1893.
39. *West Shore* 14 (1888):619. There were alternatives. Flumes, rather than railroads, might have been built, or plants might have been moved to the new areas of production. In fact, both were being done to a degree, but neither was an answer for the industry as a whole. Plants moved into the hinterland would still have to get their cut out to market, while flumes were not practical in relatively flat areas, such as the land around Puget Sound.
40. *West Shore* 8 (1882):182; Portland *Oregonian*, 1 Jan. 1882; Meany, "History of the Lumber Industry," p. 260; *Canada Lumberman* 2 (15 Sept. 1882):278. There were earlier railroads in the Northwest, but they were not built primarily for logging.
41. *West Shore* 13 (1887):264; Berner, "Port Blakely Mill Company," pp. 158, 161–64.

With[in] two miles of the end of the road there is enough timber to pay for the road and collect stumpage to pay for all our land."[42]

In 1884 *West Shore* noted that railroads and steam engines were replacing skid roads and ox teams because logs could be hauled "perhaps ten to fifteen" miles by rail for no more than it cost to haul them one or two miles with oxen.[43] Under the circumstances, it is no wonder that the number of logging railroads increased rapidly.

Even along the lower Columbia, where timber stands near the watercourses had as yet not been so severely depleted as those near the sound and where flumes often proved less expensive than railroads as a means of tapping stands in the interior, logging railroads were putting in their appearance. The Weidler interests had a railroad up Abernethy Creek on the Washington side of the Columbia some fifty miles downstream from Portland.[44] The Vancouver, Klickitat and Yakima Railroad pushed into the mountains behind Vancouver and brought logs out to the Michigan Mill Company, a new plant erected for the express purpose of utilizing stands made accessible by the line.[45] "The dawn is breaking," the editor of the *Clarke County Register* wrote. "Vancouver has a great future." [46] The West Shore Mills in Astoria was also partially supplied with logs hauled by rail. Owner J. C. Trullinger had constructed a line up the Walluski River. This line, made of old material and using secondhand rolling stock, cost approximately $4,000 a mile to build and was considered a good investment at the price. It had a capacity of two hundred thousand feet of logs per day.[47] Other roads were soon being built in the woods along the lower Columbia.[48]

42. Quoted in Berner, "Port Blakely Mill Company," p. 163.
43. *West Shore* 10 (1884):73. Similarly, George Emerson wrote that to "log off Willapa timber by rail when the road is completed will without doubt be the most economical and satisfactory method." See Emerson to A. Simpson, 16 Mar. 1892, EL.
44. *West Shore* 14 (1888):415–16; Fred Lockley, *History of the Columbia River Valley from the Dalles to the Sea,* 3 vols. (Chicago: S. J. Clarke Publishing Co., 1928), 1:334–35; Poor, *Manual of Railroads, 1890,* p. 1176.
45. *West Shore* 9 (1883):189; 12 (1886):142; 14 (1888):96; 15 (1889):60. Vancouver *Independent,* 28 May 1890, 12 Aug. 1891.
46. Vancouver *Clarke County Register,* 22 Dec. 1887.
47. Portland *Oregonian,* 5 Aug. 1887. *West Shore* 9 (1883):124; 13 (1887):504. Meany, "History of the Lumber Industry," p. 260. Emma Gene Miller, *Clatsop County, Oregon, A History* (Portland: Binfords & Mort, 1958), pp. 212–13. A much higher cost per mile, $30,000, is reported in Poor, *Manual of Railroads, 1890,* p. 1172. This probably included, in addition to construction costs, the cost of rolling stock and miscellaneous expenses. Regarding Trullinger, see Astoria *Astorian,* 5 Jan. 1886; Miller, *Clatsop County,* p. 71; Joseph Gaston, *Centennial History of Oregon, 1811–1911,* 4 vols. (Chicago: S. J. Clarke Publishing Co., 1912), 3:292.
48. *West Shore* 9 (1883):47; 11 (1885):319; 12 (1886):142. Lockley, *History of the Columbia*

Though the redwood coast had not been reached by any of the transcontinentals and the Humboldt County area would have to wait until 1914 before it had rail connections with the outside, short local lines for use in logging put in their appearance in the redwoods during the same period that they did in the Northwest—and for the same reasons: they provided access to new stands that were desperately needed, and, as the timber was cut back farther and farther from the watercourses, they were a less expensive means of getting logs out of the woods than the ox teams and skid roads that had dominated hitherto.

By the 1880s John Vance had built a line fifteen miles long with branch tracks to tap different tracts of previously inaccessible timber. Nearby, John Kentfield, D. R. Jones, and their associates constructed the South Bay Railroad (later the Humboldt Logging Railroad) to tap other stands.[49] In 1885 the Albion mill was logging stands nineteen miles from the plant. Logs were hauled by rail to a nearby stream, floated toward the coast, and carried the rest of the way to the mill on a second company-owned railroad. The sawlogs were transported by rail for four and a half of the nineteen miles between logging camp and mill.[50] Others, including the Caspar mill, also had their own rail lines, while, farther south, the completion of the Santa Cruz and Felton Railroad led to a rejuvenation of the declining lumber industry of the Santa Cruz redwoods. Lumber began once again to flow with regularity from the forests of Santa Cruz to San Francisco. By 1880 there were feeder lines up many of the creeks, and it was estimated that Santa Cruz County was turning out as much as fifty million board feet of redwood annually.[51]

River Valley, 1:335; Leslie M. Scott, "History of Astoria Railroad," *OHQ* 15 (1914):221–40; Henry V. Poor, *Manual of the Railroads of the U.S., 1892* (New York, 1892), pp. 11–12, 15. For a further discussion of the coming of logging railroads to the Northwest, see Meany, "History of the Lumber Industry," pp. 259–64.

49. Edgar Cherry & Co., *Redwood and Lumbering*, pp. 50–56. Weidberg, "History of John Kentfield & Co.," pp. 18–20, 28. C. Nelson to J. Kentfield, 5 Oct. 1875, JKCP, 1:7. J. W. Henderson to C. Nelson, 27 June and 11 July 1875, JKCP, 1:6. J. W. Henderson to J. Kentfield & Co., 8 June 1876, JKCP, 2:1; 13 July 1877, JKCP, 2:6.

50. [California] State Board of Forestry, *First Biennial Report, 1885–1886* (Sacramento, 1886), p. 140. See also A. T. Hawley, *The Climate, Resources and Advantages of Humboldt County, California,* . . . (Eureka, Calif., 1879).

51. San Jose *Mercury*, reprinted in *Canada Lumberman* 6 (15 July 1886):8; *Canada Lumberman* 13 (July 1892):7; Sanford, "Short History of California Lumbering," pp. 38–43; Harrison,

H. Brett Melendy has said that railroads "can be given major credit for the rapid expansion of the industry" in Humboldt County at a time when the stands near tidewater and watercourses had nearly disappeared.[52] The statement would appear to have validity when applied to the rest of the redwood coast and to the established areas of lumber production in the Pacific Northwest as well. One tabulation of 1889 indicated that there were twenty-two logging railroads in operation in Washington, seven in Oregon, and thirty-seven in California. There may well have been more, for, according to the *Oregonian* in 1887, logging railroads were "being introduced everywhere." [53]

Although railroads aided lumbermen, they created problems for them as well. The construction of logging railroads called for large outlays of capital, which even established firms such as the Port Blakely Mill Company found difficult to accommodate. Through the years the company had been built and expanded with the private resources of its owners and with reinvested profits. The company's first railroads were financed in these ways, but the demands for capital soon began to outrun the supply available from such sources. Charles Holmes and William Renton began to investigate the possibilities of selling railroad bonds to finance further construction. By September 1888 they had so changed their financial thinking that Holmes was writing to Renton, "If you can bond the road for more than cost as they do in California it may be a good idea." [54] In Richard C. Berner's words, "Modern Puffery Capitalism was subduing Early Yankee." [55]

The owners of the Port Blakely Mill Company found that logging railroads brought other problems, too. Loggers sometimes tried to block the expansion of the company's lines by obtaining title to key

History of Santa Cruz County, pp. 192–97; W. W. Elliott & Co., *Santa Cruz County, California: . . .* (San Francisco, 1879), p. 28; Bruce A. MacGregor, *South Pacific Coast: An Illustrated History of the Narrow Gauge South Pacific Coast Railroad* (Berkeley, Calif.: Howell-North Books, 1968), pp. 22–25, 39–49, 118–38, 186–91; U.S. Forest Service, California Forest and Range Experiment Station, "Bibliography of Early California Forestry," 69 vols. (typescript; Berkeley: University of California Forestry Library, 1939–41), 52, pt. 2:133 and *passim*.

52. Melendy, "One Hundred Years of the Redwood Industry," p. 177. See also W. W. Elliott & Co., *History of Humboldt County*, p. 139; Palais and Roberts, "History of the Lumber Industry in Humboldt County," p. 15; Carranco and Fountain, "California's First Railroad," pp. 247–48, 251.

53. *West Shore* 16 (1889):461; Portland *Oregonian*, 25 Aug. 1887.

54. Renton, Holmes & Co. to Port Blakely Mill Co., 22 Sept. 1888, PBMCP. Generating the capital to build logging railroads was probably easier for firms that had liquidated their holdings in the Lake States than for old West Coast firms, such as that at Port Blakely, which had the bulk of their funds tied up in fixed investments.

55. Berner, "Port Blakely Mill Company," p. 169.

blocks of land. The Northern Pacific, which intended to build a line to Grays Harbor, sought to prevent the extension of the Port Blakely line in that direction.[56] Other builders of railroads must surely have met with similar obstacles.

The presence of railroads forced operational changes on the lumber industry. A large volume of logs was needed if logging railroads were to operate at optimum efficiency. This requisite led to larger logging camps. One observer noted that a camp with fifteen men had been considered large in the early seventies, but by the eighties thirty to sixty men were frequently employed.[57]

Prior to the advent of the iron horse, the Port Blakely Mill Company had obtained their logs from independent loggers. The company thus was able to avoid the outlay of investment capital that would have been necessary to establish logging camps of their own. Before building lines into an area, however, it seemed prudent to own major blocks of timber in the area. Since it was their own timberland that was being logged, the owners of the Port Blakely firm now felt constrained to establish camps to fell the trees and get them out to the sidings. Many of the first logs came from the railroad right of way. Since failure to clear the right of way quickly and efficiently could hold up the entire building program, company logging camps were established to do the job. Under Sol G. Simpson, in time to become a major force in the lumber industry of the Northwest in his own right, these camps accomplished the task with dispatch. The sole dependence of the Port Blakely Mill Company on private loggers quickly passed with the introduction of railroads into the woods.[58]

Other large mills were moving in the same direction, but independent loggers continued to operate. Charles Hanson's Tacoma Mill Company obtained a large percentage of its logs from independent loggers cutting along the rail lines that supplied his mill. In the past, log rafts had been scaled upon delivery at the mill so that buyers would not be paying for logs stolen or lost while being floated to the mill. With the supply of good logs limited, competition between companies for the logs from independent operations was intense. Hanson, who could get logs by rail with little danger of loss, sought a competitive advantage

56. Ibid., p. 163.
57. *Northwestern Lumberman*, reprinted in *Canada Lumberman* 3 (1 Jan. 1883):11.
58. Berner, "Port Blakely Mill Company," pp. 159–64, 166; G. Emerson to A. Simpson, 7 June 1898, EL. Sol G. Simpson was apparently not related to A. M. Simpson.

over his rivals when he tried to persuade the industry to scale logs in the woods. Firms such as the Puget Mill Company that were not serviced by rail lines saw the scheme for what it was and rejected the proposal.[59]

While rail lines helped increase production by opening up new sources of logs, to what extent railroad land grants contributed to this development is unclear. There were extensive grants in western Oregon and Washington, but without careful study of tax, property, and other records, it cannot be safely asserted that they furnished the logs transported over the various lines. These logs may well have come from nongrant tracts that were close to the new railroads, lands that could be obtained through pre-emption or private purchase. Clearly, nearly all of the largest grant, that to the Northern Pacific Railroad, remained in that firm's hands until 1900, when the Weyerhaeuser interests purchased nine hundred thousand acres. Even then, major logging of this timberland was not undertaken for several years.[60] Land grants were certainly not essential for the expansion in the eighties: production on the redwood coast, where there were no land grants, rose just as it did in the Northwest.

Railroads caused changes not just in production but in marketing patterns, too. The construction of the Santa Cruz and Felton Railroad brought a great increase in production in the area south of San Francisco. The greatly increased quantities of redwood that began to move into San Francisco from nearby forests made marketing lumber from Humboldt and Mendocino counties in the Bay Area especially difficult in the late seventies.[61] The construction of the Oregon-Pacific Railroad up the Santiam River also created new competition for the cargo mills. Long trainloads of lumber were soon moving down the line from the previously untapped Cascade Mountains for use in the Willamette Valley and in California as well.[62] On the other hand, the completion of the Canadian Pacific in the mid-eighties tended to lessen competition from British Columbia. With transcontinental connections, Vancouver boomed and the domestic demand for lumber greatly increased. In addition, lumber producers in the vicinity of Burrard

59. W. H. Talbot to C. Walker, 13 June 1887, AC, Walker, incoming corres.
60. Hidy, Hill, and Nevins, *Timber and Men*, pp. 211–47; Berner, "Port Blakely Mill Company," pp. 161–64. Cf. Meany, "History of the Lumber Industry," pp. 212–27.
61. Sanford, "Short History of California Lumbering," pp. 38–39.
62. Portland *Oregonian*, 13 Aug. 1884. *West Shore* 13 (1887):265; 16 (1890):172–73.

Inlet now turned to interior markets that had previously been beyond their reach. In time these new outlets became the primary ones for their mills. Though they continued to engage in the Pacific lumber trade following the completion of the Canadian Pacific, the percentage of the cargo trade that went to Canadians declined sharply during the 1890s.[63] It was fortunate for Puget Sound mills that Canadians turned their attention to the interior, for by this time the lumber being produced in British Columbia was of better quality than that from the sound. Australians, especially, favored the Canadian product.[64]

Railroads affected the lumber industry of the West Coast in another way during the eighties. There was much railroad construction elsewhere around the Pacific Basin, and such construction led to increased demand for lumber from the West Coast long before freight rates on the transcontinentals opened significant new markets for Douglas fir and redwood lumber in the interior of the United States. Millowners built logging railroads not just in order to obtain new sources of timber that would keep their mills running at the old levels and put them in a better position vis-à-vis the independent loggers; the increased demand for lumber in the markets of the Pacific Basin would have made tapping additional stands necessary in any case. This fact was recognized at the time, though it has been generally overlooked since. The *West Shore* stated, "To meet this demand [in the markets of the Pacific area], railroads are building into the timber. . . ." [65]

Chile was among the areas where demand was increasing. For all practical purposes the War of the Pacific came to a close with the capture of Lima by Chilean forces in 1881. The victorious Chileans turned to domestic problems and soon undertook an extensive railroad-building program. Though Chilean forces subdued the Araucanians in 1883 and thus opened southern Chile to development, the bulk of the construction materials used in Chile's railroad projects came from the west coast of the United States rather than from Chilean sources. *West Shore* noted that 1882 was a "very prosperous year for

---

63. *Canada Lumberman* 8 (May 1888):8; 9 (Jan. 1889):1; 9 (March 1889):1, 6; 9 (May 1889):1. Lawrence, "Markets and Capital," pp. 37–41, 113. Meany, "History of the Lumber Industry," pp. 338–64. Unlike its counterparts on the Canadian mainland, the mill at Chemainus on Vancouver Island continued to compete actively in the cargo trade into the 1890s. See Lawrence, "Markets and Capital," pp. 68–70.

64. *Canada Lumberman* 3 (1883):290; 9 (July 1889):1; 11 (May 1890):10.

65. *West Shore* 13 (Jan. 1887):50. Cf. *Canada Lumberman* 9 (Mar. 1889):1; Tattersall, *Importance of International Trade to Oregon*, p. 7; Meany, "History of the Lumber Industry," p. 137; Tattersall, "Economic Development of the Pacific Northwest," p. 183.

lumbermen in the Pacific Northwest." One reason for the prosperity, the journal explained, was the increased demand in Chile that came with the end of fighting.[66]

The construction of railroads in Chile continued through the rest of the decade. As the victor in the War of the Pacific, Chile had at its disposal the rich mineral deposits in the provinces of Tarapacá and Antofagasta. As the nation's mineral exports skyrocketed, the country itself prospered. Public revenues, benefiting from the customs on exports, rose from approximately 15 million pesos annually before the war to 37 million in 1886 and 58.5 million in 1890. Much of this went to finance development projects. More than a thousand kilometers of rails were laid and many other public works built during the presidency of José Manuel Balmaceda (1886–91). According to one distinguished Latin American scholar, "It was a veritable orgy of material progress and even a new department of state had to be created to look after it—the ministry of public works in 1887. . . ." [67] As the chief suppliers of the forest products needed in this building program, lumbermen along the west coast of the United States benefited accordingly. But when warfare again broke out in Chile in 1891, lumber importation came to a halt. The Port Discovery mill closed, blaming the war and the resulting dullness in the foreign trade.[68]

As one of the vanquished in the War of the Pacific, Peru was at first in no position to undertake public works such as those of Chile, no matter how badly they were needed. The nation was in serious financial straits. However, in 1886 President Andrés A. Cáceres reached an agreement with Michael A. Grace, representative of a group of British capitalists who held Peruvian government bonds and were anxious to reap some return from their investments. Peru was in no position to amortize the debt, but in the so-called Grace Contract agreed, in return for the cancellation of most of it, to turn over control of the Peruvian railroads to the bondholders for a period of sixty-six years, to give them the first three million tons of guano exported each year, and to pay eighty thousand pounds sterling to the bondholders during each of the next thirty-three years. For their part, the bondholders agreed to invest in the railroads of Peru, so as to facilitate

66. *West Shore* 8 (1882):182. See also Claudio Véliz, *Historia de la Marina Mercante de Chile* (Santiago: University of Chile, 1961), pp. 254–62; Eureka *Humboldt Times*, 18 June 1881.
67. Galdames, *History of Chile*, p. 343.
68. *Canada Lumberman*, 12 (June 1891): 13. Other factors undoubtedly were also involved.

repair and extension of the lines, and to make new loans of up to six million pounds sterling to the Peruvian government.

To many Peruvians it seemed a harsh settlement, but it was probably as good as any representative of the nation could have obtained under the circumstances. After strenuous objections, the congress of Peru finally agreed to the contract in 1889. Almost immediately the Peruvian Corporation, Ltd., which had been established by the bondholders, began to repair and extend the country's rail network. This activity, together with the increasing production of silver between 1886 and 1895 (an increase encouraged by railroads to the mining areas), helped return prosperity to the land.[69] Like rail construction in Chile, this construction increased demand for forest products from Puget Sound, Oregon, and California.

A third long-time market for lumber from the West Coast also grew during the 1880s as a result of a burgeoning program of railroad building. Shipments of lumber to Australia had been infrequent during the 1860s and only somewhat more common during the 1870s.[70] By November 1883, however, shipments were being dispatched to ports in the Australian colonies with greater frequency, and *West Shore* observed that the Australian market was becoming important to producers on Puget Sound and in adjacent areas.[71]

Railroad construction had commenced in Australia as early as the late 1850s. Additional mileage was added during the 1860s, but in 1870 there were still only one thousand miles of railways on the continent. The rate of construction accelerated during the next two decades, however, and by 1890 ten thousand eight hundred miles of track had been completed. Over six thousand miles of new track were opened during the 1880s alone. This rapid expansion of the rail system was made possible by large amounts of British investment capital and by the enthusiastic manner in which the various colonial governments pushed the programs ahead. Very few of the lines were the result of private

69. Pike, *Modern History of Peru*, pp. 151–54; Levin, *The Export Economies, passim.* There was also considerable railroad construction in Peru during the 1870s. See Stewart, *Henry Meiggs*, pp. 42–224.

70. Astoria Customs House, "Marine List and Memoranda," vol. 2 *passim.* Foreseeing that the lumber trade with Australia might soon become more active, the agents of an Australian firm suggested sending a sample of spruce, as it might lead to "quite a trade." Dempster and Keys to Renton, Holmes & Co., 11 Dec. 1878, PBMCP. See also Grattan, *The United States and the Southwest Pacific*, p. 107.

71. *West Shore* 9 (1883):286. San Francisco *Journal of Commerce*, reprinted in *Canada Lumberman* 8 (July 1888):9; 8 (Aug. 1888):1.

enterprise. Needless to say, the rapid expansion of Australia's railways created a demand for Douglas fir and redwood ties and bridge timbers that was without precedent in that quarter.[72]

Yet the construction of railroads does not wholly explain the great increase in forest products shipped from the west coast of North America to Australia during the 1880s. Other factors were also at work. Australia's population grew continually from 1860 to 1900, the most rapid increase occurring in the years 1881–89. The discovery in 1883 of one of the world's richest mineral deposits, the Broken Hill silver-lead-zinc complex, together with other less important mineral finds during that decade, spurred economic activity in Australia. In addition, agricultural output was rising. Railroads made it economically feasible for farmers to raise wheat farther and farther inland. The production of sugar and wool also rose during the 1880s. Increasing agricultural surpluses meant increased agrarian building plus increasing trade, which in turn meant a greater need for building materials and for forest products to service vessels calling in Australian waters.[73]

Collectively, these developments made the Australian colonies much more attractive as markets than they had been in the sixties and seventies. Consular records of the United States show that American lumber dominated in Australia, New Zealand lumber was second, and British Columbia lumber third. The Puget, Washington, Port Blakely, Tacoma, and Port Discovery mill companies were all involved in the trade, sending cargoes primarily of rough lumber, timbers, or ties. Norway furnished most of the finished lumber. In 1886 shipments from the United States totaled 31.8 million feet; in 1888, 55.4 million feet. In 1892, after Australia's expansion had halted, they dropped to 12 million feet.[74]

Southern California was yet another area of burgeoning demand

72. Grattan, *The Southwest Pacific to 1900*, pp. 283–85, 316–17; Herbert Burton, "The Growth of the Australian Economy," in *Australia*, ed. C. Hartley Grattan (Berkeley: University of California Press, 1947), pp. 158–59; Brian Fitzpatrick, *The British Empire in Australia: An Economic History, 1834–1939* (Melbourne: Melbourne University Press, 1941), pp. 222–30; R. M. Crawford, *Australia*, rev. ed. (London: Hutchinson's University Library, 1960), pp. 131–32.
73. Grattan, *The Southwest Pacific to 1900*, pp. 261–63, 269–80, 311–16; Fitzpatrick, *The British Empire in Australia*, pp. 244–55; Burton, "Growth of the Australian Economy," pp. 158–61; Crawford, *Australia*, pp. 131–33; *West Shore* 10 (1883):39.
74. USBFC, *Consular Reports*, no. 65 (July 1886), pp. 244–45; no. 86 (Nov. 1887), pp. 377–78; no. 151 (Apr. 1893), pp. 134–37. As had been the case earlier, shipments often outstripped demand, and demand itself often fluctuated sharply. For examples, see the frequent reports on the Melbourne lumber market published in *Canada Lumberman* beginning with the issue of September 1881. See, especially, 3 (1883):214; 18 (May 1897):6.

during the 1880s. Once again, the motivating force behind the rise of this market was the construction of railroads. During the 1860s tiny vessels—such as the schooners *Mose*, *Susie*, and *Twin Sisters*, which plied on an irregular schedule between San Francisco and the Bolsa de San Joaquín (later Newport)—were able to fill the needs of southern California. John Kentfield and others dispatched increased quantities in the 1870s, but it still was far from being a major market. It became one in the 1880s.[75]

Many had been anticipating a boom in the southern part of the state for some time. With the completion of the Santa Fe railroad in 1885 and the rate wars that soon followed, the long-anticipated boom became a reality. Thousands of new settlers flocked into the area, setting off a surge of real estate and building activity.[76]

West Coast lumbermen quickly moved to take advantage of the development. Indeed, the Washington Mill Company, anticipating the boom that would follow the arrival of the railroad, actively began seeking business in southern California as early as 1883. Its efforts were not without success, and cargoes had been dispatched to Wilmington, San Diego, San Luis Obispo, and Los Angeles well before the Santa Fe drove its golden spike at Cajón Pass.[77] For the Port Blakely Mill Company, the effects of the boom were limited until 1886. Then the effect of rising sales at its yard in San Buenaventura began to be felt.[78]

The Pope and Talbot interests were especially aggressive in moving to take advantage of the new developments in southern California. Joining with the Hooper brothers, who had lumber interests in San Francisco and on the redwood coast, they soon developed chains of

75. Harris Newmark, *Sixty Years in Southern California, 1853–1913*, 4th ed. (Los Angeles: Zeitlin & Ver Brugge, 1970), pp. 274, 380; Duncan Gleason, *The Islands and Ports of California* (New York: Devin-Adair Co., 1958), pp. 88–89; W. F. Montgomery, "Pioneer Lumber Dealers in Los Angeles," *Historical Society of Southern California Quarterly* 24 (1942):66–68; Aubrey Drury, "John Albert Hooper," *CHSQ* 31 (1952):293. For examples of the earlier trade, see schooner *California*, trip records, 1869–70, GMC journals, 13:2–13; 26:326–29 and *passim*.
76. For a thorough study of southern California during this period, see Glenn S. Dumke, *The Boom of the Eighties in Southern California* (San Marino, Calif.: The Huntington Library, 1944). See also Larry Booth, Roger Olmsted, and Richard Pourade, "Portrait of a Boom Town: San Diego in the 1880's," *CHSQ* 50 (1971):363–94.
77. Washington Mill Co. to W. Adams, 28 Apr. 1883, 26 June 1884; Washington Mill Co. to San Diego Mill Co., 20 Nov. 1884; Washington Mill Co. to Swartz & Becker, 20 Nov. 1884; R. Kendrick to W. Adams, 26 Feb. 1884, WMCP. See also Phineas Banning, "The Settlement of Wilmington" (MS, Bancroft Library, Berkeley, Calif.), pp. 5, 10, 13; Richard Webster Barsness, "The Maritime Development of San Pedro Bay, California, 1821–1921" (Ph.D. diss., University of Minnesota, 1963), pp. 178–277 *passim*.
78. Berner, "Port Blakely Mill Company," p. 168.

lumberyards in southern California and in the San Joaquin Valley. In the Russ Mill and Lumber Company in San Diego and the San Pedro Lumber Company they acquired strategically placed manufacturing facilities for planing and resawing. The yards and mills associated with the Hooper–Pope and Talbot combine were able to offer both redwood and Douglas fir products to prospective customers, since the operations of the one firm centered on Puget Sound and the other on the redwood coast.[79]

Just how large the market in southern California was at its height is not clear. Extant records indicate that some 72 million board feet of redwood were shipped there in 1883, about 35 million in 1884, and about 30 million—approximately 10 percent of the total cut of the redwood coast—during the year ending 10 October 1885. Figures are not available for Puget Sound, but, though the percentage of the total cut was probably not as great, the quantities were clearly quite large.[80]

Then in the spring of 1888 the real estate boom in southern California came to an abrupt end. By March demand for lumber was in sharp decline and, though it was not realized at the time, the prosperity that West Coast lumbermen had known through most of the decade had begun to trickle away. In less spectacular fashion demand was falling elsewhere. By October W. H. Talbot was describing the lumber market as "completely demoralized." The downward trend continued in the months that followed. By the time the nation as a whole entered a major depression in 1893, the lumber industry on the Pacific Coast had already been in a depressed state for at least two years.[81]

79. Coman and Gibbs, *Time, Tide and Timber*, pp. 204–9. For an account of the Hooper interests, see Drury, "John Albert Hooper," pp. 289–305. The acquisition of yards in southern California by major manufacturers demonstrates how far the industry had come since the gold rush. Then lumber merchants had built sawmills to insure sources of lumber; by the eighties lumber manufacturers had come to dominate the industry. However, unlike in many sectors of industry in the United States, the line between merchants and manufacturers was never clear-cut in the Pacific Coast lumber industry. Pope & Talbot, Renton, Holmes & Co., and other firms continued to serve in a mercantile capacity and to charge commissions for their services to manufacturing enterprises that were primarily owned by the same people.
80. [California] State Board of Forestry, *First Biennial Report, 1885–1886*, pp. 145, 168. Cf. Eureka *Humboldt Times*, 18 Aug. 1887, 20 Apr. 1888. By contrast, little lumber appears to have been shipped to southern California by mills located in Oregon. See Portland *Oregonian*, 1 Jan. 1885.
81. Berner, "Port Blakely Mill Company," p. 168; *West Shore* 14 (1888):412; Talbot to C. Walker, 17 Oct. 1888, AC, Walker, incoming corres.; Gleason, *Islands and Ports of California*, pp. 90–92; *Pacific Coast Wood and Iron* 11 (1889):94; San Francisco *Alta California*, 1 Jan. 1889, 1 Jan. 1890.

But in spite of its spectacular collapse in 1888, southern California never completely disappeared as a market for lumber. As the 1890s commenced, the Puget Mill Company was still sending cargoes to its outlets in southern California.[82] In 1892, the Kerkhoff-Cuzner Mill and Lumber Company, which had eight lumberyards in the Los Angeles area, purchased lumber from the Gardiner Mill Company, Stimson Mill Company, Noyo Lumber Company, Caspar Lumber Company, Sierra Lumber Company, and Elk River Mill and Lumber Company. Sales from its yards totaled over $138,000. Other outlets undoubtedly did as well.[83]

Unless a mill owned lumberyards, shipping to such economically volatile areas as southern California was fraught with risks. The operations of Los Angeles lumber dealer Perry Whiting well illustrate the relationship between producers and lumberyard operators. Having started from scratch, Whiting by the 1890s had built to the point where he was submitting orders for one hundred thousand feet of lumber, worth some $2,000. At first he was surprised that he did not have to pay in advance. Soon Clarence DeCamp of the Caspar Lumber Company, from which Whiting was buying his lumber, approached him with an offer of additional credit. Thereafter Whiting bought on six-months notes at 6 percent interest. Within a year Whiting was approaching the $25,000 limit on credit he and DeCamp had agreed on, and the president of the Caspar mill began to worry. The owner asked for early payment, saying the money was needed badly for extending their logging railroad. Whiting agreed to pay early if the mill would discount his notes 1 percent per month. Whiting borrowed from a bank and, to the surprise of the mill company executive, paid in full the next day. Impressed, the lumber company reversed its tack and offered Whiting further credit.

But Whiting was disgusted with the firm's duplicity and transferred his business to the Hammond Lumber Company. Within two years Whiting owed Hammond $75,000; concerned, the millowner came to Los Angeles to investigate. When he found that Whiting owed a total of $100,000, Hammond said, "Well, you may not realize it, but you

82. Puget Mill Co. to Russ Lumber and Manufacturing Co., 17 June, 27 July 1891; Puget Mill Co. to San Pedro Lumber Co., 25 July 1891, AC, Walker corres.

83. Kerkhoff-Cuzner Mill & Lumber Co., Ledger, Kerkhoff-Cuzner Papers, Huntington Library, San Marino, Calif. See also Whiting, *Autobiography*, pp. 143–58, 203, 215; Newmark, *Sixty Years in Southern California*, p. 515; Ann M. Connor, *Caspar Calling* (Santa Rosa, Calif.: by the author, 1967), p. 3.

are broke." Whiting demurred, saying his assets were worth twice as much as his debts. Hammond noted that if he were to demand payment, Whiting would have to close his doors; but Whiting replied that Hammond could not demand a cent until the notes were due. As Whiting tells it: "He looked at me with his piercing gray eyes for at least five seconds, then said, 'You are a smart young man, and I hope you come out all right; as long as you pay your notes when they are due, we will give you credit.' " It was a wise decision. Not only did Whiting's firm survive, it prospered in the years to come.[84]

In addition to growth of demand in long-established markets, railroad construction during the 1880s also led to consumption in new ones. Though lumber had been sent to Chile, Peru, and southern California for a number of years, little had ever been dispatched to ports on the west coast of Mexico. However, during the 1880s railroad construction was undertaken at various points along Mexico's western shore. A line was constructed from Guaymas to Nogales and shorter ones from Altata to Culiacán, from San Blas to Huaristemba, from Manzanillo to Armería, and from Salina Cruz to Ixtepec. Shipments of ties moved southward to fill the demand created by this construction. The market proved transient: little further railroad building took place on Mexico's west coast until after 1898; but while it lasted, many cargoes of ties and construction materials were dispatched there from the lumber coast.[85]

It must have been a colorful trade. Captain W. Colby of the schooner *Courser* has left a vivid account of the problems he faced getting his vessel into the "harbor" at Altata. Once there, Colby complained, a passenger on his vessel told the Mexican tallyman that the cargo of ties had been condemned at the mill. The passenger apparently hoped that the Mexican consignees would refuse the cargo and that he could then pick it up for himself at bargain rates. The scheme nearly worked. The consignees refused to accept the ties until the captain used "a little Policy." He judiciously distributed potatoes

84. Whiting, *Autobiography*, pp. 165–67. Perhaps the difficulties Whiting experienced with the Caspar Mill Co. sprang from the fact that his competitors, Kerkhoff and Cuzner, were half-owners of the sawmill firm. See Connor, *Caspar Calling*, p. 3.

85. Daniel Cosío Villegas, *Historia Moderna de México*, 8 vols. (Mexico City: Editorial Hermes, 1955–70), 7:539–42, 566–70; Washington Mill Co. to W. J. Adams, 19, 24 Apr., 1, 4, 6 May 1884, WMCP; Edgar Cherry & Co., *Redwood and Lumbering*, p. 30; Eureka *Humboldt Times*, 18 June 1881. Even during the slack years of 1884–98, occasional shipments were sent. For example, see E. G. Ames to C. Walker, 9 Feb. 1892, AC, Walker, private corres.

and other items from the ship's stores that were not available in the Mexican port. The buyer soon decided that the ties would be satisfactory. When officials of the Port Blakely Mill Company, who had been complaining of the captain's misuse of ship's stores, heard his account of what had taken place, they promptly dropped their complaints.[86]

Larger and more lasting than those in Mexico were markets that developed during this same period in North China and were reached through the port of Tientsin. In 1425 the Ming emperor Yung Lo had established Tientsin as a walled city forty miles from the sea at the north end of the Grand Canal. The first Dutch embassy to Peking passed through Tientsin in 1855 and described it as an active commercial city, "very populous and so full of trade, that hardly the like . . . is to be found in any other city in China." [87]

But the city was important as a regional trading center, not as an international port. Although Tientsin was occupied during the second Anglo-Chinese war, it was not opened to foreign commerce until the additional conventions of 1860. Once opened, Tientsin's primary importance still was local; in 1868 the American consul there described Tientsin as "a small outport of Shanghae." [88] The silt-choked Pei-ho on which the city is located tended to discourage all except shallow-draft coastal steamers and junks from calling there.[89] Even the shortage of houses in Tientsin in the 1870s was insufficient inducement to attract lumber-laden vessels from abroad.

Then, belated modernization, including extensive railroad construction, abruptly changed the picture in North China. Previously, what construction materials had been needed in Tientsin had been transshipped from Shanghai or Hong Kong or obtained from domestic sources. These sources were incapable of furnishing the quantities the new projects required.[90] In spite of the limitations of Tientsin as a port,

86. W. Colby to Renton, Holmes & Co., 4 Mar., 26 May, 1 June, 29 July 1882, PBMCP. For other problems associated with the trade with Latin America, see Seattle *Daily Intelligencer*, 18 Aug. 1877; *Canada Lumberman* 13 (Jan. 1892):11.

87. Quoted in O. D. Rasmussen, *Tientsin: An Illustrated Outline History* (Tientsin: Tientsin Press, 1925), p. 9.

88. Consular despatches, Tientsin, 6 Jan. 1868.

89. Ibid., 16 Oct. 1872; Tientsin *Chinese Times*, 28 Feb. 1891 and other dates; Jones, *Shanghai and Tientsin*, pp. 117–19.

90. Dealers in Shanghai hoped to continue to dominate the lumber trade of China even after the coming of railroads in the north. See Farnham & Co. to Renton, Holmes & Co., 10 Aug. 1885, PBMCP; Wright, ed., *Twentieth Century Impressions of Hong Kong, Shanghai, and Other Treaty*

shipments of forest products began to come directly to it from western North America. According to the American consul in the city, Sheridan P. Read, the first shipments arrived from the Northwest in 1883. Other shipments followed, and an active trade quickly developed that was to continue through the rest of the nineteenth century and into the twentieth.[91]

The dominant position that Douglas fir timber held in the Tientsin market by the 1890s was not obtained without competition. Almost every lumber-producing area around the Pacific sought to capture a portion of the burgeoning market in North China. A steam sawmill was erected in Tientsin to saw logs imported from Korea. Other logs were sawn in Korea and the cut shipped to Tientsin. Because of poor transportation facilities in Korea, it was necessary to cut the logs into short lengths to get them out to market. Thus, however inexpensive, Korean lumber could not compete with American for orders for larger dimensions.[92] Timber products were also imported from Asiatic Russia, Southeast Asia, and the forests of South China. William Forbes and Company, described by Consul Read as the "firm most extensively interested in lumber here," advertised that it had sawmills in Sandakan (Borneo) and Hong Kong. The firm also imported from British Columbia and Korea.[93] Redwood entered the competition late, probably because of its relatively high price. Apparently it was not until 1896 that redwood was shipped in significant quantities to North China. In that year thirty thousand redwood ties, described as "a trial shipment to see how that particular kind of wood will last in China," were dispatched to Tientsin.[94]

Ports, pp. 578, 588. Hong Kong and Shanghai continued to import forest products in the 1880s, but few railroads were constructed in the Yangtze Basin or South China, and when they were built, they were usually supplied with ties from local sources. As a result, the quantity of imports did not skyrocket as it did in the north. See USBFC, *Commercial Relations of the United States with Foreign Countries during the Years 1894 and 1895*, 2 vols. (Washington, D.C., 1896), 1:591; Tientsin *Chinese Times*, 23 June 1888; Wright, ed., *Twentieth Century Impressions of Hong Kong, Shanghai, and Other Treaty Ports*, pp. 578, 837.

91. Tientsin *Peking and Tientsin Times*, 15 Sept. 1894. George Emerson, for one, was eager to sell ties to buyers in China. Such orders, he explained, "take a class of material worth very little in San Francisco. . . ." Even logs often left in the woods could sometimes be used. See Emerson to A. Simpson, 15 Dec. 1897, EL.

92. Tientsin *Chinese Times*, 22 Oct. 1887, 14 Apr., 2 June 1888; Tientsin *Peking and Tientsin Times*, 15 Sept. 1894.

93. Tientsin *Chinese Times*, 11 June, 22 Oct. 1887, 17 and 24 Mar., 21 Apr., 5, 12, 13, 19, 26 May, 2, 9 and 16 June, 20 Oct. 1888, 3 Aug. 1889, 19 July, 4 Oct. 1890, 7 and 14 Mar. 1891; Tientsin *Peking and Tientsin Times*, 15 Sept. 1894; USBFC, *American Lumber in Foreign Markets*, pp. 91, 93; Lawrence, "Markets and Capital," p. 70.

94. Tientsin *Peking and Tientsin Times*, 10 Oct. 1896.

Of all the sources of competition, the most important was Japan, one of the most extensively forested nations in the world. The various *daimyo* of Tokugawa Japan had carefully husbanded the forests of their domains so that the nation entered upon the hectic modernization that followed the Meiji Restoration of 1868 with abundant forest resources upon which to draw. The feudal restrictions on the use of Japan's forests were withdrawn as a result of the political changes of the restoration; timber felling was pushed forward with vigor, and the exportation of Japanese forest products soon began.[95] As a part of its efforts to open up the previously undeveloped northern island of Hokkaido and to gain foreign exchange, the Meiji government itself encouraged the growth there of a forest industry that could sell part of its cut abroad.[96] During the 1880s Japanese were offering pine planking and railroad ties in the markets of China at prices North Americans could not match. Vessels arrived regularly from Otaru and other ports in Japan with cargoes of forest products. The amount involved may have been even greater than the tonnage of vessels arriving from Japan would seem to indicate, for, according to the *Peking and Tientsin Times*, much of the lumber represented as coming to North China from ports further down the coast was in fact originally from Japan and was merely being transshipped to Tientsin.[97] As the Japanese overcut their forests, however, exports declined. By 1894 Consul Read was reporting, "Oregon pine is safe from rivalry."[98]

Some boosters of the development of China were disenchanted to see the large quantities of forest products that were being imported.

95. Imperial Japanese Commission to the Louisiana Purchase Exposition, *Japan in the Beginning of the 20th Century* (Tokyo: for Imperial Japanese Commission to the Louisiana Purchase Exposition by Japan Times, 1904), pp. 224–25, 237, 241, 250–51, 274, 282; Tsuneaki Sato, "Japanese Industries: Agriculture and Forestry," in *Fifty Years of New Japan*, ed. Count Shigenobu Okuma (London: Smith, Elder & Co., 1909), pp. 583–93; Johannes Hirschmeier, *The Origins of Entrepreneurship in Meiji Japan* (Cambridge, Mass.: Harvard University Press, 1964), p. 276. See also Hattori Kintaro, "Nihon ringyō hattatsu shi jo ronsetsu" [Introductory remarks on a history of the development of the forest industry of Japan], *Mita gakkai zasshi* 45 (1952):46–52.
96. Thomas C. Smith, *Political Change and Industrial Development in Japan* (Stanford, Calif.: Stanford University Press, 1955), pp. 113–14; Mary Harbert, "The Open Door Policy: The Means of Attaining Nineteenth Century American Objectives in Japan" (Ph.D. diss., University of Oregon, 1967), pp. 273–76; John A. Harrison, *Japan's Northern Frontier* (Gainesville: University of Florida Press, 1953), pp. 106–8.
97. Tientsin *Chinese Times*, 11 June, 22 Oct. 1887, 2, 16 June 1888, 13 Apr. 1889; Tientsin *Peking and Tientsin Times*, 15 Sept. 1894.
98. Quoted in Tientsin *Peking and Tientsin Times*, 15 Sept. 1894. See also USBFC, *Commercial Relations of the United States with Foreign Countries during the Year 1898*, 2 vols. (Washington, D.C., 1899), 1:1096.

"In Manchuria," *The Chinese Times* insisted, "there are immense stands of the very finest timber, very much superior in size and fineness to the wood of Oregon" which should be supplying the country's needs.[99] The paper's editor overestimated both the size and the quality of the Manchurian stands, but he understood very well why timber from there was in fact not competitive with imported woods. First, the editor explained, "Chinese timber is so loaded with local exactions in addition to the regular Custom duties that the Japanese product is laid down much cheaper, and so the Railway Company must pay foreigners for material grown abundantly in their own country." [100] Second, transporting of logs or lumber from the forests of Manchuria "is impracticable for lack of railways" or other inexpensive means of handling such bulky cargo. Under the circumstances, the cost in Peking and Tientsin of domestically produced timber was so high that only the rich could afford to use it.[101] Such prices hardly served to recommend its utilization in the building projects that accompanied China's attempts at modernization.

In the final analysis, it seems clear that railroads played a major role in helping to bring on the great boom that the lumber industry of the West Coast experienced during the 1880s. But this boom did not result primarily from the opening of new markets to the east via the transcontinentals or from the demand for forest products that the construction of America's transcontinentals created. The gains from both were limited: markets to the east proved elusive, and the demand for building material transient and often filled by transmontane competitors. The boom cannot even be attributed primarily to the creation of a rail system that tied the forested hinterlands to the mills and ports, thus making possible increased exportation by sea. The impact of railroads on the lumber industry during the 1880s was multifaceted, but one key element clearly was the greatly increased demand for forest products that railroad construction throughout the entire Pacific Basin helped to create. Along the lumber coast railroads tapped previously inaccessible forests; around the rim of the Pacific the building of thousands upon thousands of miles of railroads created

99. Tientsin *Chinese Times*, 8 Oct. 1887.
100. Ibid., 11 June 1887. See also Imperial Japanese Commission to Louisiana Purchase Exposition, *Japan in the Beginning of the 20th Century*, pp. 224–25.
101. Tientsin *Chinese Times*, 8, 29 Oct. 1887, 14 Apr. 1888; USBFC, *Commercial Relations, 1894–95*, 1:612; USBFC, *Commercial Relations of the United States with Foreign Countries during the Years 1895 and 1896*, 2 vols. (Washington, D.C., 1897), 1:801–6.

markets of unprecedented size. An increased supply filled an increased demand. The end result was a degree of prosperity for the lumber industry of the Pacific slope that had not been known since the days of the Gold Rush, and a level of production undreamed of even then.

Changing conditions in various markets served by the cargo mills gradually diminished the prosperity they had enjoyed through most of the eighties. The real estate boom in southern California collapsed, demand dried up in Australia, and war reopened in Chile. The 1890s commenced with Pacific Coast lumbermen more concerned over retrenchment and consolidation than expansion. By the end of 1893 retrenchment had given way to a desperate struggle for survival.

But if the eighties closed with confidence and prosperity muted, they also closed with the industry larger and stronger than ever before. The trickle of investments from the Lake States had not yet turned into a flood that threatened to sweep the cargo mills before it. The depression of the nineties, which was to undermine and weaken them, had not yet struck. The decade of the eighties was their Golden Age.

CHAPTER X

Technological Changes Ashore and at Sea

INNOVATIONS and improvements have come to the forest products industries of the Pacific slope in an almost constant stream from the time of the area's first sawpits up to the present. In the expansive, prosperous 1880s the pace of change was especially rapid. Numerous new devices and techniques came to be utilized in the logging, milling, and transporting of the forest products of the Far West. The railroad was one such innovation, but there were many others. Their collective impact was great: they not only changed production and transportation but, through the financial demands they made on lumbermen, helped bring structural changes to the industry as well.

One of the most basic changes in logging was the adoption of new methods of felling timber. Crosscut saws were in use in the forests of the Far West as early as the 1840s, but down to the 1880s axes alone were used to fell trees. The crosscut was used only in cutting trees into lengths once they were on the ground. Then, somewhere in the redwoods, lumbermen discovered that by using crosscuts in combination with axes—which were still used in clearing the undercut—the time required to fell a tree was reduced by four-fifths. Moreover, because saws cut so much faster than axes, a sawn tree would topple from its stump quickly and neatly, while a chopped tree would often tear slowly away, causing unpleasant consequences. With the uncut portion holding fast to the stump, the tree would sometimes split as it fell or at other times be pulled so that it would fall not where intended, but across rocks or other impediments that would shatter the trunk and render it useless.

This new method of felling timber was in general use in the redwoods by the early 1880s, had been introduced into the Grays Harbor area, and was being adopted elsewhere. Unfortunately, the name of the discoverer of this improved method of felling timber has been lost with the passage of time.[1]

The ax itself was undergoing change. Originally West Coast lumbermen used a single-bitted ax, but during the early eighties the double-bitted ax replaced it. In 1878 the Washington Mill Company's store at Seabeck began carrying a new brand of ax. The company informed the manufacturer that their loggers preferred the new brand not because it was 25 percent cheaper than the brand in use earlier, but because its steel held an edge better. To a timber faller such considerations were more important than the lower price.[2]

Changes in methods of getting logs out of the woods were as important as those in felling timber, and more widely noticed. Most of the first sawmills were supplied with logs from trees that had grown at waterside. The trees were felled directly into the water, or were simply rolled in, and the logs were then floated to the mill.[3] However, as trees on the banks of the streams, bays, and sloughs of the Far West were cut, new methods had to be adopted. Because of the size of logs in the Far West and the very different terrain and climate, methods that had been successfully used in the woods of the Lake States and the East were not practical.

Western lumbermen developed the skid road system to solve the problem. By clearing paths through the forest, putting logs crossways on these paths at close intervals, greasing these skids, and attaching six to ten span of oxen to a string of logs, it was feasible to tap stands as far

1. Ingersoll, "In a Redwood Logging Camp," p. 202; Edgar Cherry & Co., *Redwood and Lumbering*, p. 44; Hugh Bower, "The Stand" (typescript account, Bancroft Library, Berkeley, Calif.); Bilsland, "An Historical Comparative Analysis," p. 47; Brown, *Sawpits in the Spanish Redwoods*, p. 11; Stewart H. Holbrook, *The American Lumberjack*, rev. ed. of *Holy Old Mackinaw* (New York: Collier Books, 1962), pp. 169–71; California State Agricultural Society *Transactions* (1877), p. 147; E. J. Stewart, transcript of interview, Mar. 1953 (Bancroft Library, Berkeley, Calif.).
2. Bower, "The Stand"; Washington Mill Co. to Adams & Taylor, 23 Jan., 6 Feb. 1878, WMCP. See also Washington Mill Co. to William Mann, Jr. & Co., 6 Feb. 1878, WMCP. As George Emerson put it, "Loggers are very particular about the axe they buy." See Emerson to H. Pennel, 20 Feb. 1892, EL.
3. For descriptions of the equipment and techniques used, see D. O. L. Schon, "The Handlogger: Unique British Columbia Pioneer," *Forest History* 14 (1971):18–20.

as two miles from the water. Bull teams and bullwhackers quickly became common sights in the woods of the region. But by the 1880s this method, too, was beginning to prove inadequate. In 1881 a Seattle paper reported that the best timber had been cut from the shore of Hood Canal one and one-half miles back for its entire length.[4] The same was true of most other readily accessible areas. Some millmen were able to obtain sawlogs by relogging cutover tracts: Charles D. Stimson's mill in Seattle was kept busy for two years on logs obtained from previously logged land near Appletree Cove.[5] But such operations were exceptions to the general rule. If milling were to be continued, much less expanded, methods would have to be devised to tap timber stands beyond the reach of skid roads.[6]

Logging railroads were one answer, but there were several others. Loggers constantly sought out streams along which the timber had not yet been cut. If a stream were large enough to float logs, it was soon in use. The *West Shore* announced in 1883 that in Columbia County, Oregon, every "stream of any size has been cleared of obstructions, so that logs can be run down them in high water season."[7] By the end of the 1880s, the same was true of almost any county along the lower Columbia, around Puget Sound, or along the lumber coast.

Yet, still more logs were needed. One answer was the splash dam, a device for turning tiny streams into torrents large enough to float logs. A dam would be built on a stream and water stored behind it. When a large head of water had been accumulated, it would be released and would quickly sluice logs that had been dumped into the pond behind the dam, together with others collected along the watercourse below the dam, out to where they could be handled by conventional means.[8]

4. Seattle *Tri-Weekly Fin Back*, 31 Aug. 1881, cited in Buchanan, "Economic History of Kitsap County," p. 214. See also Portland *Oregonian*, 1 Jan. 1882. Change apparently came more slowly to coastal British Columbia. See *Canada Lumberman* 3 (16 Apr. 1883):113.
5. Thomas D. Stimson, comp., "A Record of the Family, Life and Activities of Charles Douglas Stimson" (photocopy, University of Washington library, Seattle), p. 11.
6. Among the many descriptions of skid road logging are Holbrook, *The American Lumberjack*, pp. 163–65; Meany, "History of the Lumber Industry," pp. 245–46; Edgar Cherry & Co., *Redwood and Lumbering*, pp. 34–50; *West Shore* 2 (1877):68; Stewart, interview. Innovations came even to the skid roads in the eighties. Petroleum-based lubricants began to be used in place of dogfish oil as skid grease. See Thomas F. Gedosch, "A Note on the Dogfish Oil Industry of Washington Territory," *PNQ* 59 (1968):100–2.
7. *West Shore* 9 (1883):128.
8. Nelson C. Brown, *Logging: The Principles and Methods of Harvesting Timber in the U.S. and Canada* (New York: J. Wiley & Sons, 1949), pp. 356–63; Meany, "History of the Lumber Industry," pp. 256–69; Peter J. Rutledge and Richard H. Tooker, "Steam Power for Loggers: Two Views of the Dolbeer Donkey," *Forest History* 14 (1970):20.

Alex Polson built a splash dam in Pacific County, Washington, in 1881. The next year either he or Cyrus Blackwell introduced the technique into the area of Grays Harbor. Use of the splash dam spread quickly. In 1888 the main logging operations supplying George Weidler's Willamette Steam Mills, Portland's largest sawmill, were using splash dams. Weidler's crews floated some ten million feet of logs a year out of tiny Abernethy Creek by this means.[9] Elsewhere flumes and chutes were being put to use to get logs out from otherwise inaccessible stands.[10]

A different approach, but one accomplishing the same thing in the long run, was to build mills in the vicinity of timber stands and then float the lumber out of the woods by flumes. The Garcia Mill, near Point Arena on the redwood coast, was one such operation. Lumber passed directly from the saws or planer into a box flume that carried it four miles to a hoisting works, a waterwheel-powered system of rollers, that lifted the lumber to the top of the riverbank. There it was loaded onto railroad cars to be hauled the final two and one-half miles to the coast and put aboard schooners for shipment to San Francisco. Not all operations were so complex, but flume arrangements of one type or another rapidly increased in number during the 1880s.[11]

Another reflection of the shifting sources of log supply was to be seen on the major waterways. As mill operators came to obtain their logs from sources farther and farther from the millsites, powerful new tugs for towing log rafts were often added to the holdings of mill companies, and booming and rafting techniques reached a high level of development. On the lower Columbia and Puget Sound during the eighties logs were rafted fifty miles and more to get them to the sawmills. Vessels such as Weidler's steam tug *Wonder* did yeoman service in this work.[12]

9. *West Shore* 14 (1888):415–16; Bilsland, "An Historical Comparative Analysis," pp. 46–47. See also G. Emerson to A. Simpson, 16 Mar. 1892, EL.

10. Peterson and Powers, *A Century of Coos and Curry*, p. 488; J. E. Robertson, "The Log Flume As a Means of Transporting Logs," *The Timberman* 10 (1909):45–46; Brown, *Logging*, pp. 262–74, 387–98; Eugene S. Bruce, *Flumes and Fluming*, U.S. Department of Agriculture Bulletin no. 87 (Washington, D.C., 1914), *passim*; G. Emerson to H. Pennel, 11 Aug. 1891, EL.

11. Melendy, "One Hundred Years of the Redwood Industry," pp. 231–32. Brown, *Logging*, pp. 387–98. Bruce, *Flumes and Fluming, passim. West Shore* 9 (1883):13; 13 (1887):337; 15 (1889):151. Hittell, *Commerce and Industries*, pp. 580, 585. Evans, *History of the Pacific Northwest*, 2:216. California State Agricultural Society *Transactions* (1877), pp. 146–47.

12. Wright, ed., *Lewis and Dryden's Marine History*, p. 252; Newell, ed., *McCurdy Marine History*, p. 26; McNairn and MacMullen, *Ships of the Redwood Coast*, pp. 110–20; Brown, *Logging*, pp. 370–78; Buchanan, "Economic History of Kitsap County," pp. 156–61, 228;

Perhaps the most important advance in getting out logs was the adaptation of steam power to woods operations. The breakthrough came with the introduction of the steam donkey—like the skid road, a West Coast invention. Basically the steam donkey was little more than a power-driven spool to which rope was attached for pulling in logs, but with modifications and improvements it quickly revolutionized western logging. John Dolbeer, the inventor of the device, first tried it in yarding logs on his Salmon Creek operations near Eureka in August 1881. The power of this first steam donkey was small, so several blocks had to be employed to increase the pull; moreover, the hemp rope used was not overly strong and tended to stretch, especially when wet. These factors kept down the distance logs could be pulled. In addition, the design of this first model was such that the directions from which logs could be pulled were limited. Nevertheless, Dolbeer considered the tests a success and patented his invention soon after.[13]

At first, power equipment was used only in the yarding of logs, an operation for which oxen were poorly suited because they could not be used on steep ground and were difficult to maneuver in tight quarters or where the footing was poor. Once logs had been brought to the yards by Dolbeer's steam donkey, ox teams took over and pulled them to streamside so they could be floated out of the woods. But as modifications made steam donkeys more powerful, efficient, and versatile, their role expanded. Dolbeer himself developed and patented improvements. Others added contributions as well. About 1891, David Evans designed and put into use a modified version called the bull donkey. Larger and more powerful than Dolbeer's machine, it took the place of oxen on the skid roads. Logs could now be brought to landings beside a stream or railroad by power equipment alone. The days of the ox teams and bullwhackers were clearly numbered.[14]

Meany, "History of the Lumber Industry," pp. 251–53, 255–56; Clark, "Analysis of Forest Utilization," p. 62.

13. Edgar Cherry & Co., *Redwood and Lumbering*, p. 43; John Hittell, *Commerce and Industries*, pp. 425–27; Wilde, "Chronology of the Pacific Lumber Company," pp. 19–20; Rutledge and Tooker, "Steam Power for Loggers," pp. 20–21, 24, 27; Melendy, "One Hundred Years of the Redwood Industry," pp. 28–48; Bower, "The Stand"; Stewart, interview. Cf. Asa S. Williams, "Logging by Steam," *Forestry Quarterly* 6 (1908):2. Prior to the adoption of the steam donkey it was necessary to leave the larger logs in the woods or split them before hauling them out. See Melendy, "One Hundred Years of the Redwood Industry," p. 30; [California] State Board of Forestry, *First Biennial Report, 1885–1886*, p. 139; *West Shore* 8 (July 1882):136; Clark, "Analysis of Forest Utilization," pp. 26–27; G. Emerson to H. Pennel, 9 Feb. 1892, El.

14. Rutledge and Tooker, "Steam Power for Loggers," pp. 20–28; Meany, "History of the

The adoption of power logging was speeded by the early abandonment of manila rope for pulling logs. Its main shortcomings were its
bulkiness, which made it difficult to handle and limited the quantity
that could be stored on a drum, and its relatively short life when used
in the woods. The wire rope available in the 1880s left much to be
desired, but it was an improvement nonetheless. It not only proved
more durable, but also, because it had a greater tensile strength and was
less bulky, increased the distance that logs could be pulled.[15]

The use of power logging equipment quickly spread beyond the
redwood coast. In 1887 Cyrus Blackwell had a steam donkey in
operation near Grays Harbor. Simon Benson, a logging magnate from
Portland who was associated with George Weidler, soon was using
one in the forests along the Columbia. Others followed suit.[16]

At first only the larger, more adequately financed logging operations
used steam donkeys and similar power equipment but by the late
nineties even the operators of small shows were coming over. Included
among them were William Kyle on the Siuslaw and Harrison and
Diven on the Alsea. The reasons are plain. Benson reported that he
reduced the cost of getting out logs from $4.50 per thousand with bull
teams to $2.10 per thousand when he changed over to steam donkeys.[17]
Kyle put it less precisely, but more colorfully. Hauling logs to the river
with bull teams, he said, "is out of date, and is costly, the proper way is
with a logging engine and wire rope, when the machine dont work it
dont cost anything to keep it and you dont have to feed it when it is
not earning anything."[18] Steam donkeys not only got logs out of the

Lumber Industry," pp. 246–49.
15. Rutledge and Tooker, "Steam Power for Loggers," pp. 21, 27–28; Meany, "History of the
Lumber Industry," p. 247; William H. Gibbons, *Logging in the Douglas Fir Region*, United
States Department of Agriculture Bulletin no. 711 (Washington, D.C., 1918), p. 57; *Timberman*
10 (Aug. 1909):52. Gradually wire rope was improved, although it was not until after the turn of
the century that really satisfactory steel cable appeared. See C. D. Meals, "Wire Rope and Its
Application to the Logging Industry," *West Coast Lumberman* 54 (1928):146–47; James
O'Hearne, "Care and Choice of Wire Rope in Logging," *Pioneer Western Lumberman* 60
(1913):27–28; American Wire & Steel Co., *American Wire Rope: Catalogue and Handbook* (New
York, 1913), pp. i–ix, 129–34, 237. The first wire rope advertisement in the *Canada Lumberman*
appeared in October 1894.
16. Bilsland, "An Historical Comparative Analysis," p. 47; Meany, "History of the Lumber
Industry," pp. 246–49; G. Emerson to H. Pennel, 11 Aug. 1892, EL; Emerson to A. Simpson, 31
Mar. 1892, EL; *Canada Lumberman* 15 (Jan. 1894):10.
17. G. Diven to Meyer and Kyle, 30 May 1899, WKSP, 154; Kyle to P. Snodgrass, 22 Aug.
1899, WKSP, 174; Lockley, *History of the Columbia River Valley*, 1:629.
18. Kyle to Snodgrass, 22 Aug. 1899, WKSP, 174.

woods more cheaply, they also made it possible to log when it was too muddy for bull teams to work.[19] By the turn of the century, there were 293 donkey engines in use in Washington, 35 in Oregon, and 61 in California. Thirty-five of the latter were in Humboldt and Mendocino counties.[20]

While the crosscut saw, the double-bitted ax, the steam donkey, and a number of lesser innovations were changing the pattern of woods operations, equally drastic changes were being brought to the manufacturing processes. Early mills were simple affairs in which little, if anything, other than the saw itself was machine-powered. Invention after invention changed this. By the 1880s, there was hardly an operation in the larger, newer mills that had not been mechanized. A correspondent for the *Oregonian* described the "typical" mill on Puget Sound at the beginning of the decade.

The logs are now drawn up a slip from the water into the mill upon an endless chain armed with grappling dogs, and as they one by one fall upon the bed are seized by steel teeth of live rolls, which at the motion of brakes toss backward and forward, like playthings the largest, leaving them just right for the cross cut saws which governed by brakes dart upward from their hiding places and in a jiffy the log is ready for the rotaries. From the rotary the lumber, save square timber, passes to the gang edgers, drawn along by live rolls. Beyond the edgers are stationed the markers, one to each saw, who cull and mark the lumber, from whom it is hurried along the live rolls to the sluices, down which it slides to the wharf, where one or two vessels are constantly loading. Not the least interesting features of these mills are their admirable arrangements of endless chains for carrying off the bark, tops, buts [*sic*], slabs, edgings, sawdust, etc.[21]

Elsewhere along the coast the larger mills were moving in the same direction.

More and more steps in the manufacturing of lumber were being

---

19. G. Emerson to A. Simpson, 2 May 1892, 15 Oct. 1895, EL.
20. United States Census Bureau, *Twelfth Census of the United States Taken in the Year 1900*, 10 vols. (Washington, D.C., 1902), 9:821–22. Legislative restrictions slowed adoption of steam loggers in British Columbia. See Schon, "The Handlogger," p. 18.
21. Portland *Oregonian*, 5 Jan. 1880. See also Melendy, "One Hundred Years of the Redwood Industry," p. 61; Edgar Cherry & Co., *Redwood and Lumbering*, pp. 56, 66–68; Compton, *Organization of the Lumber Industry*, pp. 36–40; *Canada Lumberman* 2 (15 Feb. 1882):1. Mechanized as these mills were in comparison to their predecessors, they still were rather crude by modern standards. One laborer later recalled that pullers on the green chains often went home evenings black and blue from shoulders to elbows. From his point of view, there "was no convenience in the mill whatever at the time." A. P. Alexander, transcript of interview, 23 Aug. 1944 (Bancroft Library, Berkeley, Calif.). Unlike many of the innovations in woods operations, most of those in milling were first worked out in the Lake States and then adopted, sometimes with modifications, by millmen of the Far West.

mechanized. At the same time, steps long mechanized were undergoing change. Head rigs, the saws that did the basic cutting in a mill, were being improved. The old up-and-down sash and muley saws of pioneer times were in the 1860s largely replaced by circular saws. The latter had the advantage of continual cutting and were more efficient because they did not have to repeatedly overcome inertia as did reciprocal saws. The absence of reciprocal motion made possible the great speeds that made the circular saw practical. Since the cutting edge of a circular saw was at some distance from the source of power at its center, the amount of power at the outer rim was reduced. As a result, the depth of the bite of the saw had to be reduced to keep it from lugging down or binding. However, by increasing the speed of rotation, circular saws could more than make up for the reduction in cut due to their small bite.

Still, there were disadvantages to the circular saw. The size of log that could be cut with one was limited to slightly less than half the diameter of the saw. Moreover, as saw size was increased to handle larger logs, tension on the saw increased. Normally, saw speed also increased with size. Increased tension and heat from high-speed cutting meant that thicker steel had to be used to manufacture a large circular saw that would not crack or break. Thicker saws meant wider kerfs. As saws grew bigger, lumbermen saw a larger and larger portion of their logs being turned into sawdust.[22]

By the 1870s the double circular saw had replaced the old single circular as the head rig in nearly all the larger mills. This was nothing more than two circular saws mounted one above the other and, to prevent their choking with sawdust, set so their blades rotated in opposite directions. Using two saws in this manner, it was possible to construct head rigs that could handle logs too large for any single saw. Since the individual saws in a double circular head rig tended to be smaller than that in a single circular head rig, this new arrangement also made it practical to use thinner steel in manufacturing saws so that, sometimes at least, the kerf and, thereby, waste were reduced. Unfortunately, in sawing, the top saw often was only partially immersed in the log, while the bottom saw was wholly engaged in

22. *The Saw in History*, 2nd ed. (Philadelphia, Pa.: Henry Disston & Sons, 1916), pp. 13–14, 30–33; Stanger, *Sawmills in the Redwoods*, pp. 137, 139; W. J. Blackmur, "The Circular Saw: Its Possibilities," *West Coast Lumberman* 48 (1925):75–76, 81; Meany, "History of the Lumber Industry," pp. 308–11.

cutting. The top saw therefore ran cooler than the bottom and the resulting different rates of bending under the tension of sawing often produced a lateral gap between the two blades. This gap left a step on the face of the cut that had to be removed in the planer. In the end, the double circular saw generally produced about as much waste as its forerunner, so its primary advantage was its capacity for handling larger logs.[23]

Circular saws, both single and double, took a giant step forward when Nathan W. Spaulding developed a practical method for equipping them with replaceable teeth. While engaged in operating a series of small sawmills in the Sierra, Spaulding became convinced that there were profits to be made in manufacturing saws for lumbermen. In 1859 he moved to Sacramento and began producing saws. He soon discovered that if replaceable teeth were inserted into rounded sockets around the outer edge of a circular saw, not only could they be made to hold securely, but the saw itself was less apt to crack or break than if sockets with square corners were utilized. He patented his discovery and soon was busy trying to keep up with demand.

Replaceable teeth solved a problem that had long plagued mill operators. Previously, when a saw struck a spike or rock and lost teeth, the saw was either discarded or, if salvageable, new teeth were cut into it. But as new teeth were cut or old ones sharpened, the diameter of the saw was reduced and cutting capacity fell. After a few repetitions of this process, otherwise reparable saws had grown so small that they had to be abandoned. Spaulding's insertable teeth made it possible to replace or sharpen teeth without reducing a saw's size. This was such a significant improvement that Spaulding, while busily trying to fill all his orders, also found himself engaged in struggles against those who infringed on his patents. To better supply potential customers, Spaulding moved his operations to San Francisco in 1861. He later added a plant in Chicago. By the 1880s insertable teeth, marketed by Spaulding and various competitors, had become standard throughout the lumber industry.[24]

23. *The Saw in History*, pp. 31–32, 37; Stanger, *Sawmills in the Redwoods*, pp. 137, 139; Rodney C. Loehr, "Saving the Kerf: The Introduction of the Band Saw Mill," *Agricultural History* 23 (1949):169; J. Richards, "Selections from *A Treatise on the Construction and Operation of Woodworking Machines*," *Forest History* 9 (1966):22; Hidy, Hill, and Nevins, *Timber and Men*, pp. 165–66; Melendy, "One Hundred Years of the Redwood Industry," pp. 58–59.
24. *The Bay of San Francisco*, 2:477–79; *The Saw in History*, pp. 14, 40–45; Stanger, *Sawmills in the Redwoods*, pp. 136, 139; Hittell, *Commerce and Industries*, pp. 425–27; *Canada Lumberman* 1 (1881):15, 16.

Even after double circular saws came into use, the largest logs still had to be split before they could be utilized. Some were blasted with dynamite, others were cut by giant sash saws before being sent to the head rig. Many were simply left in the woods. In 1869 David Evans of Russ and Evans in Eureka patented a device that added a third circular saw to the head rig and thus made it possible to handle the largest sawlogs.

Evans' Third Saw was adopted in mills along the redwood coast, but not until the widespread adoption of band saws in the 1880s was a really satisfactory solution found for the problems lumbermen faced as a result of the size of the logs available along the Pacific Coast.[25] Band saws were huge loops of steel into which teeth had been cut and which were stretched tight between flywheels set above and below the log carriage. They presented a cutting edge long enough to carve through the largest log. Moreover, the band saw was faster, required less power, and took a smaller kerf, thus wasting less of the log, than circular saws.[26]

Apparently the first successful band saw on the Pacific Coast was installed at the Dolbeer and Carson plant in 1885. It was not used as the mill's head rig, but it was not long before band saws were put to that use. The Elk River Mill and Lumber Company of Humboldt County installed a band saw head rig in 1888, the band of which was fifty-one feet in circumference and could cut a plank eighty inches wide. In 1889, when the St. Paul and Tacoma Lumber Company opened their Plant A on Puget Sound, it had two band saw head rigs, which could cut two hundred fifty thousand feet of lumber a day. Such a capacity, undreamed of but a few years before, was in large part the result of the adoption of the band saw, an innovation as important in the milling of lumber as the steam donkey and logging railroad were in getting logs out of the woods.[27] In 1891 the owner of the mill at Port

25. San Francisco *Journal of Commerce*, 23 Aug. 1883; Edgar Cherry & Co., *Redwood and Lumbering*, p. 56; Melendy, "One Hundred Years of the Redwood Industry," pp. 59–60.

26. Loehr, "Saving the Kerf," pp. 168–72. Melendy, "One Hundred Years of the Redwood Industry," pp. 56–62, 157, 162–63. Wilde, "Chronology of the Pacific Lumber Company," pp. 15, 47–55. Meany, "History of the Lumber Industry," pp. 309–11. G. Emerson to A. Simpson, 6 June, 5 Dec. 1893, 25 Mar., 15 Oct. 1895, 12 July 1897, EL. *Canada Lumberman* 1 (15 Aug. 1881):2; 3 (15 Jan. 1883):27; 5 (1 Feb. 1885):42. *The Saw in History*, pp. 14–16, 37, 45–49. *Timberman* 29 (Dec. 1927):104.

27. Eureka *Humboldt Times*, 31 Oct. 1885; Melendy, "One Hundred Years of the Redwood Industry," pp. 61–62, 165, 167; George Douglas, "History of the Pacific Lumber Co. . . . As Told . . . to D[erby] Bendorf" (typescript copy, Bancroft Library, Berkeley, Calif.); B. S. Allen,

The San Diego waterfront during the boom of the eighties. In the foreground is the yard of the Southern California Lumber Company. In spite of the presence of wharfage, lumber is being lightered ashore. Courtesy of the Historical Collection, Title Insurance and Trust Company, San Diego

As donkey engines became larger and more powerful, they came to be used for skidding as well as yarding logs. Here a bull donkey skids redwood logs near Fort Bragg, California, around 1899. Courtesy of the U.S. Forest Service

The two-masted schooner *Big River*, shown here under sail, was typical of the vast fleet of little vessels that carried lumber from the bar harbors and outports to San Francisco. These broad-beamed, single-deck vessels carried prodigious loads in their holds and on deck, yet required only a small crew. Courtesy of the Bancroft Library, Berkeley, California

The adoption of bandsaws for the head rigs in mills, such as this one in the Union Lumber Company's plant in Scotia, California, greatly increased the capacity of mills to handle huge logs. Courtesy of the U.S. Forest Service

Discovery decided to replace his single and double circular saws with two band saws. As one competitor said in reporting on the decision: "He thinks that his mill will then be perfect."[28] Such confidence may have been unwarranted, but it serves to illustrate the importance millowners attached to this new device.

Other changes designed to bring greater efficiency to milling were also adopted during the period. Blowers were added to speed the disposal of sawdust at the Port Discovery and Port Hadlock mills, while the operations of Pope and Talbot—and probably other firms as well—underwent organizational changes and modified their methods of handling lumber in an effort to reduce overhead and improve service to their customers.[29]

At the same time mills were also becoming increasingly diversified. Initially practically all the millstuff shipped in the lumber trade of the Pacific, whether destined for domestic or foreign markets, had been rough, green, unfinished products. But as San Francisco and the other markets served by the cargo mills matured, a demand developed for dried and planed lumber, for mouldings, and other more highly manufactured products. Millmen would probably have moved to fill the demand in any case, but they were hastened in that direction by the rise of small, marginal mills in the hinterlands and on out-of-the-way ports that were moving into the market for rough lumber and, indeed, taking much of the profit out of that trade.

Bit by bit the established mills acquired new equipment to meet the rising demand for specialized products. Devices were added for manufacturing box shooks, mouldings, sashes and doors, and various kinds of dressed lumber. E. G. Ames urged such diversification on the Puget Mill Company. We "must be on the look out to save every thing we can," he wrote, adding that "some day & perhaps not far

---

reply to questionnaire on lumber industry (typescript copy, Bancroft Library); Coman and Gibbs, *Time, Tide and Timber*, pp. 214–15; A. T. Furlong, transcript of interview, Mar. 1953 (Bancroft Library). Band saws were in use in the Lake States at an earlier date. For other claims to the first band saw on the West Coast, see Ryder, *Memories of the Mendocino Coast*, pp. 35–36; Wilde, "Chronology of the Pacific Lumber Company," p. 15.

28. Pope & Talbot to Puget Mill Co., 14 May 1891, AC, incoming corres. Similarly, George Emerson wrote Asa Simpson: "Band saws should be installed in all our mills . . . [for] no investment of money can pay larger returns." Emerson to Simpson [Sept. 1891], EL.

29. W. Walker to W. H. Talbot, 6 Mar. 1888, PTP, Walker corres.; Ames to W. H. Talbot, 24 Nov. 1885, 4 Mar., 1 Apr. 1888, PTP, Ames corres.; Coman and Gibbs, *Time, Tide and Timber*, pp. 158–61, 372, 383–91, 398–401. On the role of organizational change in business growth, see Alfred D. Chandler, Jr., *Strategy and Structure: Chapters in the History of Industrial Enterprise* (Cambridge, Mass.: M.I.T. Press, 1962), esp. pp. 283–323.

distant . . . the material now burned up will be a source of revenue to us that will be worth looking after."[30] Another firm moving toward diversification was the Willamette Steam Mills which, the *Oregonian* declared, was able to turn out "every style of lumber available anywhere."[31] Most mills of any size were equipped with planing mills, and many tried adding dry kilns as well. The dry kilns were not a marked success at first. As one observer put it: "They collectively demonstrated how not to dry lumber." Most of the new equipment was more immediately successful, however.[32]

Increasingly sophisticated milling operations added, through manufacture, greater value to the raw materials being handled. As the *West Shore* observed in 1884, rough lumber hardly paid its cost, but the application of skilled labor and machinery in the process of planing increased the value of this same lumber by 100 percent, in the manufacture of doors by 200 percent, in the manufacture of sash 300 percent, and in the production of mouldings 500 percent.[33] These manufactured items increased the opportunity for profits on sales in established markets and at the same time opened more distant markets because, having greater value, they could bear greater transportation costs. Millmen began to reach out to those markets that had been too remote to penetrate with rough, green lumber. Shipments to the East Coast (via Cape Horn rather than the transcontinentals), England, continental Europe, and South Africa began to be made. As yet these markets were of minor significance, but in time they were to be of great importance to the lumber operations of the West Coast.[34]

30. Ames to W. H. Talbot, 11 Apr. 1888, PTP, Ames corres.

31. Portland *Oregonian*, 1 Jan. 1882.

32. Allen, reply to questionnaire. See also Melendy, "One Hundred Years of the Redwood Industry," pp. 284–85; *The Bay of San Francisco*, 1:301; Wilde, "Chronology of the Pacific Lumber Company," p. 34; Eureka *Humboldt Times*, 3 Apr. 1886; Furlong, interview; Weidberg, "History of John Kentfield & Co.," p. 35.

33. *West Shore* 10 (1884):136. Many planing mills were erected in San Francisco as well as in the forested areas. See *The Bay of San Francisco*, 1:301–2; [W. H. Murray], *The Builders of a Great City: San Francisco's Representative Men* . . . (San Francisco, 1891), pp. 59, 84.

34. *Canada Lumberman* 3 (1 Mar. 1883):70; 3 (15 May 1883):146, 155; 5 (1 Oct. 1885):330; 5 (15 Oct. 1885):342. Melendy, "One Hundred Years of the Redwood Industry," pp. 168–69. Coman and Gibbs, *Time, Tide and Timber*, pp. 181–83. Wilde, "Chronology of the Pacific Lumber Company," p. 18. Edgar Cherry & Co., *Redwood and Lumbering*, pp. 13–14, 33–34, 75. Also helping make possible the penetration of more distant markets was the fact that lumber prices were increasing relative to wholesale prices as a whole because production in areas near the primary American markets was declining. When more distant sources were drawn upon, prices went up. See Compton, *Organization of the Lumber Industry*, pp. 2, 76; G. F. Warren and F. A. Pearson, *Wholesale Prices for 213 Years, 1720 to 1932* (Ithaca, N.Y.: Cornell University, 1932), pp. 81, 116–19; Tattersall, "Economic Development of the Pacific Northwest," p. 185.

Mechanization and diversification were costly. Cyrus Walker complained that "they . . . make improvements so fast and so many of them that it cost big money to keep up with them all."[35] Such sums were not always available to the area's older firms. These had been established for the most part with investment capital from the San Francisco mercantile community and, though additional funds had been added from the same source from time to time, the bulk of the subsequent expansion of the cargo mills had been financed with internally generated capital. As Elisha Douglass has observed, lumbermen tended to be "suspicious of outsiders and . . . reluctant to share power with boards of directors or stockholders they did not know."[36] This suspicion led many Pacific Coast mill operators to strain their own financial resources rather than seek funds from independent sources to make improvements. In other times or places, this might have proven a wise decision, but during the eighties huge amounts of capital accumulated in the Lake States were being reinvested not in the declining lumber industry of that area, but in the Far West. The reluctance of the operators of the cargo mills to draw on others for their capital hindered them in meeting the challenge of the newcomers from the Lake States.

Newspapers and journals carried frequent reports of firms that were moving, or contemplating moving, to the West Coast.[37] Enough new mills appeared on the Pacific shore to lend credence to the wildest rumors. A group of lumbermen from Michigan established the Michigan Mill Company and erected a plant in Vancouver, Washington. J. W. McDonnell and others from the same state built a mill at Ballard, only to sell it to another millman from Michigan, Charles Stimson. Chauncey and Everett Griggs arrived from Minnesota and established the St. Paul and Tacoma Lumber Company. They had behind them one and one-half million dollars of paid-up capital. Robert

35. Walker to W. H. Talbot, 8 Oct. 1885, PTP, Walker corres.
36. Douglass, *The Coming of Age of American Business*, p. 176. Cf. Frederick W. Kohlmeyer, "Northern Pine Lumbermen: A Study in Origins and Migrations," *Journal of Economic History* 16 (1956):531–34, 538; *West Shore* 8 (July 1882):126.
37. Compton, *Organization of the Lumber Industry*, p. 2. *West Shore* 9 (1883):79; 13 (1887):757; 15 (1889):222; 17 (1891), p. II. See also T. Burke to C. Walker, 8 Nov. 1887, AC, Walker, personal corres.; G. Emerson to A. Simpson, 23 Feb. 1895, EL. Usually this meant that management, manpower, and capital were being moved to the Far West, but sometimes equipment was also transferred. More frequently, old plants were simply sold for whatever they would bring, and new equipment was purchased for use in the mills being erected in the Far West.

Dollar and C. R. Johnson appeared on the redwood coast and began investing in milling operations and laying the groundwork for their later empires in lumbering.[38]

In addition, many lumbermen from the Lake States were buying up large tracts of timberland in preparation for opening sawmills at a later date. At least twenty-one timber-holding companies acquired tracts in the redwoods during the period from 1880 to 1902. Ten were from the Lake States. The giant Weyerhaeuser operation was beginning to consider acquiring timberlands in the Northwest, and others were already doing so. Capitalists from the Lake States, and from Ottawa as well, were also investing in the forests of British Columbia.[39]

If the established firms on the West Coast wished to compete with these well-financed new ventures, they would have to invest in mill improvements and, in order to assure continuing supplies of sawlogs, acquire large stands of timber—something most of them had not previously done. Some of the older firms accomplished this without too much difficulty, but others found it hard to obtain the necessary capital from internal sources and hesitated, or were unable to draw from external sources. The Washington Mill Company was in financial straits as a result of the cost of mill improvements even before the mill burned in 1886. The fire marked the beginning of the end: though the firm continued for some time, it never really recovered. The decline of John Kentfield and Company is partially attributable to the same sort of financial dilemma.[40]

The Port Blakely Mill Company and Pope and Talbot were more

38. Stimson, comp., "A Record of the Family, Life and Activities of Charles Douglas Stimson," p. 11; *West Shore* 15 (1889):280; Cox, "Organizations of the Lumber Industry," pp. 4–6; Ryder, *Memories of the Mendocino Coast*, pp. 14ff.; Melendy, "One Hundred Years of the Redwood Industry," pp. 23–24, 149; *Pacific Coast Wood and Iron* 25 (1896):1; Dollar, *Memoirs*, 1:13–27; O'Brien, "Life of Robert Dollar," pp. 28–47.

39. Edgar Cherry & Co., *Redwood and Lumbering*, pp. 18–21. Melendy, "One Hundred Years of the Redwood Industry," pp. 21, 24. Humboldt Times, *Souvenir of Humboldt* (Eureka, Calif., 1903), pp. 44–46. Palais and Roberts, "History of the Lumber Industry in Humboldt County," pp. 9–10. *West Shore* 16 (1890):138, 942. Hidy, Hill, and Nevins, *Timber and Men*, pp. 211–12. Lawrence, "Markets and Capital," pp. 37, 65–81. *Canada Lumberman* 8 (Feb. 1888):10; 12 (Feb. 1891):8, 11. This movement began in the eighties and continued at an accelerated pace through the next two decades.

40. Gedosch, "Seabeck," pp. 29–34. Rather than rebuild at Seabeck, the firm purchased a plant at Port Hadlock, which E. G. Ames described as a fine, well-arranged mill that ought to make money. The reasons for the firm's failure to prosper following the transfer are not clear, but it appears to have resulted from the financial embarrassments under which the fire of 1886 forced it to operate. See Ames to W. H. Talbot, 1 Apr. 1888, PTP, Ames. corres. On the Kentfield failure see Weidberg, "History of John Kentfield & Co.," pp. 37–38, 71–72.

fortunate. They not only kept pace, they continued to be leaders. During one fifteen-month period in the 1890s, for instance, Pope and Talbot filled orders for 304 different dimensions of lumber, some in more than one grade or special lengths.[41] The newcomers could hardly have done more, and most of them that prospered did so by catering to markets reached by transcontinental railroads, outlets that the Pope and Talbot and Port Blakely mills could service only with difficulty because of their locations on the west side of Puget Sound. The cargo markets were secondary concerns to most of the new mills; trade there continued to be dominated by the same firms as before.[42]

Technological changes in the woods and mills were paralleled by changes in the cargo trade itself. As time passed, the demand for rough construction materials in California tended to give way to demand for more finished products. In the cities, for example, frame construction became less common, but demand grew for planed lumber, mouldings, and similar items for use in brick and stone buildings. Even the quality of frame buildings was changing: instead of the hastily constructed temporary structures of the Gold Rush era, they were now permanent dwellings that used large amounts of finished lumber.

But in manufacturing these specialized products, large quantities of rough, low-grade items were also produced. Much of what came from even the finest of timber stands was too low in quality to be used in any other way. Buyers for the finished goods of the large mills were plentiful; however, the low-grade items proved more difficult to move. If these goods were not sold, the finished items, those for which there was a good demand, would have to carry the entire cost of production by commanding higher prices. An increase in prices might well have meant the loss of the margin of advantage that millmen of the Far West held over producers elsewhere. Fortunately for the lumbermen of the Pacific Coast, the various markets around the Pacific Basin complemented those in California. China, Australia, and Latin America took large quantities of the rough, inexpensive grades. The demand for railroad ties and mining timbers in both domestic and foreign

41. Berner, "Port Blakely Mill Company," pp. 161–62; Fred Talbot to Walker, 29 May 1899, AC, Walker, personal corres. However, Pope & Talbot officials sometimes complained of the necessity of plowing profits back into their business in order to meet the competition. See C. F. A. Talbot to Walker, 8 Nov. 1887, W. Talbot to Walker, 8 May, 15 Dec. 1888, AC, Walker, incoming corres.
42. Coman and Gibbs, *Time, Tide and Timber*, pp. 210–20, 446–47; Meany, "History of the Lumber Industry," *passim*; *The Bay of San Francisco*, 1:301.

markets was especially welcome, for large portions of the poorer parts of logs were used in manufacturing such items.

Even with these outlets for the lower grades, mill managers often found themselves burdened with an excess of low-quality lumber. Shippers frequently resorted to overgrading cargoes in an attempt to move some of this poorer material, but the amount that could be disposed of in this fashion was limited. A typical reaction was that of a buyer in Hawaii who angrily wrote to J. A. Campbell, manager of the Port Blakely Mill Company, that a shipment of lumber Campbell had sent to the islands was "rubbish. . . . You had better sell this kind of lumber to the Chinese Lumber yard[s]—we won't have it." [43] Without the sort of market for low-quality products that China represented, the marketing problems faced by the managers of the large mills on the Pacific Coast would have been even more difficult to overcome than they were. To an increasing degree the scattered markets of the Pacific lumber trade were becoming an interdependent whole.[44]

Changes in the industry also had a great impact on small lumber mills. Small firms, unable to keep pace in the adoption of technological improvements and the diversification of production, found themselves in an increasingly disadvantageous competitive position. While their larger competitors, with their specialized equipment, were able to put the various parts and grades of sawlogs to those uses that would give optimum returns, small mills had a more restricted range of choice and thus tended to profit less even when their log supply was of a quality equal to that of the larger firms. They also had fewer choices in marketing. Most were too small to furnish quickly sufficient lumber to fill vessels that were large enough to pay when put on the long runs to China, Australia, or South America. The markets for low-grade products in those places were thus effectively closed to the smaller mills. With a narrower range of products to sell and a narrower range of markets in which to sell them, the small mills were more vulnerable when demand for standard products slumped in California or the Northwest. During the recession in the lumber industry of 1884, the larger mills with diversified businesses did the best, the *Oregonian*

43. J. C. Allen to J. A. Campbell, 22 Nov. 1898, PBMCP, box 35. Cf. *Canada Lumberman* 18 (May 1897):6.
44. Redwood lumber was higher priced than Douglas fir, and thus mills that produced it were unable to benefit from these markets for inexpensive materials to the same degree as mills on Puget Sound. See Edgar Cherry & Co., *Redwood and Lumbering*, p. 74.

noted. When the much more severe depression of the nineties struck, many of the small mills were forced out of business altogether.[45]

The case of Eugene Semple's Lucia Mill Company in Vancouver, Washington, illustrates the situation of the smaller mills. The mill had a daily capacity of twenty-five thousand board feet. In addition to the basic equipment, the mill also had a planer. Semple could find markets for his planed lumber, especially in a number of hamlets east of the Cascade Mountains, but he had great difficulties finding buyers for his low-grade cut. His customers to the east of the Cascades could obtain rough lumber from small mills nearby. The cost of transportation priced Semple's rough out of that market. Moreover, he could obtain no relief through sales in Portland, for the Willamette Steam Mills and other plants in that city could undersell him there. By 1884 Semple was in trouble with his creditors. In desperation he tried to get his customers to take some of his rough lumber by warning them that future orders for planed lumber would not be filled unless rough were also purchased. This threat generated little new demand, however, so he found it necessary to raise the prices of his planed lumber in order to cover the cost of manufacturing the lower grades. He undoubtedly lost sales as a result.

Unbalanced demand also restricted him in buying logs. He could afford to purchase only those that would yield a high percentage of lumber worth planing. He had to pass up mixed lots, even when they were offered at bargain prices. Semple claimed that it was better for him to buy quality logs at $7.00 a thousand than poor- to middle-grade logs at $4.50. Since even the best logs available on the Columbia River yielded only 25 percent clear lumber, Semple's predicament would appear to have been insoluble unless he found new markets.[46]

He set out to do just that. Tiny as his operation was, he began to explore the possibility of shipping to San Francisco, where there was some demand for rough lumber. He tried to induce Asa Simpson to market his cut at "extremely close figures." Simpson showed no interest. Semple then tried to make arrangements to market lumber through relatives in California. Again he was unsuccessful.[47] Only his

45. Portland *Oregonian*, 1 Jan. 1885; Melendy, "One Hundred Years of the Redwood Industry," p. 289.
46. *West Shore* 9 (1883):189; Semple to Norman Merrill, 22 Apr. 1884; Semple to A. L. Stokes, 28 Apr. 1884; Semple to Proebstel Bros., 23 June 1884; Semple to Reese & Redman, 7 July 1884, Semple Papers, Lucia Mill, outgoing corres.
47. Semple to G. Flavel, 2 May 1884; Semple to [?], 7 May 1884; Semple to "nephew," 3, 4, 18 June 1884: Semple to [L. B. Mizner], 12, 18, 22 June 1884, Semple Papers, Lucia Mill, outgoing

winning of contracts to supply lumber for construction at Vancouver Barracks and for a causeway on the south bank of the Columbia saved Semple from failure. For the time being, he prospered. There was obviously little future in the operation, however, and Semple soon turned the mill over to others and left Vancouver for activities that seemed to offer greater opportunity. Within three years Grover Cleveland had named him governor of Washington Territory.[48]

The situation faced by other small mills was generally quite similar to that faced by the Lucia Mills. Only in isolated, outlying areas, where logs might be obtained at prices considerably below those being paid in the major areas of production, were the small mills able to stay competitive. In the main centers of production, small mills declined in number and eventually disappeared almost entirely.[49]

While technological developments were bringing change to the woods and mills, other changes were taking place at sea. The two were not unconnected: the increasing volume of lumber that was shipped as mills became larger and more mechanized made improvements in transportation virtually mandatory.

In the first years of the Pacific lumber trade, whatever vessels were available were utilized. On both coasts and in Europe new vessels were constructed in order to take advantage of the high freights offered during the Gold Rush. The supply of sailing vessels had been more than ample before the excitement of the rush; the tonnage added during the rush made it inevitable that there would be a tremendous surplus when demand returned to normal. As this surplus materialized, many owners who would have disdained such employment for their vessels but a short time before put them to work as lumber carriers. As

corres. Semple later bitterly complained that the mills of the Portland and Vancouver area were "shut out from the ocean trade by the San Francisco combine . . . [and] restricted to supplying a country within the sound of their steam whistles." The failure of Semple's attempts to enter the cargo trade had apparently run into organizational as well as natural obstacles. See L. C. Palmer and Semple, open letter to lumbermen, Portland *Oregonian*, 28 Apr. 1886.

48. Vancouver *Clarke County Register*, 27 Nov. 1884; Evans, *History of the Pacific Northwest*, 2:551; *West Shore* 15 (1889):38; Alan Andrew Hynding, "A Biography of Governor Eugene Semple" (Ph.D. diss., University of Washington, 1965), *passim*; Bancroft, *History of Washington, Idaho and Montana*, p. 298.

49. For example, see J. B. Montgomery to Walker, 10 Jan. 1894 [dated 1893], AC, Walker, personal corres. The decline of small mills in the areas of the cargo mills was not paralleled by a decline of small mills elsewhere. As population grew in the interior, many small, localized mills sprang up; the number of small mills in both Oregon and Washington actually increased sharply during the 1880s. See Meany, "History of the Lumber Industry," pp. 128ff.

a result, through the 1850s and 1860s most of the vessels that made up the fleet of lumber transports working in the Pacific were vessels originally built either on the East Coast or abroad for hauling mixed, general cargoes.[50]

Skepticism about the durability of ships built of Douglas fir encouraged this dependence on outside sources. But by the 1870s the production of sailing vessels was well under way on the Pacific Coast. It had become clear that properly cut and treated Douglas fir was a fine shipbuilding material, even possessing some advantages over the traditional oak. In 1875 the Board of Marine Underwriters of San Francisco published specifications for the construction of vessels from West Coast timber which, when followed, resulted in vessels that won first-class ratings and the lowest insurance rates.[51]

When lumbermen began to contract for new tonnage, rather than buy secondhand ships built on the East Coast or charter the sailing equivalent of the modern-day tramp steamer to haul the cut of their mills, it was only a matter of time before they began to seek modifications designed to improve the effectiveness of these vessels as transporters of lumber. What resulted was not a single type, but a variety of types, each specially suited for the role it was expected to play in the lumber trade of the Pacific.[52]

The first of the specially designed lumber droghers to be widely adopted on the West Coast was the single-decked, two- or three-masted schooner. The origins of the design used on the West Coast are obscure. The three-masted schooner had appeared on the Atlantic

50. Cutler, *Greyhounds of the Sea*, pp. 274ff.; Fairburn, *Merchant Sail*, 4:2605 and *passim*; Caspar T. Hopkins, *Ship Building on the Pacific Coast (No. 2)* (San Francisco, 1874), pp. 10–11; Hopkins, *Report on Port Charges, Shipping, and Shipbuilding* . . . (San Francisco, 1885), p. 41. These vessels often left much to be desired. According to one observer, "In the early sixties [they were] almost all . . . old vessels, built in eastern yards, poorly adapted to the lumber trade and undergoing constant repairs." Even as early as the mid-fifties, however, efforts were being made to obtain vessels particularly fitted to the needs of the Pacific lumber trade. See Caspar T. Hopkins, "The California Reminiscences of Caspar T. Hopkins," *CHSQ* 27 (1948):65–66; A. J. Pope to Wm. Pope & Sons, 19 Sept., 18 Nov. 1855, 2 July 1857, PTP, Pope corres.

51. Hopkins, *Ship Building*, pp. 9, 16–18; Hopkins, *Report*, pp. 41–59. For an account of some of the more successful shipbuilding enterprises on the West Coast, see Newell, ed., *McCurdy Marine History*, p. 21; Buchanan, "Economic History of Kitsap County," pp. 172–89.

52. Shipbuilding on the West Coast has been little studied. In 1935 Howard I. Chapelle called for a study of the contributions of builders on the Pacific shore, but the call has gone largely unanswered. Among the few works are Robert Johnson's "Schooners Out of Coos Bay" and David W. Dickie's "The Pacific Coast Steam Schooner," in *Historical Transactions, 1893–1943*, of the Society of Naval Architects and Marine Engineers. More are needed. See Chapelle, *The History of American Sailing Ships* (New York: W. W. Norton & Co., 1935), pp. 264–67.

Coast as early as 1801, but its popularity had remained rather limited until mid-century. A few of the vessels that rounded Cape Horn during the Gold Rush were schooners. Perhaps some of these proved successful when put to work hauling lumber along the coast. Some of the men who came to California with experience in shipbuilding behind them, such as Asa Simpson, may have recognized that the schooner was the type of vessel best suited for use along such a coast. From whatever source, the schooner quickly became the most commonly used vessel in the coastal lumber trade.[53]

The schooner rig had decided advantages for coasting. A schooner-rigged vessel could sail more nearly into the wind than a square-rigger and could maneuver more handily in tight sailing quarters, an ability that was to prove vital on the Mendocino coast. A schooner had relatively little weight in the overhead rigging and when built with a heavy, fairly flat bottom could sail while empty without having to take on ballast. In a trade such as that in lumber, where cargo was generally hauled only one way, this was a valuable asset. Moreover, a schooner-rigged vessel required a much smaller crew than a square-rigged ship of the same tonnage, so by using schooners lumbermen could keep down the cost of transportation. Where low-priced bulk cargoes such as lumber were involved, this was vital.[54]

But the lumber schooners built on the West Coast were not ordinary schooners. Henry Hall described them as "flat, one-decked vessels, with long bows, handsome square sterns, and broad beam . . . excellent sea boats." [55] William Armstrong Fairburn was less complimentary. Though he admired the deeper-bodied, two-decked schooners that were in use on the Atlantic Coast, he was scornful of the unorthodox Western vessels. The flat, single-decked design was an adaptation to the special demands of the lumber trade. Officials in the

53. Hutchins, *American Maritime Industries and Public Policy*, pp. 546–52; Cutler, *Greyhounds of the Sea*, pp. 37–38; Carl G. Cutler, *Queens of the Western Ocean* (Annapolis, Md.: U.S. Naval Institute, 1961), pp. 335–38; Kortum and Olmsted, "A Dangerous-Looking Place," pp. 49, 51. The Astoria Customs House Records indicate that schooners visited the Columbia River more frequently than any other kind of vessel until at least the early eighties. If the Washington Mill Co. papers are an accurate indicator, they disappeared on Puget Sound somewhat earlier. Though eight schooners called at Seabeck in 1858 and thirty-two in 1859, there were none arriving by 1868. Larger types of vessels were being used.

54. Hutchins, *American Maritime Industries and Public Policy*, pp. 549–52, 554–65; Cutler, *Queens of the Western Ocean*, pp. 335–38; B. R. Crowninshield, "Wooden Sailing Ships," *Transactions of the Society of Naval Architects and Marine Engineers* 15 (1907):196; Fairburn, *Merchant Sail*, 4:2616–18.

55. Hall, *Report on the Shipbuilding Industry*, p. 133. Cf. Hopkins, *Report*, p. 42.

Pope and Talbot organization believed it the best design for coastwise lumber shipments on the Pacific shore.[56]

Vessels with two decks created problems when put to hauling lumber. Loading the lower hold with lumber through hatches on deck was a difficult process. Heavy pieces and long timbers, which ideally should have been stowed in the lower hold, often had to be carried in the upper hold or on deck. The vessel's sailing qualities ofttimes suffered as a result. If nothing else, loading a two-decked vessel through hatches on deck was time-consuming and costly. To avoid such difficulties, large ports were frequently cut through the hull either in the bow or the stern so as to give direct access to the lower hold. Once a ship was equipped with such ports, a cargo of lumber could be more quickly and properly stowed. These ports were not without drawbacks, however. If they were cut too small, they were of no great help; if they were made too large, the vessel was apt to ship water through them in rough weather. The ship *Florence* was lost on a run from Puget Sound to Honolulu when the covers to large ports that had been cut to facilitate the loading of lumber apparently worked loose during a gale, causing the ship to take on water and founder. Loading ports might make a two-decker a better lumber carrier, but they could not make it an ideal one.[57]

Thus, when lumbermen on the West Coast began building their own vessels, they quickly abandoned the multidecked design that was almost universal among merchant sail and began building vessels with but a single deck. To give access to the large, single hold, oversize deck hatches were added. Stowing lumber on such a vessel was a relatively easy task. Since there was no second deck contributing to the structural strength of these schooners, designers compensated by making the vessels rather flat. The reduced distance between the keel and deck helped assure adequate structural rigidity. However, the size of the hold was also decreased, and upward of half of the cargo was therefore

56. Fairburn, *Merchant Sail*, 4:2605–6, 2609; C. Walker to W. H. Talbot, 9 Jan. 1888, PTP, Walker corres. According to Fairburn, the *Cassandra Adams* was "the best product of West Coast designers and builders"; but he hastens to point out that it was copied from an East Coast model.
57. C. M. Scammon, "Lumbering in Washington Territory," *Overland Monthly* 5 (1870):59–60; Portland *Oregonian*, 12 Sept. 1897; Seymore, "Port Orchard Fifty Years Ago," p. 259; Meany, "History of the Lumber Industry," pp. 312–13; Fairburn, *Merchant Sail*, 3:1750. For examples, see Tientsin *Chinese Times*, 23 Apr. 1887; [Adams, Blinn & Co.] to Washington Mill Co., 12, 20 Mar. 1858, WMCP; E. Mallett to G. Ainsworth, 30 Apr., 21 Aug. 1882; Mallett to L. Hawkins, 3, 13 Aug. 1882; Mallett to J. Ainsworth, 23 Jan. 1882, APSA, box 3, *Kate Davenport* 3.

stowed on deck. Lumbermen were happy with this arrangement, as it was even easier to put a deckload on a lumber schooner than it was to load its hold.

There are many stories of the huge deckloads carried on these vessels. Old-timers claimed that loads were piled onto lumber schooners until the deck itself was sometimes under four feet of water. To captains' protests over the amount of lumber being put aboard, a standard reply of millmen is said to have been, "Hell, its just lumber. You've never seen wood sink have you?" [58]

Though the stories are surely exaggerated, the deckloads were of prodigious size. On one of the first voyages of Kyle's schooner *Bella*, she sailed from Tillamook Bay in mid-winter with only sixteen inches of freeboard, and A. W. Beadle complained that the captain "will have to load the vessel deeper if we are going to make any money with her. She will carry a big load if he will only put it on." [59] Similarly, when the marine surveyor at Port Townsend refused to issue Pope and Talbot vessels the certificates necessary for insurance unless ample freeboard were left, company officials were angered and threatened to have others inspect their vessels unless the surveyor changed his policy. To leave such freeboard, they insisted, "would be ruinous to any single deck vessel. . . . They are built to take deck loads, and if they are not allowed to take same, their profits are necessarily lost." [60] Perhaps they were right. The *Bella* lost money when Captain Smith took her out of Tillamook Bay with sixteen inches of freeboard.

Suited though the three-masted, single-decked schooner was for the coastal lumber trade, it was not entirely adequate for shipping to the distant markets of China, Australia, or South America. The result was the appearance of the so-called great schooners. These four-, five-, and six-masted vessels could carry far larger cargoes than the three-masters, while keeping the economical schooner rig, and thus could turn profits on long voyages. Indeed, the larger the schooner, the fewer crewmen per ton were required. These added savings helped to compensate for the two-deck construction that was necessary in order to build

58. McNairn and MacMullen, *Ships of the Redwood Coast*, p. 25; Coman and Gibbs, *Time, Tide and Timber*, p. 180.
59. Beadle to Meyer and Kyle, 9 Feb. 1898, WKSP, 145. See also Beadle to Meyer and Kyle, 1 Oct. 1898, WKSP, 150; Smith to Meyer and Kyle, 23 Dec. 1898, WKSP, 148; Meyer and Kyle to P. Ellinsen, 1 Jan. 1898, WKSP, 170.
60. Pope & Talbot to Puget Mill Co., 23 May 1891, AC, incoming corres. See also Ames to W. H. Talbot, 5 July 1885, PTP, Ames corres.

structurally sound vessels of the size of the great schooners. The first four-master was constructed in 1880. The type quickly became popular with lumbermen shipping to distant ports. During the 1890s forty-eight four-masted schooners were built in the Northwest. By this time five-masted schooners were beginning to appear as well, and six-masted schooners followed soon after. The first five-master built for service at sea was *Louis*, launched from Asa Simpson's yard on Coos Bay in 1888. Great schooners proved well suited for the Pacific lumber trade. Westerners testified that no foreign sail were able to compete in the trade with these economical giants. They continued to haul lumber on transpacific runs well into the twentieth century.[61]

But the great schooners had one distinct disadvantage: they were not as fast in following winds as square-rigged vessels of the same dimensions. On the broad Pacific, where long legs of a voyage might be made with the trade winds at one's back, this was a real handicap. A number of adaptations were tried in an attempt to add speed without losing the economies of the schooner rig. Yards were added on the foremast of many four- and five-masted schooners so that a square sail could be set extending out from the windward side of the foremast to help catch following breezes. Other means of adding extra sails to the basic schooner sail plan were also tried.[62] Many shipowners went a step further and rigged their vessels as barkentines rather than pure schooners. With this sail plan—square-rigged on the foremast and fore-and-aft rigged on the other masts—greater speed could be achieved, while much of the economy of the schooner was maintained. The barkentine became one of the most popular types of lumber vessel; on the long transpacific runs, it was without equal.[63]

61. Fairburn, *Merchant Sail*, 2:1582; 4:2606–23. Hutchins, *American Maritime Industries and Public Policy*, pp. 545–56. Coman and Gibbs, *Time, Tide and Timber*, pp. 178–80. Johnson, "Schooners Out of Coos Bay," pp. 29–31 and *passim*. Matthews, "Shipbuilding History of the Matthews Family," 1:10–13. John Lyman, "Five-masted Schooners," *American Neptune* 5 (1945):137–41. See also Murphey, *Shanghai*, pp. 73–75.
62. P. A. McDonald, "Square Sails and Raffees," *American Neptune* 5 (1945):142–45; McDonald, "Mutton Spankers and Ringtail Topsails," *American Neptune* 5 (1945):235–39; D. L. Dennis, "Square Sails of American Schooners," *Mariner's Mirror* 49 (1963):226–27; Fairburn, *Merchant Sail*, 4:2621–22. Some owners were still moving in the direction of economy of sail, however. If schooner rigs were more economical than square, reduced schooner rigs should be even more economical, they believed. This led to the bald-headed schooner, a vessel that kept the schooner mainsails, but eliminated all sail carried high. See Fairburn, *Merchant Sail*, 3:1654; Coman and Gibbs, *Time, Tide and Timber*, pp. 179–80.
63. Fairburn, *Merchant Sail*, 4:2621; Cutler, *Greyhounds of the Sea*, p. 246; Francis E. Clark, "An Early Four-masted Barkentine," *American Neptune* 25 (1965):112–15; John Lyman, "Two Masts Square-, Two Masts Schooner-Rigged," *American Neptune* 4 (1944):237–38; *West Shore* 9 (1883):274.

The quest for economy led in other directions, too. In 1879 the *Olympus* was launched at Seabeck for the Washington Mill Company. This vessel was designed to maintain the economies of the single-deck plan even at the expense of sacrificing the schooner rig. Though the *Olympus* was a full-rigged ship in its sail plan, its hull was that of a huge, single-decked lumber schooner. The beamy, flat-hulled ship had a tremendous cargo capacity, reportedly being capable of carrying 1.3 million feet of lumber. It was the largest single-decker ever built, but "could sail like a yacht." In *Lewis and Dryden's Marine History of the Pacific Northwest* it is referred to as "the finest sailing vessel ever built on Puget Sound." William Armstrong Fairburn was less impressed. He called the ship "somewhat freaky." [64]

While some lumber schooners were being built larger and given modified sail plans in order to meet the special demands of the long-distance shipping of lumber, others were being modified in different ways in order to fit them for use in carrying lumber from the smaller harbors of the Pacific Coast. Relatively flat-hulled though the basic lumber schooner was, it still drew too much water to make it practical for use at some of the less hospitable bar harbors. To meet the needs of lumbermen, shipbuilders sought a means of further reducing a vessel's draft without sacrificing too much carrying capacity. They found it in the retractable centerboard.

In most schooners a degree of dead rise was maintained in order to prevent the vessel from making too much leeway and being too unmanageable when under sail. This dead rise increased the draft of the vessels and kept them from crossing the shallower bars. To reduce draft, shipbuilders began to construct flat-bottomed schooners that were kept from drifting sideways before the wind by a centerboard that projected down from the hull. These centerboards were built so they could be raised while crossing shallow bars. With the appearance of the centerboard schooner on the West Coast, the size of the lumber carriers calling at places such as Tillamook Bay and the Nehalem, Alsea, and Siuslaw rivers was suddenly increased. One of the many centerboard schooners built to service these ports was William Kyle's *Bella*. The centerboard schooner was not an invention of the shipbuilders on the Pacific Coast; it had been in use on the Hudson

---

64. Wright, ed., *Lewis and Dryden's Marine History*, pp. 271–72; Fairburn, *Merchant Sail*, 4:2606. Fairburn's comment hardly seems just. True, the vessel was not typical, but it did well the task for which it was designed. Performance ought to be the test of a vessel.

River and various sounds along the Atlantic Coast since about 1830, but the type proved well suited for the use to which Westerners now put it.[65]

Though centerboard schooners did much to provide improved transportation at the smaller harbors, they presented problems, too. They were difficult to sail, and finding an adequate captain for one was not easy. One reason for the problems of the *Bella* may have been Captain Smith's lack of experience in handling a centerboard craft when he became her master. As A. W. Beadle put it, "We are all aware of the fact that it takes a man who is used to these 'Flat Bottomed' Boats, to make time with them." [66] The advantages must have outweighed the disadvantages, however, for the design remained popular through the nineties. As late as 1898 one shipbuilder reported from San Francisco, "I have never seen so many Flat Bottom Boats building." [67]

The centerboard schooner was a response to the special challenges of the bar harbors; the steam schooner evolved in response to those of the redwood coast. In the tiny dogholes marking that shore a vessel dependent upon wind for its motive power had a narrow and undependable margin of safety. About 1880 someone attached a steam engine to the after end of a lumber schooner, thus freeing the vessel from dependence on the vagaries of the winds. Authorities disagree over which vessel was the first to be converted. The *Beda, Laguna, Surprise, Newport,* and *Alex Duncan* have all been given the honor. The number of claimants serves to indicate how quickly the idea caught on. Shipyards on San Francisco Bay were soon busy with conversion orders. The Fulton, Risdon and Deacon Iron Works were kept busy building engines and boilers. Soon the construction of steam schooners, in contrast to the mere conversion of ordinary schooners to steam, was underway. Robert Dollar's *Newsboy*, launched in 1888, is often credited with being the first vessel built specifically for service as a steam schooner, but others apparently preceded it.[68]

---

65. Hutchins, *American Maritime Industries and Public Policy*, p. 555; Johnson, "Schooners Out of Coos Bay," pp. 27–28; Bandon (Ore.) *Recorder*, 23 Sept. 1901. For details of two centerboard schooners intended for use as lumber carriers, see specification folders, schooners *Oakland* and *Hugh Hogan*, Kruse & Banks Papers, University of Oregon library, Eugene.
66. Beadle to Meyer and Kyle, 9 Nov. 1898, WKSP, 153. See also F. H. Rogers to Kyle, 3 Feb. 1897, WKSP, 148.
67. H. W. Hansen to Kyle, 24 Feb. 1898, WKSP, 146.
68. McNairn and MacMullen, *Ships of the Redwood Coast*, pp. 14, 17, 129–37; Matthews, "Shipbuilding History of the Matthews Family," 1:5, 2:1–14, and *passim*; Johnson, "Schooners

As conversion gave way to construction, steam schooners came to resemble their sailing predecessors less and less. The hull was lengthened to increase the carrying capacity, and superstructures were added aft to house officers and passengers above deck. Later, oil replaced coal as fuel; engines were moved amidships, thus making possible the addition of more loading gear; the square stern of the sailing schooners gave way to a rounded stern; and steel hulls replaced wooden. All these adaptations were designed to fit the steam schooner more adequately to its assigned tasks. It was to be well into the twentieth century before the basic idea of the steam schooner as it had emerged in the 1880s was abandoned.[69]

The carrying capacity of the early steam schooners was not large. Most could haul between two hundred and four hundred thousand feet of lumber. In view of this, it was often predicted that the steam schooner would never be able to haul lumber to San Francisco from points more distant than Humboldt Bay. Perhaps the predictions were influenced by the fact that neither coastal nor deep-sea steamers, both of which had been operating in the waters of the Pacific for years, had been used to haul lumber. Freight rates on these vessels were too high for them to serve as lumber carriers. Few foresaw how much more economically a steam-powered vessel might be run if it were designed specifically for use as a lumber drogher. However, before the end of the eighties Cyrus Walker was investigating the possibility of running steam schooners to the Puget Mill Company sawmills on Puget Sound, and not long afterward steam schooners were operating all along the coast from Puget Sound in the north to San Pedro in the south.[70]

This development was a direct challenge to the small sailing vessels that had for so long dominated the trade of the bar harbors and outports along the Pacific Coast. To meet this new competition, many owners began to have their sailing craft towed up the coast by tug. The savings in time and wages were sufficient to more than make up for the $300 or more that the towing cost. By the end of the nineties, the days of the sailing schooner were nearly over on the redwood coast. A. W. Beadle

Out of Coos Bay," p. 29; Dickie, "The Pacific Coast Steam Schooner," p. 39; Hopkins, *Report*, pp. 60–72; O'Brien, "Life of Robert Dollar," pp. 46–47, 389.

69. McNairn and MacMullen, *Ships of the Redwood Coast*, pp. 18–20, 101–5; Johnson, "Schooners Out of Coos Bay," pp. 38–59; Newell, ed., *McCurdy Marine History, passim*; Dickie, "Pacific Coast Steam Schooner," pp. 39–48.

70. Walker to W. H. Talbot, 9 Jan. 1888, PTP, Walker corres.; Drury, "John Albert Hooper," p. 295.

stated that by then towing was "the only way to make these Boats pay."[71]

The new types of lumber carriers developed in the 1880s had the combined effect of driving down the cost of transporting lumber to market. Other things were pushing in the same direction. Expenditures on river and harbor improvements made many ports easier, and thus cheaper, to service.[72] Huge booms and cable loaders replaced apron chutes along the Mendocino coast, thus making loading both faster and safer.[73] Down-Easters replaced clippers as the major square-rigged merchant vessels. They carried far more cargo for their size than their famous predecessors, thus providing an economical means of transportation for men such as John Ainsworth who could use two-decked vessels to advantage in their trade in lumber. Down-Easters also provided reasonably priced charters for many a millowner when the usual fleet of lumber vessels failed to furnish sufficient tonnage to meet demand. Perhaps even more important was the worldwide decline in charter rates, from which lumbermen, along with other shippers, profited.[74]

The 1880s were, indeed, an epochal period for the lumber industry. Changes in logging techniques, in milling, and in the shipping of forest products had taken place before and were to continue to do so in the years to come; but the innovations of the eighties were especially important because they came at a time when the lumber industry of the Far West was undergoing a basic reorientation. These innovations made possible or, at the very least, speeded the transition. In turn, these innovations were encouraged by the tremendous increase in capitaliza-

71. Beadle to Meyer and Kyle, 6 Oct. 1899, WKSP, 154.
72. Coy, *Humboldt Bay Region*, pp. 130–31, 223–24, 277–78; Lawrence, "Markets and Capital," p. 66; Gibbs, *Pacific Graveyard*, pp. 41–46; Portland Board of Trade, *Annual Report* (1882), pp. 14–15; Ryder, *Memories of the Mendocino Coast*, pp. 36–37; *Canada Lumberman* 8 (Apr. 1888):7.
73. Allen, reply to questionnaire; McNairn and MacMullen, *Ships of the Redwood Coast*, pp. 21–32; Ryder, *Memories of the Mendocino Coast*, p. 22; Kortum and Olmsted, "A Dangerous-Looking Place," pp. 53–55; Connor, *Caspar Calling*, pp. 75–76.
74. Robert Carse, *The Twilight of Sailing Ships* (New York: Grosset & Dunlap, 1965), pp. 42–43, 46–47; Fairburn, *Merchant Sail*, 2:1577–82 and *passim*; MacArthur, *Last Days of Sail*, p. 35; Hutchins, *American Maritime Industries and Public Policy*, pp. 372–73; J. d'A. Samuda, "On the Influence of the Suez Canal on Ocean Navigation," *Transactions of the Institute of Naval Architects* 11 (1870):1–2; David A. Wells, *Recent Economic Changes* (New York, 1890), pp. 29–30; Douglass C. North, "Ocean Freight Rates and Economic Development," *Journal of Economic History* 18 (1958):537–55; G. S. Graham, "The Ascendancy of the Sailing Ship, 1850–85," *Economic History Review*, 2nd ser., 9 (1956):74–88; Coman and Gibbs, *Time, Tide and Timber*, p. 176. Analysts do not concur on the cause of the decline of freight rates, but do agree that there was a significant drop.

tion that took place during the period. Though the number of sawmills in the nation declined by 12 percent during the eighties, capital invested in the lumber industry increased by over 200 percent. The industry in the Far West got more than its share of the increase. Great quantities of capital were transferred from the declining lumber industry of the Lake States to the still largely untapped forests on the Pacific Coast. The result was that lumbering along the western shore moved rapidly away from the relatively simple sort of resource exploitation that it had long been and toward the more complex world of modern industry.[75]

75. Compton, *Organization of the Lumber Industry*, p. 38; Melendy, "One Hundred Years of the Redwood Industry," pp. 160–73.

CHAPTER XI

The Quest
for Order

FROM an early date millmen on the Pacific Coast resorted to
cooperative action in an attempt to reduce the uncertainties of the
lumber trade. Such action must have seemed desperately needed to
those interested in lumbering as a long-range investment. Productive
capacity had outstripped demand by the 1850s and continued to run
ahead of it during the decades that followed. Even during good years,
mills were often unable to run at full capacity for prolonged periods.
Whenever the market turned downward, running time had to be
reduced, profits dropped, and the less stable mills had to struggle to
stave off bankruptcy. Many millmen resorted to price cutting, rebates,
and other adjuncts of unrestrained competition in their efforts to
remain solvent. Even those who doubted the wisdom of such practices
for the industry as a whole felt obliged to adopt them so as to prevent
other millmen from gaining a competitive advantage. Cooperative
action appealed to many as preferable to such unrestrained competition.

Before the 1880s intermill cooperation was sporadic and halfhearted;
but during the eighties, when fixed investments were growing and, at
the same time, competition was increasing, such action became a
regular feature of the cargo trade. However, it remained for the
depression of the 1890s to bring a level of desperation that drove even
the most independent of millmen to aid in erecting combinations that
were virtually all-inclusive and thus had some real hope of success. In
the end, even these were to fail.

Intermill cooperation was first undertaken in the Humboldt Bay area in 1854. The effort was short-lived. The owners of nine firms joined to form the Humboldt Lumber Manufacturing Company.[1] They hoped that together they could make sales and penetrate markets not open to them individually. They anticipated that additional sales in these new markets would reduce the oversupply of redwood in San Francisco as well as make possible fuller use of the productive capacity of their mills. The combine enjoyed initial success, exporting 20.5 million feet of redwood during its first year of operations. However, a number of the firm's customers, caught in the depression of the mid-fifties, failed to meet their obligations. The company soon found it necessary to close down. In April 1855 the company's chief creditors took over the mill properties.[2]

The fate of the Humboldt Lumber Manufacturing Company suggested to lumbermen on the Pacific Coast that it was less risky to depend upon one's own resources than to join in cooperative ventures.[3] No new arrangements of this sort appeared until the 1870s, when the financial stringencies of the times led to cutthroat business practices that reduced the lumber industry on the Pacific Coast to near anarchy. By that time, the small pioneer mills, built as speculative ventures or to fill the extraordinary demands of the Gold Rush, had given way to larger mills, the owners of which viewed their plants not so much as speculations or temporary operations as long-range investments. To such men and under such circumstances cartels had much to offer. A number of the industry's leaders, including Charles Holmes, C. L. Dingley, William Adams, and S. H. Hammond began meeting daily for lunch.[4] Formal combinations soon resulted, both for the redwood

1. Porter and Livesay, *Merchants and Manufacturers*, p. 119, argue that in the 1850s merchants took the lead in erecting agencies for price-fixing and market control, in contrast to the 1870s and after when manufacturers took the lead. But in the Pacific Coast lumber industry the distinction between the role of merchant and manufacturer became blurred after the former erected sawmills to supply their needs during the gold rush. The cargo trade's major merchants and major manufacturers were inextricably intertwined from that time on.

2. The new management that took over after the failure instituted a cash and carry policy which, though it failed to generate the volume of business done previously, kept the mills from further financial disaster. Arcata *Humboldt Times*, 9 June 1855.

3. The lack of action on a proposal in 1855 for a similar combination in the Northwest may well be attributed to the fate that befell the Humboldt County organization; so, too, might be Andrew Pope's negative reaction to similar proposals in 1859–60. See Olympia *Pioneer and Democrat*, 3 Mar. 1855; Coman and Gibbs, *Time, Tide and Timber*, p. 409.

4. Ames, "Story of Pope & Talbot as told by Mr. E. G. Ames," mimeographed copy, PTP. Informal cooperation seems to have occurred in other times and places as well. Cartels included a number of forms of horizontal combination between independent firms that were aimed at

coast and the Northwest. These were only the first of a series of attempts to bring order into the lumber trade of the area. Millmen were still seeking stability through cooperative arrangements as the twentieth century began.

In the Northwest, the first effort appears to have been merely a gentlemen's agreement. Near the beginning of the 1870s, lumber dealers in San Francisco agreed to increase prices, and mill operators from the area around Puget Sound joined them. Just who was involved in this attempted price-fixing is not clear, but in view of the close ties most of the mills on the sound had with San Francisco, it seems likely that more than sympathy for the purposes of the Bay Area's dealers led Northwesterners to join the effort.[5]

According to Edward Clayson, the Washington Mill Company frequently violated this agreement. The company would sell cargoes at the stipulated price and then discount the bills. Other parties to the agreement consequently chastised the Seabeck firm: they bought all the logs they could on Hood Canal and hauled them out past Seabeck to their own plants, thus forcing the Washington Mill Company to pay in cash to obtain logs and to haul long distances.[6]

This combination was active during 1871 and 1872, but how long it remained so cannot be determined. In 1876 it was reported that the operators of the mills on Puget Sound had agreed to reduce their running time to three-quarters time for sixty days.[7] This decision may well have been a product of the same combination, though evidence of activity during the intervening years has not been uncovered.

In 1877 the Port Blakely, Port Madison, Port Gamble, Seabeck, and Tacoma mills formed a new combination. They adopted common

---

limiting competition. Though they were often successful in Europe, American law and public opinion combined to make them less so here. They were, in the words of John D. Rockefeller, "ropes of sand." See *Canada Lumberman* 16 (Aug. 1895):10; Porter, *Rise of Big Business*, pp. 54–56.

5. Seattle *Weekly Intelligencer*, 24 Apr. 1871; Buchanan, "Economic History of Kitsap County," p. 223. Early in 1872 lumber dealers in San Francisco established an exchange to promote their interests. W. H. Gawley was elected president. This exchange apparently was formed by the same group that had been attempting to increase lumber prices. See Seattle *Weekly Intelligencer*, 22 Apr. 1872; Buchanan, "Economic History of Kitsap County," pp. 223–24.

6. Clayson, *Historical Narratives of Puget Sound*, p. 66. According to Clayson, the action of the other mills crushed the prestige of the Washington Mill Co., but it is difficult to see how this activity could have hurt that firm financially any more than it did those engaged in bidding against it for logs. When the latter made a purchase on Hood Canal, they had to haul it even farther than the Washington Mill Co. did when it made one.

7. Seattle *Weekly Intelligencer*, 25 Mar. 1876.

price lists and sought to reduce the supply of lumber by limiting running time and paying subsidies to enable some mills to close altogether. They also adopted a system of fines to punish violators, and an assessment of ten cents on each thousand feet of lumber produced by member firms to finance the association's operations.[8] At first it worked satisfactorily. In January 1878 Charles Holmes exulted, "We have the best of the coast trade now and most of it. . . ."[9] However, by July the combine had lost control of the market, and it was soon abandoned. Members, as well as nonmembers, apparently contributed to this result. Officials of the Port Blakely Mill Company claimed that both the Port Madison and Seabeck mills violated the agreements they had helped formulate; and many lumber dealers in San Francisco had never been parties to the pact.[10]

In December 1880 a new combination, the Pacific Pine Manufacturers' Association, was formed. The eleven mills that made up this new cartel included the major mills on Puget Sound as well as plants on Coos and Winchester bays and Aaron Sutro's mill at Aaronville in California. The association had greater powers than its predecessors, regulating both prices and running time and carrying out inspections to insure compliance. In addition, the association collected one dollar a thousand on all lumber its members shipped to the Bay Area. It used this money to buy up cargoes bound for San Francisco when the market seemed in danger of becoming overstocked. These cargoes were deposited in a company "boneyard" and later sold for whatever they would bring. In an attempt to give their organization greater effect, the members agreed that any retailer who sold lumber for less than the association's retail price schedule would in the future have to pay the full retail price for any stock he purchased from association members. The mills that made up the association were the major

8. Ibid., 7 and 10 Jan., 7 Feb., 6 Apr. 1878; California State Agricultural Society *Transactions* (1877), pp. 148–49; Renton, Holmes & Co. to Port Blakely Mill Co., 9 Jan., 4 Apr. 1878, PBMCP; Gedosch, "Seabeck," pp. 20–21; Berner, "Port Blakely Mill Company," p. 165; Coman and Gibbs, *Time, Tide and Timber*, p. 409. The combination was referred to under different names: the Pacific Pine Manufacturers' Association, the Pine Lumber Association, and the Pine Manufacturers' Association. Available documents do not make it clear which was the official appellation.

9. Renton, Holmes & Co. to Port Blakely Mill Co., 7 Jan. 1878, PBMCP. The combination's control may have reached well beyond Puget Sound, for in 1879 it was rumored that Asa Simpson was being subsidized by the combine to keep his mill near the mouth of the Umpqua idle. See Roseburg *Plaindealer*, 20 Dec. 1879.

10. Renton, Holmes & Co. to Port Blakely Mill Co., 4 Apr., 2 Aug., 28 Sept. 1878, PBMCP; Gedosch, "Seabeck," pp. 20–21, 29; Berner, "Port Blakely Mill Company," pp. 164–65.

suppliers to the city's wholesale lumber market, so this threat undoubtedly carried a good deal of force. In spite of the extra powers given to it, this combination also failed. By 1882 the Tacoma Mill Company was apparently running nights and underselling the schedules, as was the Port Discovery Mill. The Port Blakely Mill Company, with more orders in hand than it could fill in the allowable running time, justified its decision to run full-time by the fact that others already were doing so. The agreement that had led to the creation of the Pacific Pine Manufacturers' Association was soon allowed to lapse.[11]

In spite of the failure of the Pacific Pine Manufacturers' Association, intermill cooperation did not come to an end in 1882. Talk of reviving the combination was in the air during the latter part of 1883, but nothing came of it. However, cooperative action was still being taken, without formal organization. Correspondence in the Washington Mill Company and Pope and Talbot papers indicates that collective efforts to control production were still being made two years after the Pacific Pine Manufacturers' Association came to an end.[12]

In spite of their repeated failures, lumbermen in the Northwest continued to seek stability through cooperation. In 1886 a group of them joined to form the Pacific Pine Lumber Company. Unlike its predecessors, this new organization was to have more than a fleeting existence. As before, the combination regulated production and enforced quotas through a system of fines. In an attempt to control price-cutting, the founders of the cartel adopted a new method for handling sales. All sales in San Francisco and its hinterland were handled through a central yard in the bay city. Members purchased lumber from this outlet to stock their own yards. Profits earned by Pacific Pine's yard were then passed on to members as dividends. The Pacific Pine Lumber Company also obtained foreign orders and distributed them among the member mills on the basis of capacity, though most mills continued to sell their lumber directly to their

11. Pacific Pine Manufacturers' Association, Constitution and By-Laws, WMCP; *West Shore* 8 (1882):183; Gedosch, "Seabeck," pp. 28–29; Berner, "Port Blakely Mill Company," p. 165; Buchanan, "Economic History of Kitsap County," p. 226. The combination may actually have been functioning before its formal organization in December 1880. See *West Shore* 6 (1880):166.
12. Renton, Holmes & Co. to Port Blakely Mill Co., 25 Dec. 1883, PBMCP; Gedosch, "Seabeck," p. 29; Washington Mill Co. to W. J. Adams, 22, 27 Jan., 8 July 1884; R. Kendrick to R. Holyoke, 1 Aug. 1884, WMCP; C. Walker to W. H. Talbot, 14 Sept. and 12 Oct. 1885, PTP, Walker corres. See also *West Shore* 10 (1884):73; *Canada Lumberman* 4 (1 June 1885):93.

established customers both overseas and in the Northwest.[13]

In addition to limitations on running time and controls over lumber prices, Pacific Pine undertook other activities designed to benefit its members. A committee was formed to investigate the possibility of some form of mutual fire insurance.[14] When the Washington Mill Company plant at Seabeck burned in 1886, the association seized the opportunity to reduce the productive capacity of the overexpanded industry. It paid Adams and his associates $750 a month in return for a promise not to rebuild for at least a year. Rather than rebuild, the Washington Mill Company purchased the mill at Port Hadlock. Overall production was thus reduced and the plant at Port Hadlock passed from owners who refused to cooperate with the combination, and who were considered "a bad lot" by their competitors, to ones amenable to collective action. Contemporary reports indicate that the association also arranged for the closure of some mills of nonmembers by paying them subsidies.[15] In addition to all this, Pacific Pine encouraged other cooperative efforts, including intermill activity to drive down the price of logs on the open market.

At first the Pacific Pine Lumber Company was quite successful. W. H. Talbot observed that "its hard work doing anything with men who will *not stick to an agreement*"; but the protections built into the new combination made it possible to do just that. He added, "This new scheme cannot be beaten & has worked well up to date." [16] After the cartel had been in operation for about ten months, A. W. Jackson of Pope and Talbot reported its successes were "up to the most liberal estimates." Profits were running over 30 percent, and Jackson

13. C. Walker to W. H. Talbot, 9 Dec. 1885, PTP, Walker corres.; Renton, Holmes & Co. to Port Blakely Mill Co., 14 Feb., 17, 24 Sept. 1886, PBMCP; Pacific Pine Lumber Co., price schedules, various dates, and resolution of board of directors, 15 Mar. 1888, PBMCP, box 51; Pacific Pine Lumber Co. to contracting parties, 11 June, 5 Aug. 1889; E. M. Herrick [president Pacific Pine Lumber Co.] to Renton, Holmes & Co., 6 Dec. 1888, PBMCP; Pope & Talbot to Puget Mill Co., 26 Sept. 1891, AC, Puget Mill Co., incoming corres.; *Canada Lumberman* 7 (Mar. 1887):8; Berner, "Port Blakely Mill Company," p. 166; Gedosch, "Seabeck," p. 29; Coman and Gibbs, *Time, Tide and Timber*, pp. 409–11. The combination's yard was located on land leased from the Oregon Improvement Co. and formerly occupied by the A. M. Simpson interests. See Oregon Improvement Co. and Pacific Pine Lumber Co., agreement, 13 May 1889, OICP.

14. Insurance Committee to Pacific Pine Lumber Co., 6 Dec. 1888, PBMCP, box 51.

15. Seattle *Post-Intelligencer*, 7, 12, 17 Sept. 1886; *West Shore* 16 (1890):311; Renton, Holmes & Co. to Port Blakely Mill Co., 7, 9 Sept. 1886, PBMCP; Gedosch, "Seabeck," p. 34. Whether the new management at Port Hadlock was much of an improvement over the old seems open to question. For example, see E. Ames to C. Walker, 16 Apr. 1890, AC, Walker, personal corres.

16. W. H. Talbot to C. Walker, 22 Oct. 1887, AC, Walker, incoming corres.

estimated that they would be between 20 and 25 percent during the coming year—"about all that can be expected in 1888." [17]

The initial success of the combination attracted new members. By 1889 some thirteen firms belonged to Pacific Pine, though by 1892 three of these—the West Shore Mills in Astoria, the Marshfield Mills on Coos Bay, and the Stetson and Post Mill Company in Seattle—had dropped out. Other changes in membership appear to have occurred between the formation of Pacific Pine in 1886 and its dissolution in 1897, but the details are not evident from the available records. [18]

Though the Pacific Pine Lumber Company started off well, it soon encountered difficulties. The company found it necessary to complain to members about the quality of lumber and pilings being dispatched to the central yard in San Francisco. Moreover, the company had difficulty getting the member mills to ship the kinds of stock needed in San Francisco, even though mills were fined one dollar a thousand for failure to fill "mandatory" orders. The management of the company also drew up "restricted" lists enumerating the types of lumber, already in oversupply, that should not be sent. W. H. Talbot found it necessary to remind Cyrus Walker, superintendent of the Puget Mill Company's plant at Port Gamble, that "the mills must cut on P.P.L. Co. memos when cutting Frisco lumber." [19] In spite of fines and directives, unwanted sizes accumulated in the yards in San Francisco while others remained in short supply. In February 1891 there were some eleven million feet of restricted sizes choking the available storage space in the bay city. [20]

But the problems connected with stocking the San Francisco yard were minor compared with others that beset the combination. When the southern California real estate bubble burst in the spring of 1888, lumber prices dropped sharply. Reductions in the number of hours that Pacific Pine's member mills were allowed to run failed to stem the

---

17. Jackson to C. Walker, 26 Nov. 1887, ibid.
18. A. D. Moore to C. Walker, 12 July 1887, ibid.; Pacific Pine Lumber Co. to contracting parties, 22 Mar. 1889, PBMCP, box 51; *West Shore* 16 (1890):311; Berner, "Port Blakely Mill Company," p. 166; Coman and Gibbs, *Time, Tide and Timber*, pp. 409–13. Most of the changes in membership seem to have been among the mills from outside the Puget Sound area.
19. Quoted in Coman and Gibbs, *Time, Tide and Timber*, pp. 410–11.
20. Pacific Pine Lumber Co., resolution of board of directors, 9 July 1888, PBMCP, box 51; Pacific Pine Lumber Co. to contracting parties, 29 Jan., 8 June 1889, ibid.; Pacific Pine Lumber Co. to Renton, Holmes & Co., 7 Nov., 17 Dec. 1894, ibid.; E. Herrick to Renton, Holmes & Co., 16 Mar., 18 May 1888, ibid.; E. Herrick to members of Pacific Pine Lumber Co., 15 Aug. 1888, ibid.; Coman and Gibbs, *Time, Tide and Timber*, pp. 410–11.

decline. New mills had appeared on the scene, and established ones had increased their capacities through remodeling and additions. The Pacific Mill Company, which had constructed a large plant at Tacoma, joined the combine, but other newcomers did not. Thus, while capacity was up, the percentage of the market controlled by the Pacific Pine Lumber Company was down and so was its ability to control the market.[21]

A temporary reprieve came as a result of a fire that burned much of Seattle in 1889. The Stetson and Post Mill Company was the leading lumber producer in the city, but that firm alone could not satisfy the demand for construction materials that rebuilding the city soon created. The Seattle firm purchased large quantities of lumber from Pacific Pine. Orders were distributed among the member firms on a pro rata basis. In view of the depressed nature of the lumber market in other quarters, they welcomed the business even though they received only $10 a thousand.[22]

But this windfall was only temporary. Lumber cargoes shipped from Port Gamble fell from seventy-eight in 1888 to sixty-nine in 1889 to fifty-five the following year. At the end of 1890 the Puget Mill Company closed its plants at Utsalady and Port Ludlow. Others, including the mills at Port Discovery and Port Madison, followed suit.[23] Avoiding complete shutdown, Asa Simpson ordered production at his plant on Grays Harbor (and perhaps his other plants as well) reduced by 25 percent. The mill ran only intermittently, and George Emerson, the mill superintendent, observed that it was an "expensive and awkward" arrangement made possible only because there was so little work elsewhere that crews would remain with the company to pick up whatever employment they could.[24]

Closures and reductions in running time failed to bring relief. Charles Moore used the occasion of the shutting down of his mill at

21. San Francisco *Alta California*, 1 Jan. 1889; Pacific Pine Lumber Co. to Puget Sound Brokerage Co., 24 May 1888, PBMCP, box 51; W. H. Talbot to C. Walker, 17 Oct. 1888, AC, Walker, incoming corres.; Berner, "Port Blakely Mill Company," p. 166. For details on the collapse of the boom in southern California, see Dumke, *Boom of the Eighties*, pp. 259ff.
22. Pacific Pine Lumber Co. to Renton, Holmes & Co., 15 June 1889, PBMCP, box 51; MacDonald, "Seattle's Economic Development," p. 176. For accounts of the fire and rebuilding, see *West Shore* 15 (1890):37, 759; William Gilpin, *The Cosmopolitan Railway* (San Francisco, 1890), p. 240.
23. T. O'Hara to C. Walker [May 1891], AC, Walker, incoming corres.; A. Wiley to C. Walker, 9 Feb. 1892, ibid.; Coman and Gibbs, *Time, Tide and Timber*, pp. 178–79.
24. Emerson to Simpson, 20 Apr., 6 May 1891, EL.

Port Discovery to install new equipment.[25] The mill at Cosmopolis was kept running on a full-time basis, as was a new plant that had been erected at Bellingham Bay. To move their cut, the mill at Bellingham sold at drastically reduced prices. Those connected with other mills found it difficult to imagine how the firm could afford to sell so cheaply. E. G. Ames of Pope and Talbot suggested that the new mill might have an especially inexpensive source of logs "similar to our Utsalady Log proposition," or that perhaps they were "running a big bluff in Order to get a subsidy from PPLCo not to Run." [26] Foreign orders, including one for twenty-one million feet of mining timbers for use in Australia, took up some of the slack, but they failed to provide overall relief.[27] By June 1891 the San Francisco office of Pope and Talbot was reporting that the Pacific Pine Lumber Company was "in a hole" and "we presume we will have to see them through even if it costs us big money." [28]

The flush times of the eighties were over, but the depths of the depression had not yet been reached. Australians, retaliating against tariffs levied by the United States on wool, instituted duties that virtually shut out American lumber.[29] By 1893 Pope and Talbot was laying up vessels for lack of work. Frederic C. Talbot, W. C. Talbot's younger son, reported in July: "Biz gets duller daily, lumber is being sold for most any price by all, except the P.P.L. Co[;] they have a price low enough, & if they cannot get it, then let the other fellow sell." Three months later he added, "We do not see how it can be much worse but it continues growing worse daily." [30] Talbot had no way of knowing it, but relief was nearly five years away, and by that time the

25. Emerson to Simpson, 8 Feb. 1892, EL; Pope & Talbot to Puget Mill Co., 14 May 1891, AC, Puget Mill Co., incoming corres.; E. Ames to C. Walker, 10 June 1891, AC, Walker, incoming corres.

26. E. Ames to C. Walker, 10 June 1891, AC, Walker, incoming corres. The latter suggestion seems the more likely since the mill had been having difficulty making ends meet only a few months earlier. See P. Atkinson to C. Walker, 15 Apr. 1891, ibid.

27. Pope & Talbot to Puget Mill Co., 26 Sept. 1891, AC, Puget Mill Co., incoming corres. The order for mining timbers was welcome not only for its size, but because "the roughest kind of lumber will do"; Pope & Talbot instructed the mill to "work in a lot of your poor logs which would otherwise go on the slab fire." Cf. *Canada Lumberman* 12 (Sept. 1891):17; 13 (Jan. 1892):14.

28. Pope & Talbot to Puget Mill Co., 1 June 1891, AC, Puget Mill Co., incoming corres.

29. Producers in British Columbia quickly took up the slack. *Canada Lumberman* 13 (Nov. 1892):10.

30. Talbot to C. Walker, 24 July, 6 Oct. 1893, AC, Walker, incoming corres. See also Talbot to Walker, 23 Aug., 6 Oct. 1893, ibid.; Coman and Gibbs, *Time, Tide and Timber*, p. 184; Clark, *Mill Town*, pp. 29–42.

old cargo trade in which his firm had engaged for over forty years would be almost gone. Mills of a new type, serving new markets, would be rapidly coming to the fore.

Lumbermen did not sit back and wait for the depression to pass. With the Pacific Pine Lumber Company unable to stem the decline in prices, member firms began to seek new means of dealing with the crisis. In 1892 representatives of the combination visited "every mill" in Oregon and Washington to learn if there was interest in an enlarged plan of subsidized closures or, failing that, if their owners could be induced to stay out of the markets of California for certain sums.[31] Asa Simpson was not interested. As George Emerson explained: "Our fleet is adapted to cheap transportation. Owning our own tugs should give us the cheapest towage. It would seem therefore, we should be able to deliver our lumber in San F[rancisco] in competition with anything on the Pacific Coast and should have no need for a combination . . . that by tempting to raise prices beyond a legitimate profit [only] but bids for the building of more mills."[32] Apparently owners of other key mills were also uninterested, for nothing came of this effort to broaden the scope of the combination. The Pacific Pine Lumber Company struggled on as best it could.

New sources of competition, as well as declining prices and the overproduction of member mills, plagued Pacific Pine. Since the market in San Francisco was their prime concern, members were greatly disturbed when some of the rising young firms in the Seattle area began shipping lumber to the Bay Area in order to move that part of their cut they could not sell locally. In 1892, when Seattle area mills joined with local retailers to maintain the price of lumber in their home market, Pope and Talbot officials saw an opportunity to strike back. Not only could they sell some of their own cut in Seattle at higher prices than prevailed in California, but, as one put it, by threatening to undersell the schedules of the Seattleites, "we may be able to get them to stay out" of the San Francisco market.[33]

It was a vain hope. One of the most aggressive of the Seattle producers, the Stimson Mill Company, had grown too large for the

---

31. Pope & Talbot to C. Walker, 7 June 1892, AC, Walker, incoming corres. *Canada Lumberman* 12 (Dec. 1891):9; 13 (Oct. 1892):8.
32. Emerson to S. Perkins, 25 July 1892, EL.
33. Pope & Talbot to C. Walker and E. Ames, 19 May 1892, AC, Walker, incoming corres. See also Pope & Talbot to C. Walker, 23 May, 15 June 1892, ibid.; MacDonald, "Seattle's Economic Development," pp. 146, 170–87.

markets available to it in the Northwest. Charles Stimson, president and general manager of the firm, felt penetration of the markets to the south to be essential. When agents failed to make sales for him in the Bay Area, he went to California and began seeking customers himself. At every turn he found his efforts blocked by the Pacific Pine Lumber Company. To break this impasse, Stimson sought membership in the combination. On being refused, Stimson declared economic war on Pacific Pine. With a modern plant plus an ample store of experience and capital gained in the woods of Michigan, Stimson was well equipped for the struggle. As his son later recounted, Stimson offered lumber for sale so cheaply as "actually to give away the lumber and pay the freight on it from Seattle to Oakland." The Pacific Pine Lumber Company's carefully protected price structure quickly collapsed. Similarly, unable to find buyers in Los Angeles, Stimson opened his own yard there and sold through it at less than market price. Though the market in both areas was plunged into "a chaotic state," Stimson got the new outlets he needed for his expanding operations.[34]

Apparently their encounter with the Stimson Mill Company, and similar episodes involving other mills, did not lead the managers of the Pacific Pine Lumber Company to change their ways.[35] In 1897 George Emerson wrote that the "condition of the Coastwise Lumber market is simply disgusting." He denounced the " 'dog in the manger' spirit" of the combination and complained that though conditions warranted an increase in prices, Pacific Pine refused to adjust its schedules upward. To do so would not hurt sales, Emerson argued, adding, "We have a very poor opinion of the business sagasity" of the combination's management.[36]

In the late nineties millowners in the Portland area, who had developed a pool of their own, also began to ship lumber to San

34. Stimson, "A Record of Charles Douglas Stimson," pp. 11–12. One must doubt that Stimson actually sold his lumber as cheaply as his son recalled, but he clearly did offer it at prices low enough to break the market. See also MacDonald, "Seattle's Economic Development," pp. 146, 179–80.

35. The details of these other encounters are not clear, but from evidence in George Emerson's correspondence, it appears that both the Slade and E. K. Wood mills on Grays Harbor were successful in attempts to wrest part of the California market from Pacific Pine Lumber Co. The Simpson interests, though not in the combine, appear generally to have abided by its schedules while continuing to hold their established place in the lumber trade of the Bay Area. Though Emerson held a low opinion of the combine, the firm which he helped to direct engaged in no overt struggles with it. See Emerson to S. Perkins, 15 Jan., 19 Feb. 1894, EL; Northwestern Lumber Co. to A. Simpson, 21 Jan. 1897, EL.

36. Northwestern Lumber Co. to A. Simpson, 28 May 1897, EL.

Francisco. They had a tremendous advantage over the mills of the larger combination. Thanks to extensive work on navigational improvements, the Columbia was ceasing to be the handicap it once had been. More important, these mills enjoyed a sizable local demand for their slabs, sawdust, and shavings. The receipts from these sales and their sales of lath (which was manufactured from what would otherwise have been waste), when coupled with their gain on scale, were sufficient to cover the cost of manufacturing.[37] Thus, George Emerson estimated that lumber produced at mills in Portland cost only $4 a thousand, that is, the cost of the logs themselves. Moreover, these mills had increasingly large markets to the east, serviced by rail, that constituted their main sources of income. In view of this, Emerson doubted whether there was any cost whatever to the lumber they shipped to California. The firms shipped their surplus and grades for which there was little demand in the rail markets; they could afford to sell these at "auction prices." When Pacific Pine engaged in a price war in an attempt to drive the Portlanders from the markets of the Bay Area, prices plummeted; but the Portland firms were not to be routed.[38]

Ineffective in the face of its many problems, the Pacific Pine Lumber Company was dissolved at the end of 1897. It was replaced by the Pacific Pine Company, a marketing agency that served only a few

37. The log scale in use then as now viewed logs as perfect cylinders, causing well known and predictable underestimates of volume. However, the scale then in use had been adopted when most mills still used circular saws for their head rigs and reflected production expectations based on that mode of milling. The Portland mills were modernized plants with band saws to do their main cutting. Since these took a much narrower kerf than circular saws, it was possible to cut substantially more from a log than the scale indicated or competitors with circular saw head rigs were able to do. See M. Bradner and P. Neff, "Log Scale versus Lumber Tally: A Discussion of Overrun and the Factors Affecting It," *Timberman* 27 (Sept. 1926):46–52; United States Census Bureau, *Twelfth Census of the United States, 1900,* 9:814.

38. Northwestern Lumber Co. to A. Simpson, 21 Jan., 2, 17 Mar. 1897, EL. *Pacific Lumber Trade Journal* 1 (1895):8; 4 (1899):19. Cox, "Organizations of the Lumber Industry," pp. 25–26. Cf. Seattle *Daily Intelligencer,* 3 Dec. 1877. Emerson had little sympathy for the Pacific Pine Lumber Co.: "Were they to do business on the principal of allowing others to live, as well as living themselves, and on the principal that anyone has a right to come into the San Francisco market and make sale of lumber if they see fit, there would probably be concessions made to the Portland people, that would allow prices to be reestablished. We are inclined to think the Pine Co. have been the 'dog in the manger' long enough and we are inclined to think the little lesson they are getting from the Portland mills will be a wholesome one, still better than the one they got a few years ago from the Slade Mill, and that they are still getting from both the Slade and Wood mills. . . . If these men were possessed of a little more brains, and less bull-dog tenacity, more philanthropy, and less of unalloyed selfishness, we think prices of lumber could be restored to about cost of product and freight. . . ." See Northwestern Lumber Co. to A. Simpson, 17 Mar. 1897, EL.

mills and made no effort to force up prices or control production.[39] Lumbermen expressed sorrow at seeing the older combination abandoned, fearing that "unhealthy competition will follow." [40] However, new agencies for bringing stability to the beleaguered lumber industry had already been adopted when the eleven-year-old cartel ceased to operate.

During 1892 Asa Simpson and others had rebuffed overtures to join in transforming the Pacific Pine Lumber Company into an all-inclusive combination of Northwestern millmen. However, as conditions continued to grow worse through 1893 and 1894, many former opponents of combination began to have second thoughts. In San Francisco retailers joined to support prices. Many may have felt as did the official of Pope and Talbot who wrote, "We have little faith in it ourselves but at the same time think that it is a very necessary matter and is our only hope of coming out even." At about the same time, a combination of all the deepwater mills on Puget Sound and in British Columbia was proposed in order to improve prices in the overseas markets. Stimson and the Willamette Steam Mills in Portland refused to join unless all the mills on the coast were taken in. Since not all millmen were ready for such a step, the plan was dropped. Again, a Pope and Talbot official expressed what was probably a widespread feeling. Perhaps it was just as well that the agreement had not been signed, he said, for it would surely have been broken shortly and in the price wars that would follow, "things would be worse than ever." [41]

The failure to effect the combination had its own negative effects. Proponents of the plan had anticipated setting the basic overseas price at $9.00 a thousand. Since this was considerably higher than the existing rates, many held off in their pursuit of orders in hope of being able to book them at the higher figure once the combination was formed. When it became clear that there would be no agreement, millmen who had been holding back hastened to book orders at lower figures. The Hastings Mill on Burrard Inlet made a number of sales, some for as low as $7.25 a thousand. Eight vessels were soon loading at its docks and the mill was preparing to run night and day. The mill was

---

39. Coman and Gibbs, *Time, Tide and Timber*, p. 413; *Canada Lumberman* 18 (Mar. 1897):13.
40. *Canada Lumberman, Weekly Edition*, 3 (27 Oct. 1897):1.
41. Pope & Talbot to C. Walker, 25 Apr., 1 June 1894, AC, Walker, incoming corres. See also Pope & Talbot to C. Walker, 25 Apr., 3 May, 12 June 1894, ibid.; R. P. Rithet to C. Walker, 7 May 1894, ibid.

so busy that it had to give up at least one cargo to the neighboring Moodyville mill.[42] The management of the Hastings Mill was engaged in a desperate gamble. Of its owner one lumber buyer wrote the superintendent of the Port Blakely mill: "You understand the *Millionere*[.] What a *farse* the fellow is almost Burst." [43] In his attempts to keep his firm solvent, the owner of the Hastings Mill helped pull down the prices for all.

The depressing effect which such sales had on the foreign market was soon matched in San Francisco. Under the Wilson-Gorman tariff of 1894, lumber was admitted to the United States duty-free. Prior to this time Canadian lumber had carried a one dollar a thousand duty, and, even though stumpage and labor were cheaper north of the international boundary, this had been sufficient to bar nearly all Canadian millstuff from the markets in California. When the new tariff schedules went into effect, lumber shipments from British Columbia to the Bay Area increased markedly, thus complicating further the already difficult situation faced by millmen on the Pacific Coast of the United States.[44] At last the situation was so desperate that the cry for an all-inclusive combine was nearly unanimous. In 1895 even Asa Simpson came to support combination.[45]

Demands for a more broadly based combination led to the formation of the Central Lumber Company. Stock in this new agency for intermill cooperation was apportioned among member firms on the

42. Pope & Talbot to C. Walker, 3 May, 12 June, 11 July 1894, ibid.

43. A. Sutherland to J. Campbell, 17 Jan. 1894, PBMCP, box 69. See also Sutherland to Campbell, 5 June 1895, 19 Aug. 1898, ibid. In his letter of 19 Aug. 1898 Sutherland observed that the owner of the Hastings Mill "has been well served as his crooked biz deserves" and added that the mill only had "a lot of bad debts to its credit."

44. W. Turner to C. Walker, 24 Nov. 1893, AC, Walker, incoming corres. C. Walker to F. Talbot, 30 Nov. 1893, AC, Walker letter books, private. F. Drew to Puget Lumber Co., Port Gamble, 20 Aug. 1895, AC, interoffice corres. Northwestern Lumber Co. to A. Simpson, 23 Aug. 1894, EL. G. Soule, undated statement [Oct. 1896], EL. G. Emerson, undated statement [Oct. 1896], EL. *Canada Lumberman* 15 (Jan. 1894):5; (Nov. 1894):3, 10; (Dec. 1894):10. At least one Canadian observer was not sympathetic. "There are only four mills in this whole province that are equipped to do this [California]trade," he wrote, "and if they are going to knock the whole of the Washington and Oregon mills out, they must be in bad shape indeed." See H. G. Ross to ed., in *Canada Lumberman* 15 (Dec. 1894):14. Cf. ibid. 16 (Nov. 1895):5.

45. According to John Cox, the plan for a combination of all the cargo mills originated with Simpson, Pope & Talbot, and Renton, Holmes & Co. Simpson was the first to sign the agreement to form such a combine, doing so on 12 Oct. 1895. D. H. Bibb of Golden Gate Lumber Co. was apparently also a prime mover in forming the organization. See Cox, "Organizations of the Lumber Industry," p. 28; Pope & Talbot to C. Walker, 12 June 1894, AC, Walker, incoming corres.; Florence *West*, 29 Nov. 1895; *Canada Lumberman* 16 (Dec. 1895):12. See also Defebaugh, *History of the Lumber Industry*, 1:453–57.

The yard of the Mendocino Lumber Company on the San Francisco waterfront, piled high with lumber in 1865. Courtesy of the San Francisco Maritime Museum, Union Lumber Company Collection

Unloading lumber at Pope & Talbot's yard at Third and Berry streets in San Francisco. In the background is the company's schooner *Okanogan*, launched in 1895. Courtesy of the San Francisco Maritime Museum

Henry L. Yesler. Courtesy of the Photography Collection, Suzzallo Library, University of Washington

Captain William Renton. Courtesy of the Photography Collection, Suzzallo Library, University of Washington

Asa M. Simpson. Courtesy of the Oregon Historical Society

John Dolbeer. Courtesy of the Bancroft Library, Berkeley, California

basis of their productive capacity. Each was to supply lumber to the central office at an established price and to receive dividends based on profits. Every effort was made to make the combination broadly based. Retailers had to join to obtain lumber, wholesalers in order to have retailers to whom to sell. Sixty cargo mills in British Columbia, Washington, Oregon, and California had joined by March 1896.[46] At first the owners of small mills had been hesitant about joining a combination with the giants for fear the latter would control the combination to the disadvantage of smaller firms. These millmen were won over by assurances that the large firms desperately needed the combination and would not dare to antagonize the small mills, for "everyone knows it does not take long for one disturber to demoralize the market." [47] Of the coastal mills of any size, the owners of only four were unconvinced. Remaining out of the combination were mills at Chemainus and Moodyville in British Columbia, Olympia on Puget Sound, and Skamokawa on the Columbia.[48]

Of the sixty firms that joined the Central Lumber Company, the Stimson Mill Company had been the most difficult to win over. At a final organizational meeting Stimson still refused to join. He pointed out that his firm had not a dollar of debt on its property, was amply financed, and had large supplies of timber. Stimson was in a position "to knock the market into a cocked hat." His opposition stemmed not from any dislike for the principle of intermill cooperation, but from pique over the earlier rejection of his request for membership in the Pacific Pine Lumber Company. He was striking back at those who had kept him out. When Stimson refused to join or to make suggestions for changes that would make the organization acceptable to him, W. H. Talbot, of the Puget Mill Company, "went out slamming the door and declaring it would be a fight to the finish." However, no fight developed. Stimson finally agreed to join, but he did so on his own terms. He signed only when the members of the combination agreed to purchase a percentage of all the lumber they sold during the coming

46. *Canada Lumberman* 17 (Jan. 1896):7; (Feb. 1896):11. W. H. Talbot to C. Walker, 16 Oct. 1895, AC, Walker, incoming corres.; *Pacific Lumber Trade Journal* 2 (1896):8; Cox, "Organizations of the Lumber Industry," pp. 27–28; Coman and Gibbs, *Time, Tide and Timber*, pp. 411–12. According to Cox the combination was incorporated in March 1896, but Coman and Gibbs state that it did not commence operations until 12 October.
47. *Pacific Lumber Trade Journal* 1 (1895):8. See also Cox, "Organizations of the Lumber Industry," p. 27.
48. *Canada Lumberman* 17 (Aug. 1896):12.

year from the Stimson Mill Company. The lumber was to be purchased at prices considerably above the market. In effect, the other members of the Central Lumber Company agreed to subsidize Stimson's operations for a year to keep him from engaging them in a price war.[49]

Stimson's was not the only mill to create difficulties for the combination. From the beginning many in British Columbia had seen the Central Lumber Company as primarily aimed against mills north of the border. The belief was given added credence when only four mills in the Canadian province joined the combination. To meet this challenge, the Victoria Lumbering and Manufacturing Company reopened its mill at Chemainus which had been closed for nearly three years. By September the mill, owned by capitalists from the eastern United States, was described as getting most of the foreign trade and being taxed to fill orders while the combination itself had few. At the same time small coastal mills selling through commission merchants reportedly gave strong competition in the San Francisco market.[50]

The Central Lumber Company soon ran into problems besides those posed by Stimson and competing mills. California retailers anticipated the increase in wholesale lumber prices that would follow the implementation of the plan for an all-inclusive combine by accumulating large stocks at the depressed prices that prevailed prior to its formation. When these were gone, many refused to do business with the Central Lumber Company. Mills in the combination were soon reduced to running only part time. Their problem was further complicated when several mills that had not previously sold in the cargo markets began to move into them and offer lumber at prices below those asked by the combine. To meet this competition, some members of the Central Lumber Company began to sell for less than the agreed schedules themselves.[51] The small mills also began to

49. Stimson, "A Record of Charles Douglas Stimson," pp. 12–13. Apparently George Emerson had the trouble with Stimson in mind when he wrote that he was sorry the Central Lumber Co. was meeting with so much opposition. "There seems to be no other hope for the business on the Coast," he added, "unless some apostle would arise among the lumbermen and show them the error of their ways." See Northwestern Lumber Co. to A. Simpson, 10 Feb. 1896, EL. Regarding Stimson's earlier position on combination, see R. Rithet to C. Walker, 7 May 1894, AC, Walker, incoming corres.

50. *Canada Lumberman* 16 (Nov. 1895):5; 17 (Apr. 1896):12; 17 (May 1896):12; 17 (Sept. 1896):14; 17 (Nov. 1896):17. *Canada Lumberman, Weekly Edition* 1 (25 Sept. 1895):1; 2 (26 Feb. 1896):2; 2 (29 July 1896):1; 2 (4 Nov. 1896):1. Jackman, *Vancouver Island,* p. 64.

51. *Pacific Lumber Trade Journal* 2 (1896):10. *Canada Lumberman* 17 (Jan. 1897):13; 17 (May 1897):6–7. Cox, "Organizations of the Lumber Industry," pp. 28–29; Coman and Gibbs, *Time,*

complain that the larger members of the combine were being given all the orders. When the agreement on which the Central Lumber Company was based came up for renewal at the end of 1896, Stimson demanded that his mill's special position be continued. This was the final straw. The agreement was not renewed.[52] The *Canada Lumberman* observed that attempts to make the combination permanent were futile "owing to the diversity of interests." [53]

With the dissolution of the Central Lumber Company and the gradual enfeeblement of the Pacific Pine Lumber Company, effectual control over prices and production disappeared during the late nineties. Some attempts were made to create new intermill combinations, but little came of them.[54] When the Pacific Export Lumber Company sought to combine secretly with the successor to the Pacific Pine Lumber Company, it was an indication of the former's aggressive merchandising, not the latter's strength. As an official of Pacific Export explained, "We believe in combination with apparent competition." [55] Lumbermen's associations had begun to appear in the Northwest as early as 1891, but as yet these had not evolved into anything that could do what combinations had failed to do. They lobbied for laws favorable to the industry and against the removal of the tariff on lumber. They also set standards for the grading of lumber; but, being voluntary organizations, they lacked the control over their members that was necessary to regulate either prices or production. Fortunately for lumbermen, both demand and prices were rising by 1897. For the moment, the need for controls was lessened.[56]

Tide and Timber, p. 412. Thomas Stimson claimed that as soon as the combination was drawn up, "numbers of those that had signed the agreement took the six o'clock trains home and began cutting prices." The Stimson Mill Co. abided by the agreement; with its special treatment, it had no need to do otherwise. See Stimson, "A Record of Charles Douglas Stimson," p. 13.

52. Cox, "Organizations of the Lumber Industry," pp. 28–29. Coman and Gibbs, *Time, Tide and Timber*, p. 412. *Canada Lumberman* 17 (Sept. 1896):14; 17 (Nov. 1896):17. *Canada Lumberman, Weekly Edition* 2 (4 Nov. 1896):1. Though he was sorry to see the Central Lumber Co. fail, George Emerson was not without hope for the future. "We think the effort made and partial success attained will lead to a new and successful effort at no very distant date," he wrote. "Meantime, I suppose all prospect of profit in the lumber business is off." See Northwestern Lumber Co. to A. Simpson, 27 Nov. 1896, EL.

53. *Canada Lumberman* 17 (Jan. 1897):13.

54. Ibid. 17 (Jan. 1897):13; 17 (Mar. 1897):13. *Canada Lumberman, Weekly Edition* 4 (7 Dec. 1898):2.

55. Pacific Export Lumber Co. to Pt. Blakely Mill Co., 20 Apr. 1898, PBMCP, box 50.

56. Cox, "Organizations of the Lumber Industry," pp. 16–17 and *passim*. Lucia, *Head Rig*, p. 9. P. Norton [president, Lumber Manufacturers' Association of the Pacific Northwest] to C. Walker, 13 Feb. 1892, AC, Walker, incoming corres. Pacific Pine Lumber Co. to contracting parties, 14 Jan., 14 Feb. 1895, PBMCP, box 51. G. Emerson to G. Stetson, 11 July 1895, EL.

While lumbermen in the Pacific Northwest were making repeated attempts at combination, millmen on the redwood coast were creating similar, and equally unsuccessful, cartels of their own. As has been shown, the first attempt at combination among the producers of redwood was the merger of nine mills to form the Humboldt Lumber Manufacturing Company in 1854. As in the Northwest, the next efforts at combination were not made until the 1870s.

The history of the first of the efforts during the seventies is obscure. In 1871 lumbermen in Mendocino County formed the Redwood Lumber Association to regulate prices and set production quotas. This pool failed to bring any real stability perhaps, as H. Brett Melendy suggests, because it failed to include the mills of Humboldt County as well as those of Mendocino. By 1877 some sort of combine, apparently rather informal, had also emerged in the area of Humboldt Bay, and by the late seventies redwood dealers in San Francisco had joined together in an attempt at regulating prices.[57]

In 1879 millmen in both counties joined to form the Redwood Lumber Company. Through pricing agreements, production quotas, and a central sales office this combine and its successor, the Redwood Manufacturers' Association, were able to bring a degree of order to the industry.[58] As was stated in a report of the California State Board of Forestry, the redwood industry had a productive capacity that was double the demand—"only through combination has a market been sustained at all."[59] In 1885 slumping demand led to a reduction of 20 percent in the quotas assigned to individual mills. The Mendocino mills refused to accept the reductions, sold under the combination's prices, and broke the market. Lumber was soon selling at less than the cost of manufacture and the market in the Bay Area was glutted.[60]

Canada Lumberman 18 (Apr. 1897):10; 18 (June 1897):15. When a degree of order finally came to the lumber industry after the turn of the century, it was the product of the lumbermen's associations rather than pools, or trusts. For discussions of these later developments, see Cox, "Organizations of the Lumber Industry"; Compton, *Organization of the Lumber Industry*; Lucia, *Head Rig*; Clark, *History of Manufactures*, 3:241–42. Cf. Clark, *Mill Town*, p. 61.

57. Melendy, "One Hundred Years of the Redwood Industry," pp. 280–82, 306; Eureka *Humboldt Times*, 21 Apr., 7 July 1877, 4 June 1878, 22 Nov. 1879.

58. J. G. Jackson et al. to lumbermen, form letter dated 1879, JKCP, 3:7. This letter outlines the organization of the combine in some detail and lists the mills that had agreed to participate. The combination was the work of Jackson (Caspar mill), Robert G. Byxbee (Navarro mill), John Kentfield, John A. Hooper (Trinidad mill), and E. C. Williams (Mendocino Lumber Co.).

59. California State Board of Forestry, *Biennial Report*, 1885–86 (Sacramento, 1887), pp. 147–48.

60. Melendy, "One Hundred Years of the Redwood Industry," pp. 282–83, 306–7; *Canada*

San Francisco's lumber dealers established a combination of their own, the California Lumber Exchange, at about the same time that lumber manufacturers were creating the Redwood Lumber Company. By February 1880, twenty-one dealers belonged. One year later, according to one authority, the exchange included all important lumber dealers in California, thirty-six in San Francisco and thirty-seven in the rest of the state. However, many of the exchange members were manufacturers as well as dealers. Through it, nonproducing dealers cooperated with the manufacturers who supplied them. It was an uneasy marriage, but until their divergent interests drove apart those who were exclusively dealers and those who were also manufacturers, the exchange served to bring a degree of stability to retail lumber prices, not only in San Francisco but in the entire market area dominated from the Bay Area.[61]

From mid-decade until June 1889 the redwood industry was without an effective intermill combination. Then the California Redwood Company was formed. Almost every mill in Humboldt, Mendocino, and Sonoma counties joined, but the organization soon ran into difficulty. The lumber produced in Humboldt County had a higher percentage of clear than that manufactured to the south. Moreover, lumber cut in the latter area tended to be hard and flinty. Because of these basic differences, members found themselves unable to agree on grading standards, quotas, or prices. Within months of its formation, the California Redwood Company foundered.[62]

Unable to maintain a combination that included both major areas of redwood production, lumbermen now moved to form separate organizations for the northern and southern forests.[63] In the north there was

Lumberman 5 (1 June 1885):93.

61. California Lumber Exchange, agreement, 24 Mar. 1882, Buhne Papers; Weidberg, "History of John Kentfield & Co.," pp. 26–29. John Kentfield belonged to both the Redwood Lumber Co. and California Lumber Exchange and also, as a seller and shipper of Douglas fir, to the Pine Manufacturers' Association. He was among those who sometimes sold below the agreed prices.

62. Melendy, "One Hundred Years of the Redwood Industry," p. 307. *Pacific Coast Wood and Iron* 11 (1889):150; 12 (1889):72. Edgar Cherry & Co., *Redwood and Lumbering*, pp. 5, 13. Wilde, "Chronology of the Pacific Lumber Company," p. 38. *Canada Lumberman* 3 (1 Nov. 1883):329. This California Redwood Co. should not be confused with the lumber manufacturing firm with the same name.

63. By 1890 production was rapidly declining in Sonoma County and the areas to the south. What little was produced was consumed locally. Thus, when one speaks of the southern area in connection with this period, Mendocino County is meant. The northern area of production included Humboldt and Del Norte counties. Even in Santa Cruz County, however, efforts were being made to regulate production and prices. See *Canada Lumberman* 8 (July 1888):1.

the Humboldt Lumber Manufacturers' Association, in the south the Mendocino Redwood Association. An attempt was made in 1892 to bring the two together, but once again millmen from the two areas could not agree on grading standards. The two organizations continued to operate separately.[64]

In spite of their repeated failures, lumbermen continued to seek means of creating a combination that would embrace the entire redwood coast. In April 1894 ten major mills joined to form anew the Redwood Manufacturers' Association. By December, twenty-one mills had joined. However, the Gualala Mill Company and the Caspar Lumber Company refused to come into the combination, thus frustrating its efforts to fix binding prices on the redwood industry.[65]

At this juncture, Edward C. Williams, president of the Mendocino Lumber Company, came forward with a plan for reconciling the interests of the northern and southern areas. Williams suggested that a quota be established for the production of redwood and that this be divided between the two regional associations, which in turn would apportion their allotment among their members. Since the mills of Humboldt and Del Norte counties produced more clear per thousand than the mills in Mendocino County, the latter were to be allowed three-fifths of the total market so as to insure them of roughly 50 percent of the income from sales. Millmen in Mendocino County were enthusiastic, those to the north cool toward Williams' proposal. In the end nothing came of it.[66]

By mid-decade, prices of lumber were so low that some firms were restricting their activity to cutting railroad ties and gathering tanbark.[67] Profits were down. As a result, millmen felt it necessary to continue efforts at forming an effective, broadly based combine. In 1896 the Caspar and Gualala mills were finally induced to join the Redwood

64. Ibid. 11 (May 1890):5; Eureka *Humboldt Times*, 6 Sept. 1890; *Pacific Coast Wood and Iron* 17 (1892):292; Melendy, "One Hundred Years of the Redwood Industry," pp. 284, 308; Wilde, "Chronology of the Pacific Lumber Company," p. 38.
65. *Pacific Coast Wood and Iron* 21 (1894):128; 22 (1894):220; 23 (1895):57. Melendy, "One Hundred Years of the Redwood Industry," p. 308; Wilde, "Chronology of the Pacific Lumber Company," p. 64.
66. *Pacific Coast Wood and Iron* 24 (1895):140; Melendy, "One Hundred Years of the Redwood Industry," pp. 287–88.
67. Little has been written on the tanbark industry. Some insight into the trade in this item can be gained through the William Kyle papers, however, for the schooner *Bella* was engaged in transporting this item for a time in the nineties. See also Bauer, "History of Lumbering and Tanning in Sonoma County"; Warren Ormsby, "Peeling the Tanoak," *Forest History* 15 (1972):6–10.

Manufacturers' Association. By April of that year, the organization included twenty-five mill companies—that is, virtually every sawmill on the redwood coast that engaged in the coastwise trade. At last it was possible to put into effect industry-wide price schedules and grading standards. The association also sought to aid its members in the opening of new markets.[68]

Though the Redwood Manufacturers' Association at last became the all-inclusive agency that the producers of redwood lumber had been seeking, it still did not succeed. Much redwood was normally marketed through independent agents. For example, J. R. Hanify and Company represented the Elk River Mill and Lumber Company as well as John Vance and Isaac Minor; Pollard and Dodge represented the Eel River Valley Lumber Company. Some redwood producers, such as Dolbeer and Carson, Navarro Mill Company, Mendocino Lumber Company, Cottoneva Lumber Company, Excelsior Redwood Company, and the Fort Bragg Redwood Company, had their own yards.[69] Since those to the east who sought to obtain redwood by rail had to obtain it from San Francisco, rather than directly from the mills, this fragmentation of marketing agencies was a very real handicap. One Chicago lumberman complained that demand for redwood in his area was not great enough to justify keeping a full line on hand; yet when he sent orders to San Francisco for those sizes and grades for which he had immediate buyers, the suppliers themselves often did not have them in stock.[70] In order to provide its members with a more effective, aggressive merchandising instrument than the yards of agents and individual members, the association opened a lumberyard of its own. However, members failed to abide by the agreement into which they had entered; both members and their agents offered lumber for less than the price schedules of the cooperative yard. As a result, by the end of 1896 the lumberyard effort was abandoned, and the close of the nineteenth century saw the lumbermen of the redwood coast still searching for an effective means of bringing stability to their industry.[71]

68. *Pacific Coast Wood and Iron* 25 (1896):140, 222; Melendy, "One Hundred Years of the Redwood Industry," pp. 308–9; Wilde, "Chronology of the Pacific Lumber Company," pp. 38, 64; Ryder, *Memories of the Mendocino Coast*, pp. 40–42.

69. *Pacific Coast Wood and Iron* 8 (1887):91; Melendy, "One Hundred Years of the Redwood Industry," pp. 304–5.

70. *Northwestern Lumberman*, reprinted in *Canada Lumberman* 5 (15 Oct. 1885):342.

71. *Pacific Coast Wood and Iron* 33 (1900):178; Melendy, "One Hundred Years of the Redwood Industry," pp. 289–91, 309–11; Eureka *Humboldt Times*, 11 Apr. 1903; Weidberg, "History of John Kentfield & Co.," pp. 36–37.

The failure of lumbermen to bring order to the redwood industry stemmed from the same basic factors that led to failure in the Pacific Northwest. In both areas productive capacity exceeded normal demand. The need for restricting production may have seemed obvious, but it was not easy to get fiercely independent millmen to agree on how to bring this about. Those combinations that were attempted were continually weakened by lumbermen who chose not to join and by the inability of the combines to keep members from violating agreements they had signed. No single lumberman emerged who was strong enough to impose order on the industry and none of the combinations was ever able to do so.

An additional cause of failure may have been a result of the effect intermill cooperation had on prices. By setting prices higher than existing economic conditions warranted, the combines may well have attracted outsiders into the cargo trade. Certainly new competitors were appearing. Sometimes these were the owners of mills already present on the West Coast, but who had been selling locally or by rail. More often, they were men who had acquired capital and experience in the lumber industry of the Lake States and were now transferring their operations to the Far West. These new competitors tended to exacerbate the very problems that the combinations had been instituted to alleviate. George Emerson, for one, felt that the fruits of combination only encouraged more lumbermen to move into the cargo trade. "Any unnatural advance in price stimulates the building of new mills that must afterwards be competitors," he argued. Emerson long felt that since entry into the area's lumber industry was relatively easy, the temporary advantages to be gained through combination would be outweighed by the additional competition that would follow.[72]

Some millmen, recognizing the dangers inherent in adding new mills to the already overexpanded cargo industry, attempted to discourage those who might build them. William Adams painted a bleak picture of the future of the trade for a would-be investor who approached him for information.[73] Similarly, when the owners of the Stetson and Post mill in Seattle found that they were unable to supply enough lumber to satisfy the demand in that city, they sought to make arrangements with the Puget Mill Company to market lumber cut at

72. Northwestern Lumber Co. to A. Simpson, 15 Oct. 1895, EL.
73. Adams to R. Maynard, 25 Dec. 1883, WMCP.

the latter's mills. By this process, it was hoped, sufficient lumber might be obtained to keep the price in Seattle low enough to discourage outsiders from building new sawmills to service the area. In 1888 the Puget Sound Cedar and Lumber Company—a joint undertaking of Stetson and Post, the Puget Mill Company, and the Port Blakely Mill Company—emerged as an instrument for accomplishing this end.[74] On another occasion, the leading mills on Puget Sound joined to purchase the Perry mill at Cosmopolis. In so doing, they were moving not only to prevent outsiders from gaining control of the operation but also to put themselves in a position to bid up the price of logs on Grays Harbor and otherwise harass the mills already in operation there, mills that were beginning to offer stiff competition to the plants on the sound.[75] Clearly, the addition of new plants and new rivals was a source of grave concern to the millmen of the Far West.

It is not readily demonstrable that the price schedules and production controls of the various intermill cartels actually caused an influx of new competitors. Schedules were quite transitory and, even while they lasted, appear to have been broken almost as often as they were observed. Moreover, it seems likely that many new operators would have appeared on the West Coast during the eighties and nineties regardless of whether there had been combinations operating in the area. Lumbermen in the Lake States were running out of timber, and many were searching for new areas in which to put their capital and knowledge of sawmilling to work. Even without artificially high prices established by intermill combines, the forests of the Pacific slope would doubtlessly have attracted a good number of them.[76]

74. C. Walker to W. H. Talbot, 12 Jan., 22 May 1888, PTP, Walker corres.; Stetson & Post to C. Walker, 27 Sept. 1888, AC, Walker, incoming corres.; W. Newell to C. Walker, 23 Dec. 1888, 3 Apr. 1889, ibid.; G. Stetson to C. Walker, 27 Dec. 1888, ibid.; E. Ames to C. Walker, 9 May 1889, ibid.; A. Jackson to C. Walker, 11 June 1889, ibid.; W. Renton to C. Walker, 12 July 1889, ibid.; Stetson & Post to Puget Mill Co., 5 Sept. 1889, ibid.; E. Ames to Pope & Talbot, 24 Jan. 1898, ibid.; Stetson & Post to Port Blakely Mill Co., 7 Jan. 1898, PBMCP, box 69; Buchanan, "Economic History of Kitsap County," pp. 227–28; Coman and Gibbs, *Time, Tide and Timber*, pp. 398–99.

75. *West Shore* 14 (1888):452; G. Stetson to C. Walker, 10 Jan., 18 May 1889, AC, Walker, incoming corres.; G. Stetson to W. H. Talbot, 24 Feb., 1 May 1888, PTP, Talbot corres.; C. Walker to W. H. Talbot, 6 and 31 Mar., 2, 5, and 22 May 1888, PTP, Walker corres.; Renton, Holmes & Co. to Port Blakely Mill Co., 6 Sept. 1901, PBMCP, box 65; Berner, "Port Blakely Mill Company," p. 166; Coman and Gibbs, *Time, Tide and Timber*, pp. 218–20, 399–401.

76. For descriptions of the migrations to the West Coast and the forces behind them, see Holbrook, *The American Lumberjack*, pp. 140ff.; Hidy, Hill, and Nevins, *Timber and Men*, pp. 160ff.

Millmen on the Pacific shore were powerless to stop this migration. The resource base on which they drew was too extensive, too widely scattered for those first present to control it to the exclusion of others. Improvements in transportation and logging technology were constantly making additional stands accessible. Therefore, it would have been necessary to acquire control over tracts that might not yield returns for years if one were to shut out potential rivals. None of the established mills was strong enough to undertake such a project, and the combines never evolved to the point where they concerned themselves with the acquisition of timberland.[77] Though sporadic efforts were made to dissuade outsiders from entering the lumber industry of the area, no combination to accomplish this end was ever devised.

Though intermill combinations failed to solve the basic problems of overproduction and unstable prices, they were sometimes successful in dealing with lesser difficulties. The various lumbermen's associations that began to appear on the West Coast as early as 1891 lobbied successfully for log-lien laws and other legislation desired by their members.[78] The Puget Sound Tug Boat Company also functioned satisfactorily. Rather than compete with one another for the towing business on the Strait of Juan de Fuca and Puget Sound, millmen in the area pooled their vessels, formed the company, and enjoyed much better service.[79] Asa Simpson made similar arrangements in connection

77. Had millmen attempted to do so, either singly or in combination, it does not seem likely that they could have acquired sufficient stands to prevent the entry of outsiders into the lumber industry of the West Coast. Federal land policies were generally ill-conceived and poorly administered, but they made it difficult enough to acquire forest tracts from the government that it was probably impossible to monopolize the stands of the Far West effectively enough to shut out determined, capable, and well-financed outsiders. The task would in all likelihood have been too difficult for existing operators even in conbination, and apparently they knew it. For a discussion of federal policy regarding forest lands, though with emphasis on inadequacies rather than how it prevented absolute monopoly, see Meany, "History of the Lumber Industry," pp. 168–236. Numerous other accounts exist. Cf. *West Shore* 8 (1882):136; Seattle *Intelligencer*, 16 Nov. 1876; Weidberg, "History of John Kentfield & Co.," p. 27.

78. See above, note 54. For evidence of earlier political activity, see C. Walker to W. H. Talbot, 9 Jan., 14 and 26 May 1888, PTP, Walker corres.; Ivan Doig, "John J. McGilvra & Timber Trespass: Seeking a Puget Sound Timber Policy, 1861–1865," *Forest History* 13 (1970):6–17.

79. W. Renton to C. Walker, 23, 29 Nov. 1888, AC, Walker, incoming corres.; C. Walker et al. to N. Hall, 16 Feb. 1889, ibid.; J. Libby to C. Walker, 18 Feb. 1891, ibid.; E. Ames to C. Walker, 10 June 1891, ibid.; Pope & Talbot to Puget Mill Co., 29 May, 28 Sept. 1891, AC, Puget Mill Co., incoming corres.; W. H. Talbot to J. Libby, 1 June 1891, ibid.; E. Ames to W. H. & F. C. Talbot, 19 Mar. 1912, PTP, Gibbs folder; Coman and Gibbs, *Time, Tide and Timber*, pp. 161–62, 249, 391–98. An earlier tugboat combine had failed when William Adams triggered an epidemic of price cutting. See C. Walker to W. H. Talbot, 16 July 1885, PTP, Walker corres.

with both the tug and steamer service on Grays Harbor and at South Bend, as did tugboat owners on Humboldt Bay.[80] On another occasion, the owners of timber stands near Grays Harbor joined to insure the Northwestern Lumber Company against losses from keeping open its hotel at Hoquiam, for all found it to their advantage to have a place where visiting businessmen could stay.[81] Mills also cooperated on orders. Indeed, smaller mills that lacked marketing facilities in California sometimes were able to run when the market was depressed only because they contracted to cut for larger operations.[82] Similarly, mills occasionally joined to operate joint lumberyards. The Puget Sound Cedar and Lumber Company in Seattle was just that. In addition, those millowners who had their own vessels cooperated at first informally and then through the Pacific Coast Shipowners' Association in combating the wage demands of the seamen's union; low-cost transportation was a key element in their success in the cargo trade, and these operators had no desire to lose the advantages it offered. Sometimes they also cooperated in the joint ownership of vessels.[83]

Perhaps the most important of these lesser attempts at intermill cooperation was the effort to regulate the price of logs purchased on the open market. As early as 1878, major producers on Puget Sound, feeling that competition for available logs had driven prices so high that mills on Grays Harbor, Coos Bay, and the Umpqua had a significant competitive advantage, agreed not to pay over $6 a thousand for logs.

80. G. Emerson to G. Stinson [*sic*], 4 Aug. 1891, EL; G. Emerson to G. Stetson, various dates, 1891, EL; Northwestern Lumber Co. to A. Simpson, 16 Nov. 1895, 21, 27 Nov. 1896, and other dates, EL; Ship Owners and Merchants Tug Boat Co., Ninth Annual Report, 8 July 1891, JKCP, 117:5. For an interesting example of the ends to which such a combine could be put, see Andrew M. Genzoli and Wallace E. Martin, *Redwood Bonanza, a Frontier's Reward: Lively Incidents in the Life of a New Empire* (Eureka, Calif.: Schooner Features, 1967), pp. 65–67.
81. G. Emerson to A. Simpson, 4 May 1891, EL.
82. C. Nelson to J. Kentfield, 21 Jan. 1873, JKCP, 1:1; Stetson & Post to J. Campbell, 28 Feb. 1898, PBMCP, box 69; Stimson Mill Co. to Port Blakely Mill Co., 18, 20 June, 6, 9, 12, 14 July 1898, ibid.; Pope & Talbot to C. Holmes, 26 Aug. 1898, PBMCP, box 51; Pope & Talbot to Puget Mill Co., 29 May 1891, AC, Puget Mill Co., incoming corres.; Pope & Talbot to Puget Lumber Co., 5 Sept. 1899, ibid.; G. Emerson to A. Simpson, 25 Nov. 1891, 18, 20 Feb., 2 June 1892, EL; G. Emerson to S. Perkins, 23 Mar. 1892, EL.
83. Adams, Blinn & Co. to Washington Mill Co., 3 Dec. 1866, WMCP; W. Talbot to C. Walker, 8 Dec. 1887, AC, Walker, incoming corres.; Pope & Talbot to Puget Mill Co., 24, 28 Sept. 1891, AC, Puget Mill Co., incoming corres.; E. Ames to W. H. Talbot, 24 May 1888, PTP, Ames corres.; Ames to W. H. & F. C. Talbot, 19 Mar. 1912, PTP, Gibbs folder; Weidberg, "History of John Kentfield & Co.," pp. 30–33, 59–70; Paul S. Taylor, *The Sailors' Union of the Pacific* (New York: Ronald Press, 1923), pp. 51–52, 66–74, 79–80, 94, 98–100, 162–63. For copies of the bylaws, shipping rules, and other documents of the Pacific Coast Shipowners' Association, see JKCP, 117:6.

There may have been informal agreements even earlier; but it is clear that efforts at keeping down the price of logs continued on an intermittent basis in the years that followed.[84] In 1887 the members of the Pacific Pine Lumber Company formed the Puget Sound Brokerage Company in an attempt to strengthen their position vis-à-vis the loggers and to protect themselves against those who might be tempted to pay more than the agreed maximum. The firm purchased logs for all cooperating mills and then distributed them on the basis of need. If the brokerage firm's executive committee failed to agree on which mills needed a group of logs the most, they made them available to the member firms on an auction basis. The company purchased logs only from loggers on an approved list.[85]

Like other cooperative efforts, the log combine was plagued by millowners who violated the terms of agreement. The owners of the Port Discovery and Port Hadlock mills were apparently the first and worst transgressors, but by February 1888 Cyrus Walker was reporting that "lots" of millmen were ignoring the brokerage firm's prices because they were short of logs and "when that is the case they dont regard any agreement." [86] As a result, the Puget Sound Brokerage Company came to an end in 1889. Nevertheless, cooperative efforts at controlling the price of logs continued.[87]

In the end, the solution of the lumber producers' problems of log supply did not come about as a result of combination. For years independent loggers had supplied most of the logs sawn at the mills. Lumbermen gave advances, generally in the form of credit at the company stores, so that the loggers could get out the necessary logs. This left the mill companies in an exposed position. Many independent loggers were irresponsible, at best, and in depression years even the most businesslike among them sometimes failed. W. H. Talbot

84. Seattle *Daily Intelligencer*, 12 Sept. 1877; Renton, Holmes & Co. to Port Blakely Mill Co., 22 Mar., 23 May 1878, PBMCP; Gedosch, "Seabeck," pp. 21, 32, 146. Even $6 per thousand seemed high to Cyrus Walker, who felt such a price made lumbering unprofitable. See Walker to W. H. Talbot, 5 Dec. 1885 and other dates, PTP, Walker corres.
85. Report of Committee on General Policy to President and Board of Directors of Pacific Pine Lumber Co., 6 Oct. 1887, AC, Walker, incoming corres.; W. H. Talbot to C. Walker, 13, 22 Oct. 1887, ibid.; Port Blakely Mill Co. to C. Walker and G. Meigs, 19 Oct. 1887, ibid.; G. Stetson to C. Walker, 20, 25 Oct. 1887, ibid.; D. Wheeler to C. Walker, 19 Nov. 1887, 8 Feb. 1888, ibid.; Pacific Pine Lumber Co. to Puget Sound Brokerage Co., 24 May 1888, PBMCP; Pacific Pine Lumber Co. to Port Blakely Mill Co., 16 July 1888, PBMCP; Pacific Pine Lumber Co. to Board of Directors of Pacific Pine Lumber Co., 4 Nov. 1889, PBMCP.
86. Walker to W. H. Talbot, 1 and 23 Feb., 6 Mar. 1888, PTP, Walker corres.
87. Berner, "Port Blakely Mill Company," p. 166; Florence *West*, 29 Nov. 1895.

asserted in 1894 that it would be better to shut down the Puget Mill Company's plants than to "tie ourselves up with irresponsible loggers and eventually get stuck." [88] Sharing Talbot's sentiments, more and more lumbermen established woods operations of their own. In 1895 loggers joined in a cooperative effort to try to force millowners to pay better prices for the logs they purchased on the open market. Their activities served only to hasten the movement toward logging camps owned and operated by the mill companies.[89] In 1896 George Emerson forecast the shape of things to come when he wrote Asa Simpson: "We think we are safe in saying that within a few years no mill can be successfully operated without owning its own timber, its own [logging] camps and its own railroad." [90]

Continuing improvement in logging equipment and techniques also served to alleviate problems of log supply. Near the end of the century, George Emerson reported that an unprecedented supply of logs was available on Grays Harbor. The reason, he explained, was that over 150 steam skidders were operating in the woods tributary to the harbor. Not surprisingly, the abundant supply had driven the price downward. Douglas fir logs were selling for $4.50 a thousand feet log scale, a rate the lumbermen operating in combination had never been able to obtain.[91]

The increased supply of logs of which Emerson spoke may have resulted from the transfer of extensive tracts of publicly owned timber to private hands through railroad land grants and subsequent logging on them, but no evidence for this has been uncovered. Indeed, the silence of Emerson, William Renton of the Port Blakely Mill Company, and E. G. Ames of the Puget Mill Company on this subject would suggest the contrary. All three were perceptive observers who reported regularly to their co-workers in San Francisco regarding developments in the Northwest and future prospects for the industry. None appears to have pinpointed land grants as a rich new source of

88. Talbot to C. Walker, 17 Jan. 1894, AC, Pope & Talbot, incoming corres. See also Pope & Talbot to Puget Mill Co., 9 July 1890, AC, Walker, incoming corres.; C. Walker to W. H. Talbot, 8 Jan. 1894, AC, Walker letter books, private; W. Adams to R. Holyoke, 10 Sept. 1883, WMCP; Seattle *Daily Intelligencer*, 12 Sept. 1877; Clark, *Mill Town*, pp. 73–74.
89. Northwestern Lumber Co. to A. Simpson, 22 Jan. 1896, EL; *Canada Lumberman* 17 (Aug. 1896):12; Cox, "Organizations of the Lumber Industry," p. 21. Cf. Northwestern Lumber Co. to A. Simpson, 12 June 1899, EL.
90. Northwestern Lumber Co. to A. Simpson, 22 Jan. 1896, EL.
91. Northwestern Lumber Co. to A. Simpson, 26 July 1899, EL. Cf. Northwestern Lumber Co. to A. Simpson, 2 June 1892, 3 Mar. 1894, EL.

timber. Renton seems to have seen them as more of a threat than an opportunity, writing in 1883 that if homestead entries on public land were halted in order to clear the way for the Northern Pacific land grant, timber tracts "will go so high no one can touch them." [92]

Land grants do not seem to have been necessary for lumbermen to acquire timberland adequate for their purposes. As pointed out earlier, during the 1880s logging increased on the redwood coast, where there were no railroad land grants, about as rapidly as in the Northwest, where such grants were extensive. Lumbermen had, of course, developed a variety of methods by which they obtained timberland besides buying it from railroads. Dummy entrymen who took up homestead claims and later turned them over to millowners in return for a small payment were a common subterfuge used from an early date to acquire forest land from a government that refused to sell it outright. Other methods were also used. These ranged all the way from timber trespass—that is, logging government land without bothering to gain title to it—to legal purchase of stumpage or land. The Cleveland administration seems to have been especially concerned with halting illegal and semilegal transfers of timberland to private parties, a fact which may help explain the statement by one of Frederick Weyerhaeuser's partners that he was "a Republican, just as every good lumberman should be." [93]

However they acquired their stands, the new system of company-owned camps and mechanized woods operations that lumbermen developed in them had numerous advantages. The mill companies now had direct control over the money they furnished to underwrite the expenses of logging. The supply of logs became more dependable, both because the mills now had greater control over logging and because mechanized operations were less seasonal than the old-fashioned method of using bull teams to get the logs to the watercourses and freshets to float them out of the woods. The quality of logs supplied also tended to be higher, for most of the company-operated camps were

---

92. Quoted in Berner, "Port Blakely Mill Company," p. 161.
93. Ibid., pp. 161–63; Meany, "History of the Lumber Industry," pp. 168–205; Buchanan, "Economic History of Kitsap County," pp. 217–18; S. A. D. Puter and Horace Stevens, *Looters of the Public Domain* (Portland: Portland Printing House, 1908), pp. 46–67 and *passim*; Doig, "John J. McGilvra & Timber Trespass," pp. 6–17; Jerry A. O'Callaghan, *The Disposition of the Public Domain in Oregon* (Washington, D.C.: U.S. Government Printing Office, 1960), pp. 71–95; Berner, "Port Blakely Mill Company," p. 162; Kohlmeyer, "Northern Pine Lumbermen," p. 537. Quotation from Kohlmeyer.

located in stands of virgin timber tapped by company railroads rather than in the picked-over forests nearer the waterways. Moreover, logs coming by train arrived more regularly and thus could be handled more systematically and economically than logs floated out on freshets by loggers operating away from the rail lines.[94]

Thus, in regulating log prices, just as surely as in regulating the production and price of lumber, the attempts of millowners at cooperative action failed to bring any lasting relief. When the sellers of lumber did gain relief from low prices, it came from the opening of new markets and the revitalizing of old ones as the depression of the nineties waned. When the buyers of logs gained relief from high prices, it came as a result of the emergence of a new pattern of woods operations.

The problem of overproduction never was solved. New mills continued to be built and older ones to expand. Only through continuous investment in improvements and continual adoption of new techniques could mills remain competitive. Those that could not or would not adjust soon declined. But, though the less competitive mills continued to close down, new capacity was added even faster than closures reduced it. Combination continued to be a will-o'-the-wisp that seemed to offer relief for the hard-pressed owners of mills that had fallen behind in the competitive struggle. However, as George Emerson pointed out, there was no way that the sort of group action the area's lumbermen were willing to take could make an old-fashioned mill the equal of a modern one.[95] So long as lumbermen on the West Coast were unable to prevent new producers from entering the area, there was no possibility for genuine relief for mills that had become submarginal operations. They were caught in a Social Darwinist world from which there was no escape.

94. Northwestern Lumber Co. to A. Simpson, 2 June 1892, 15 Oct., 3 Dec. 1895, 22 Jan. 1896; G. Emerson to S. Perkins, 25 May 1891, EL. A good, brief account of the transition from dependence on independent loggers to company-owned camps is in Berner, "Port Blakely Mill Company." See also Coman and Gibbs, *Time, Tide and Timber*, pp. 159 and *passim*.
95. Northwestern Lumber Co. to A. Simpson, 3 Mar. 1894, EL.

CHAPTER XII

The Sands
Run Out

An upturn in the business cycle, the gold rush in Alaska and the Yukon, and the Spanish-American War all led to demands that revitalized the long-depressed lumber industry of the West Coast near the end of the 1890s.[1] But the old days of the cargo trade did not return. The lumber trade of the Pacific was rapidly evolving into something new.

San Francisco had been the heart of the old trade—its main market, its source of investment capital, and the center from which, in large part, its direction had come. However, by the end of the depression of the nineties, San Francisco's importance had been permanently undermined. The city and its hinterlands continued to grow, but the rate was too slow to furnish sufficient demand for building materials to keep the coastal lumber industry prosperous, especially since much of the demand of the interior was now being filled from different sources. "A great change has taken place since the early years when country orders were of daily occurrence" in San Francisco, one commentator noted. "Almost every country town has its mill nowadays" and fills much of its own needs.[2] Moreover, millmen in the interior of southern

1. The gold rush to Alaska and the Yukon helped lumbermen by providing a new market for lumber and stimulating the growth of Seattle, Vancouver, and other cities supplying the gold fields; but it also drew away much of the work force on which the mills depended and drove up the cost of labor. See *Canada Lumberman* 18 (Dec. 1897):9, 12; 19 (Jan. 1898):3; (May 1898):3; 20 (June 1899):11. *Canada Lumberman, Weekly Edition* 4 (30 Mar. 1898):1. Emerson to Simpson, 19 July 1898, EL. MacDonald, "Seattle's Economic Development," pp. 129–45, 183–84.
2. *The Bay of San Francisco,* 1:301.

Oregon, a new group of competitors for the owners of the cargo mills, had begun to ship to California on the Southern Pacific Railroad. They filled much of the demand that had once been met by lumber entering through the Golden Gate.[3] At the same time, the increasing flow of capital into the lumber industry of the West Coast from outside the region, especially from the Lake States, lessened the importance of San Francisco as the industry's center of control.

Even those mills with both management and ownership centered in San Francisco found it necessary to adjust to the changing role the city played in the lumber trade. Some, including apparently most of the mills that belonged to the Pacific Pine Lumber Company during its last years, were able to do so successfully. They prospered during the late nineties and into the twentieth century by shipping to foreign markets, both old and new.[4] These and other firms continued to ship to San Francisco, but the old relationship was now reversed. Overseas outlets were now of primary significance, the coastwise trade to the Bay Area only complementary. San Francisco had become a place to ship one's excess cut or grades not in demand elsewhere. George Emerson summed the situation up neatly: "San Francisco stands in the same position to the lumber business that a Fire-pit stands to the old time mill. The question is into which of the two furnaces you will throw your refuse." On another occasion, with a change of metaphor, he wrote of the steps being taken by the more progressive millowners to service the demands of new markets that were developing "and thus obviate the necessity of throwing a large portion of their cut into the San Francisco lumber cesspool." [5]

The new foreign markets were widely scattered. Hamburg, Buenos Aires, Port Elizabeth (South Africa), and the major harbors of Japan were among those receiving cargoes. Producers saw great promise in the markets of Europe and in South Africa, where mining operations were being expanded rapidly. Pope and Talbot pioneered in the South African market in 1892, with others soon following.[6] By 1897 George

3. Cox, "Organizations of the Lumber Industry," pp. 59–60, 65–68; *Columbia River and Oregon Timberman* 1 (1900):8; Coman and Gibbs, *Time, Tide and Timber*, pp. 200–9.
4. Northwestern Lumber Co. to A. Simpson, 16, 21 Apr. 1897, EL; Coman and Gibbs, *Time, Tide and Timber*, pp. 185–99. See also Dollar, *Memoirs*, 3:23, 57–58. Describing the period from 1897 through 1906, Coman and Gibbs wrote, "Never before had the cargo trade been so good for so long a time" (p. 199). As shall be shown, however, not all the cargo mills prospered to the same degree as Pope & Talbot.
5. Northwestern Lumber Co. to A. Simpson, 20 Feb., 21 Apr. 1897, EL. See also E. Ames to Pope & Talbot, 4 Mar., 27 June 1898, AC, Walker, incoming, personal corres.
6. On the European markets see, for example, Pope & Talbot to C. Walker, 25, 27 Apr. 1894, AC, Walker, incoming corres. On the South African market see Pope & Talbot to Puget

Emerson was writing: "It would seem to us that there is no better place for any person to visit, who is likely to be interested in the sale of lumber, than the coast of Africa. . . . That, and not San Francisco, is the market of the future. . . ." Emerson added that even "the smallest of our Sound mills, and all the mills of Portland, have . . . established agencies over there, or have men traveling for them." [7]

Japan, too, seemed to hold out promise of becoming a major market for forest products from the West Coast. Small amounts of lumber entered the ports of the empire during the 1880s, but at first demand increased slowly since the country was able to meet most of its needs from extensive domestic stands. Though a large order had been received as early as 1886, in 1893 the total value of forest products entering Japan from the United States still amounted to less than $11,000.[8] However, by this time it was evident that the Japanese were severely overcutting their accessible stands, and increased importations were being forecast. The American consul at Osaka reported: "The trade in American lumber is increasing and bids fair to continue to do so, as it has been determined that the American production can be laid down in the open ports of the Empire more advantageously than the native product." Indeed, at the time he was writing, two lumber-laden sailing ships from Oregon were riding at anchor in the harbor at Kobe.[9] In the years that followed, Japan developed into a major outlet for lumber from the West Coast.[10]

Long-established markets overseas also expanded during this period. The rapid development of mining in the Broken Hill area of Australia led to sizable shipments of lumber to Port Pirie and Adelaide. These orders were especially welcome. As one official of Pope and Talbot

Lumber Co., 9 July 1892, ibid.; Coman and Gibbs, *Time, Tide and Timber*, pp. 182–83. See also Pope & Talbot to C. Walker, 11 Jan. 1894, AC, Walker, incoming corres. *Canada Lumberman* 7 (July 1887):8; 14 (Dec. 1893):8.

7. Northwestern Lumber Co. to A. Simpson, 12 Apr. 1897, EL. See also *Canada Lumberman* 16 (Nov. 1895):6; 16 (Dec. 1895):8; 19 (Oct. 1898):20; 20 (Nov. 1899):7.

8. USBFC, *Commercial Relations, 1894–95*, 1:612. *Canada Lumberman* 5 (2 Mar. 1885):84; 6 (1 July 1886):5; 6 (1 Oct. 1886):4; 8 (May 1888):8.

9. USBFC, *Commercial Relations, 1894–95*, 1:172. Cf. *Canada Lumberman* 19 (June 1898):4; 20 (Oct. 1899):21.

10. Dollar, *Memoirs*, 3:11–12, 196, 203; Walter A. Radius, *United States Shipping in Transpacific Trade, 1922–1928* (Stanford, Calif.: Stanford University Press, 1944), pp. 88–98 and *passim*; Ivan M. Elchibegoff, *United States International Timber Trade in the Pacific Area* (Stanford, Calif.: Stanford University Press, 1949), *passim*; E. A. Selfridge, *American Lumber in Japan* (Washington, D.C.: U.S. Government Printing Office, 1928), *passim*. By 1897 the value of lumber shipped to Japan had risen to $182,791. It continued to rise. See USBFC, *Commercial Relations, 1898*, 1:135.

explained: "With a Port Pirie order [one has] . . . one of the most desirable cargoes which any ship can take. It makes almost perfect stowage and vessels have no difficulty loading down to draught with it, and they always carry larger cargoes in feet than when taking rough lumber to any other port." [11] Since these orders were primarily for railroad ties and mining timbers, they had an added advantage: they used low-grade logs. One lumberman observed that logs unfit for manufacturing material for Port Pirie were "not fit for anything." [12]

Lumber markets in northeastern Asia also expanded during the late nineties. Increasing construction of railroads in the area furnished a demand for ties and bridge timbers even larger than that of the 1880s. Cargoes of forest products not only entered Tientsin and the other harbors of northern China with greater frequency, but began to arrive at Vladivostok and Chemulpo (Korea) as well.[13]

But, for all their importance, developments in these overseas markets had less of an impact on the lumber industry of the West Coast than did the rapid growth of demand in the interior of the North American continent. Shipments of lumber east by rail, primarily from Portland, had begun during the 1880s. The quantities grew, especially after rail rates were slashed in 1893 and the depression began to wane in 1897.[14] During 1894 some 5,200 carloads of lumber were dispatched inland from the Pacific Northwest, and by May 1896 lumbermen had begun to complain of a shortage of railroad cars for shipping lumber east. This cry was to be repeated frequently in years to come.[15] By 1899 the number of carloads of lumber shipped reached 10,000 and two years later, 24,302. During 1900, the rail markets consumed 284 million board feet of lumber from the Northwest, compared to 492 million

11. Puget Lumber Co. (Port Gamble) to Puget Lumber Co. (Port Ludlow), 25 Mar. 1892, AC, interoffice corres.

12. Puget Lumber Co. (Port Gamble) to Puget Lumber Co. (Port Ludlow), 28 Mar. 1892, ibid. See also *Canada Lumberman* 6 (Oct. 1886):4; 13 (Apr. 1892):8.

13. USBFC, *Commercial Relations, 1898*, 1:1056. USBFC, *Consular Reports*, 59:290. Consular despatches, Tientsin, 30 Jan., 5 Feb. 1897. Tientsin *Peking and Tientsin Times*, 12 June 1897. Portland *Oregonian*, 16 Nov. 1900. California State Board of Harbor Commissioners, San Francisco, *Biennial Report, 1898–1900*, pp. 13–14. *Canada Lumberman* 20 (Apr. 1899):12; 20 (Oct. 1899):21.

14. Following completion of the Great Northern Railroad, James J. Hill set the rate for lumber to the Midwest 33 percent lower than it had been previously, i.e., at forty cents per hundred pounds. To meet this competition, the Union Pacific and Northern Pacific reduced their rates. MacDonald, "Seattle's Economic Development," pp. 96–103, 180–83, assesses the impact of these reductions.

15. *Canada Lumberman* 17 (May 1896):12.

consumed by the cargo markets. By 1906 the rail trade was taking as much lumber as was the seagoing trade.[16]

Lumbermen were attracted to the rail trade not just because it was a way of marketing the excess lumber that had plagued them for so long. As a rule, it returned higher profits than did the cargo trade. For example, George Emerson reported that during 1896 the Northwestern Lumber Company's sales to customers serviced by sea brought an average price of $6.12 per thousand, local sales $7.50 per thousand, and sales to customers serviced by rail $11.00 per thousand. The figures he reported for other years were similar.[17] Under these circumstances, Emerson strove to convince Asa Simpson that their operations should be reoriented away from the San Francisco market that they had serviced for so long and toward winning and holding a place in the rail trade. As early as 1895 Emerson was writing Simpson that the "salvation of the lumber business, in our opinion, rests with the Eastern trade and we think that will be the true source of increase in Western prices. . . ." Emerson put it even more succinctly in 1899 when he wrote, "Our Eastern trade demands our full attention and we look upon all Coastwise orders as a loss." [18]

Lumbermen in coastal British Columbia, as well as those in Oregon and Washington, benefited from the burgeoning rail trade. The operators of mills on the west coast of Canada were turning their attention to the interior, where the Okanogan Valley and the prairie provinces were being rapidly developed. Mills on the coast of British Columbia shipped so much lumber to these markets that exports by sea had begun a steady decline by the end of the nineteenth century. The growth of the interior, reduced rail rates, and the higher maritime freight rates that came with prosperity all contributed to the shift.[19]

16. *Pacific Lumber Trade Journal* 4 (1899):11–12; 7 (1902):28; 9 (1904):50. Cox, "Organizations of the Lumber Industry," pp. 6–9, 59–60. Compton, *Organization of the Lumber Industry*, pp. 11, 44. Coman and Gibbs, *Time, Tide and Timber*, pp. 215–16.
17. Northwestern Lumber Co. to A. Simpson, Jan. 8, 1897, EL. In view of the higher profits to be earned in the rail trade, Emerson observed, "I do not think any of the mills prepared for Eastern shipment will be tempted to enter the [cargo] trade." See Northwestern Lumber Co. to A. Simpson, Dec. 26, 1899, EL.
18. Northwestern Lumber Co. to A. Simpson, 15 Oct. 1895, 23 Nov. 1899, EL. See also 13 Mar. 1896, 27 Feb., 16 Mar. 1897, 9, 17, 20 Aug. 1898, 15 June 1899.
19. *Western Lumberman* 11 (1914):33; 12 (1915):15; 13 (1916):13. Lawrence, "Markets and Capital," pp. 64–65. *Canada Lumberman* 19 (Jan. 1898):3; 19 (July 1898):11; 20 (Feb. 1899):5; 20 (Oct. 1899):24–26. In an attempt to circumvent high maritime freight rates some entrepreneurs tried dispatching ocean-going rafts of logs from the Northwest to California; but, with the exception of rafts sent to the Benson Lumber Co. in San Diego from 1906 to 1941, these

Similarly, redwood was being marketed increasingly by rail. In 1896 the Southern Pacific offered low rates on lumber shipped from San Francisco to the Midwest and granted a rebate to cover the cost of transporting the lumber from the millsites to San Francisco. Redwood producers had for a long time been hoping to develop significant markets for their products in the Midwest and East. The hope now began to be fulfilled.[20]

A few of the old cargo mills were fortunate enough to be located where they could continue to engage in the cargo trade while servicing rail markets as well. One was the Tacoma Mill Company. Another was the old Perry mill at Cosmopolis. By 1900 it had become the largest on Grays Harbor, with an annual cut of thirty-nine million feet, thirty-three million of which were marketed by rail. One of its owners commented in amazement that he had never thought it would get so big.[21]

Not all the cargo mills were so fortunate. As early as 1885 E. G. Ames recognized that mills such as those at Port Gamble, Port Discovery, South Bend, and Coos Bay would be operating under severe competitive handicaps when they attempted to service markets of the interior: "Those mills that are handy to the Rail Road all have the best end of the biz as they will be able to market lumber & waste that we are obliged to burn and always will be." [22] It was to prove an accurate forecast. Occasionally some of these mills did ship to rail outlets. For example, from time to time the Puget Mill Company dispatched lumber by barge to Seattle and from there east by rail, but the shipments were small.

The cost of transporting lumber to the railhead was not the only disadvantage faced by such mills when they sought entry into the rail trade. Demand in the Midwest was generally for high-grade materials, since low-quality forest products were available from sources nearer at hand. In manufacturing to meet this demand, much low-grade material was inevitably produced. Mills located on the rail lines were able to

attempts had little effect. See *Canada Lumberman* 21 (Aug. 1890):4; 29 (Nov. 1898):4–5. Kramer A. Adams, "Blue Water Rafting: The Evolution of Ocean Going Log Rafts," *Forest History* 15 (1972):16–27.

20. *Pacific Coast Wood and Iron* 26 (1896):218; Melendy, "One Hundred Years of the Redwood Industry," pp. 167–69, 284, 287, 289; Wilde, "Chronology of the Pacific Lumber Company," pp. 70–71.

21. Renton, Holmes & Co. to Pt. Blakely Mill Co., 6 Sept. 1901, PBMCP, box 65; Coman and Gibbs, *Time, Tide and Timber*, pp. 218–20, 399–401.

22. Ames to W. H. Talbot, 10 Nov. 1885, PTP, Ames corres.

dispose of much of this locally or through outlets in the Inland Empire. The St. Paul and Tacoma Lumber Company marketed quantities of low-grade lumber in the Yakima area; the mills in Portland sold much of theirs in Walla Walla. But for mills such as that at Port Gamble, these alternatives were not available. Low-grade materials brought too little to justify both the expense of barging and the cost of rail transportation to outlets east of the Cascades, and local demand was almost nil in the little company towns that had not been reached by the railroads.[23] Some of the mills located away from the rail lines sought relief by shipping their lower grades to British Columbia. Though this development led British Columbia lumbermen to demand a protective tariff, the outlet was too limited to lead to more than slight improvement in the circumstances of Puget Sound mill operators.[24]

For mills located away from rail lines and major local outlets, only the cargo markets were available to absorb the low-grade by-products that resulted from catering to the rail trade. Even buyers in the cargo markets were becoming more discriminating. These markets were taking a smaller percentage of low-grade material than they once had. According to George Emerson, the cargo trade was no longer "that of sawing lumber for the San Francisco market, to be taken away immediately and permisciously [promiscuously]. . . ." Now even the cargo mills were finding it necessary to add lumberyards so that their cut could be sorted and stored in advance of orders. "The day of sawing lumber other than to order has in the main gone by," Emerson added.[25] With such factors in mind, one official of the Puget Mill Company pointed out to the San Francisco office of Pope and Talbot that if they were to pursue the rail trade under existing circumstances, they would end up with an accumulation of much low-grade material for which markets would be difficult to find.[26]

Under these conditions, it is hardly surprising that some of the old cargo mills pushed the export trade vigorously during the late nineties and early years of the twentieth century, rather than focus their efforts

23. Northwestern Lumber Co. to A. Simpson, 27 Feb. 1896, 28 Nov. 1898, 26 Dec. 1899, and other dates, EL; Fred Hill, report on mill companies, dated 13 Sept. 1897, AC, Walker, personal corres.; Puget Mill Co. to Pope & Talbot, 6 Aug. 1898, A. Jackson to C. Walker, 17 Oct. 1899, ibid.

24. *Canada Lumberman* 19 (Mar. 1898):11; 19 (May 1898):11; 20 (May 1899):11.

25. Northwestern Lumber Co. to A. Simpson, 6 Mar. 1897, 9 Apr. 1895, EL. See also 3 Dec. 1895.

26. Puget Mill Co. to Pope & Talbot, 30 Mar. 1898, AC, Walker, personal corres. See also Pope & Talbot to C. Walker, 25 Feb. 1898, AC, Walker, incoming corres.

on the rail trade.[27] They clung doggedly to their established maritime markets. One competitor reported that Pope and Talbot "say they have had it [the Hawaiian trade] for 30 or 40 years and do not propose to give it up." To hold their own share of the Hawaiian market, Renton, Holmes and Company proposed selling lumber at a loss if necessary.[28]

The cargo mills suffered from other disadvantages besides those of location. Many were outmoded. Some of the owners of old cargo mills had changes made to permit them to supply the types of material buyers were now demanding. But modifications seldom made old-fashioned mills fully competitive with plants specially built and located for most effective participation in the new trade that was emerging.[29] Others among the owners were reluctant to expend the money necessary to make the requisite changes. Doubtless, many of these would have sympathized with William H. Talbot, who, disturbed by the continuing expense of improvements needed to stay competitive with the new mills that were being built, suggested that since Easterners were so interested in getting mills, perhaps Pope and Talbot should simply sell their mills to them and get out of the lumber business.[30]

Asa Simpson was one such operator. For years George Emerson and R. B. Dyer, superintendent of Simpson's mill at South Bend, sought to convince the captain that modernization of his plants was vital. Emerson wrote Simpson:

We think we are safe in saying that within a few years, no mill can be successfully operated without owning its own timber, its own camps and its own railroad; that it must be connected with Coastwise markets by steam transportation instead of sail, and that it must be provided with appliances to work up that portion of its product that is adapted to the Eastern market into the class of finished material required. Without all these costly appurtenances, we think, within a few years a mill plant will have fallen so far behind the march of events as to be out of the race. All this means endless thought and worry, constant investment, constant additions and new applicances.

27. For examples, see Coman and Gibbs, *Time, Tide and Timber*, pp. 185–99.
28. Renton, Holmes & Co. to Pt. Blakely Mill Co., 30 Oct. 1901, PBMCP, box 65.
29. George Emerson considered the Port Blakely mill "the best type of the old style mill," yet "out of the competition." See Northwestern Lumber Co. to A. Simpson, 3 Mar., 15 May 1894, EL. Similarly, E. G. Ames, commenting on plans to improve the mill at Port Madison, said it seemed unwise to spend much money "on such a mill in such a place"—a plant there simply could not be made competitive. See [Ames] to Pope & Talbot, 6 Apr. 1898, AC, Walker, personal corres.
30. W. H. Talbot to C. Walker, 14 May 1888, AC, Walker, personal corres. See also Coman and Gibbs, *Time, Tide and Timber*, pp. 146–47.

Today, he added, improvements must be made more frequently than ever before if one is to avoid being left behind. "We think the whole sequel of the proposition is that the old mills and the old methods are out of the competition," he wrote after describing the situation on another occasion; "the only possible method for us to meet this competition which is now in the field is to go and do as they do." [31]

Simpson's responses have not been preserved, but Emerson's own extensive correspondence indicates that the captain considered further investment in the lumber industry too speculative. He seems to have wanted a steady return on the investments made during almost half a century in the lumber trade without being forced to undertake any new risks. Only after the Northwestern Lumber Company's plant at Hoquiam burned in 1896 was Emerson given permission to make some of the changes he had been seeking. Even then Simpson was cautious. He refused to allow Emerson to install a band saw for the mill's head rig in spite of his superintendent's repeated arguments.[32] He absolutely refused to approve modernization of the plant at South Bend. In 1906, when Dyer went ahead on his own to rebuild the outmoded plant, Simpson promptly removed him from his position as mill superintendent. The captain relented only when the new plant proved more profitable than its predecessor.[33]

While the emergence of new mills and new markets was forcing basic changes upon the cargo mills, changes at sea were operating in the same direction. During the 1890s steamships replaced sailing vessels in the cargo trade.[34] By 1893, George Emerson reported, the Grays Harbor Commercial Company's mill at Cosmopolis was shipping almost all its exports by steamship. Two years later he wrote Asa Simpson: "The large tramp steamers carrying two million [board feet] and over on a ton of coal per day seem to be making quite an inroad into the Foreighn shipping business from the Sound. These we expect are iron boats of rough construction, but with the latest improvements in steam appliances." [35] The records of the Oregon Board of Pilot

31. Northwestern Lumber Co. to Simpson, 22 Jan. 1896, 3 Mar. 1894, EL.

32. Northwestern Lumber Co. to Simpson, 3 Feb., 25 Mar., 15 Oct. 1895, and other dates; George Emerson to A. Simpson, 6 Aug. 1892, EL.

33. Unidentified newspaper clipping, Dubar scrapbook 117 (University of Washington library, Seattle), pp. 46–47.

34. Among the first steamships to engage in the lumber trade was *Remus*, which loaded at Moodyville for Melbourne in 1890. The editor of *Canada Lumberman* (11 [Dec. 1890]:1) correctly predicted that this was a first step in replacing sail in the Pacific lumber trade.

35. Emerson to S. Perkins, 7 July 1893, EL; Northwestern Lumber Co. to Simpson, 11 Nov.

Commissioners show that by the period 1898–1900 the bulk of the tonnage calling at the Columbia River was steam-powered.[36] The Pacific Export Lumber Company of Portland chartered sailing vessels to carry the first cargoes it sold, but by June 1899 the firm was chartering only steam vessels.[37] Others were moving in the same direction. By the end of the century the construction of wind-powered lumber carriers had begun an inexorable decline, while age and the seas were steadily reducing the number in use. "These old vessels," complained Cyrus Walker, "are a terror in the way of repairs."[38] More and more often owners found the cost of upkeep outrunning profits and laid up their vessels to await better freight rates. Most of these vessels never went to sea again. Others sold their vessels. As Frederic C. Talbot explained, "The old vessels are gradually going[;] better take what one can get than have them lost [at sea] & be a total loss."[39]

The abandonment of sailing vessels in the lumber trade was not abrupt. Windjammers continued to transport forest products on the sea lanes of the Pacific until well into the twentieth century. Free motive power continued to attract lumbermen to sailing vessels; they found it hard to believe that ships that burned coal and required larger crews than barkentines and great schooners of similar tonnage could be a more economical means of transportation. For example, one lumber buyer wrote from Honolulu in 1898 that he could not see how steamships could be made to pay in the lumber trade. He believed that, unlike sailing vessels, steamships could be made to pay only where cargo could be carried both out and back. "However," he added, "enterprising Americans may be able to do more than ordinary people can do."[40] Of the 308 vessels that carried lumber from Puget Sound in 1903, 277 were sailing craft. As late as the 1920s Robert Dollar and a

1895, EL.

36. Oregon Board of Pilot Commissioners, "Record of Arrivals and Departures, June 1, 1898 to May 31, 1906," MS, Oregon Historical Society, Portland. Cf. F. W. Hibbs, "The Shipping and Shipbuilding of Puget Sound," Society of Naval Architects and Marine Engineers *Transactions* 12 (1904):262–65.

37. The Company, purely a marketing agency, had no mills of its own. It dispatched cargoes to China, Japan, Australia, and Ecuador. See Pacific Export Lumber Co., records of charters, MSS, Oregon Historical Society, Portland.

38. Walker to W. Talbot, 2 Oct. 1891, AC, Walker letter books, private.

39. F. Talbot to C. Walker, 5 Oct. 1894, AC, Walker, personal corres.

40. J. Allen to J. Campbell, 25 Oct. 1898, PBMCP, box 35. See also *Canada Lumberman* 2 (1882):276–77; 4 (1884):210.

few others were still shipping lumber in sailing vessels, but the dominance of such craft had long since passed. By this time steam vessels carried all but a tiny percentage of the lumber shipped by sea. In 1924 even the indomitable Dollar abandoned the use of sail. A shortage of tonnage during the early years of the new century had kept many a sailing vessel in operation, but even then the end of sail was clearly in sight. By the twenties it was clear that there was to be no resurrection.[41]

Soon after sailing vessels began to disappear from the Pacific lumber trade, steam schooners began to vanish, too. These little vessels had been perfectly suited for servicing the tiny dogholes and outports of the northern California coast and transporting lumber from there to San Francisco. But they could match the rates charged by full-sized steamships only on short routes, and by the end of the nineteenth century the main centers of production were increasingly distant from San Francisco. Some of the major redwood producers commenced operations in the forests of the Pacific Northwest. For a time modifications and improvements in design kept steam schooners competitive on these longer runs, but the larger steamers were also being improved and they finally reached a point where they could transport lumber from the Pacific Northwest to San Francisco more cheaply than steam schooners could bring it from the redwood coast. When that point was reached, steam schooners rapidly disappeared. Most of the mills they had serviced closed. The only significant lumber ports that remained could be safely reached by steamships.[42]

Changing means of transporting cargoes of lumber affected the smaller mills all along the Pacific Coast. Steamships could not enter many of the so-called harbors. Mills located on them tried shipping their cut by barge to the larger ports, but the added expense ate into their profits. Many, including William Kyle's mill at Florence, simply

41. Newell, ed., *McCurdy Marine History*, pp. 58–59, 69, 309–10; Dollar, *Memoirs*, 2:42–43, 144; O'Brien, "Life of Robert Dollar," pp. 243–51, 395–97; Hibbs, "Shipping and Shipbuilding of Puget Sound," p. 263; Pope & Talbot to C. Walker, 14 Feb. 1898, AC, Walker, incoming corres.; Portland *Oregonian*, 17 Apr. 1902.

42. McNairn and MacMullen, *Ships of the Redwood Coast*, pp. 41–42, 100–1; Dickie, "Pacific Coast Steam Schooner," pp. 39–48. Even around Humboldt Bay conditions were changing. "The golden days of the industry had passed. It had by the turn of the century become big business and most of the glamor and excitement was gone" (Melendy, "One Hundred Years of the Redwood Industry," p. 173). Large as well as small mills were disappearing. At this point, only five of the old-time firms of the county were still in operation. Cf. Wilde, "Chronology of the Pacific Lumber Company," p. 75; Defebaugh, *History of the Lumber Industry of America*, 1:272.

closed.[43] But a location accessible to steamers did not necessarily solve a small mill's transport problems. When the steamship *Inverness* cleared from Portland in June 1900 bound for Japan and China, it carried 2.6 million board feet of lumber and railroad ties.[44] Only the largest mills could produce enough lumber to fill the holds of such vessels promptly. Indeed, large as they were, even such firms as the Port Blakely Mill Company and the Puget Mill Company sometimes found it convenient to join in furnishing cargoes, rather than keep steamships waiting while the mill struggled to produce enough lumber to fill their commodious holds.[45] Thus smaller cargo mills increasingly found themselves in the position of having to market their cut through those that were larger. As in the case of barging lumber from ports too small for the steamers, this marketing arrangement ate into the profits of the smaller mills and made them less competitive. As a result, though the late nineties and the first years of the twentieth century were a prosperous period for Pope and Talbot and other giants of the cargo trade, they were years of crisis for a number of their lesser competitors. Many of the latter did not survive.[46] One observer wrote: "All things considered, the lot of the mill-man in British Columbia cannot be called an overly happy one."[47] The same thing could easily have been said about their counterparts south of the international boundary.

As the smaller cargo mills disappeared one by one, they were following a path that many of the giants would also be following in the not-too-distant future. Mills in Seattle, Tacoma, Portland, Hoquiam, and elsewhere were located where they could service both rail and cargo markets. These were in an enviable position. Their broader market base gave them a more diversified and stable demand than that enjoyed by mills which could service only markets reached by sea. In the years to come, this was to be a decisive factor in their economic struggle for survival.[48]

43. Cf. *Canada Lumberman* 3 (1883):87.
44. Portland *Oregonian*, 24 June 1900. Cf. Hibbs, "Shipping and Shipbuilding of Puget Sound," pp. 263–64.
45. Pope & Talbot to Puget Lumber Co., 4 Mar., 14 June 1899, AC, Puget Lumber Co., incoming corres.; F. Hill, report on mill companies, 13 Sept. 1897, AC, Walker, personal corres.
46. Puget Lumber Co. to Pope & Talbot, 2 Sept. 1898, AC, Walker, personal corres. Renton, Holmes & Co. to Pt. Blakely Mill Co., 6 Sept. 1901, PBMCP, box 65. Melendy, "One Hundred Years of the Redwood Industry," pp. 290–91. Wilde, "Chronology of the Pacific Lumber Company," pp. 67–68. Ryder, *Memories of the Mendocino Coast*, p. 43. *Pacific Coast Wood and Iron*, 28 (1900):55; 35 (1901):55; 37 (1902):12.
47. *Canada Lumberman* 19 (Jan. 1898):3.
48. Northwestern Lumber Co. to A. Simpson, 8 Nov. 1895, EL; Clark, "Analysis of Forest Utilization," pp. 87 and *passim*; Coman and Gibbs, *Time, Tide and Timber*, pp. 218ff. As early as

، By the beginning of the twentieth century the old cargo trade was dying. New mills, new markets, and new means of transportation were shaping a new industry. The cargo trade had played a major role in the development of the Far West. It had drawn men and capital into the region, opened areas bypassed by agriculture and mining, broadened the economic base, and, through trade with distant ports, given the region's citizenry wider horizons. In its own way, the industry helped make San Francisco the cosmopolitan center it became. The cargo trade's leaders failed in their attempts at building a highly centralized industry such as emerged in oil, steel, and tobacco, but that failure must be charged in large part to circumstances inherent to the industry rather than to failures of leadership. Their successors have been little more successful than were the pioneers of the cargo trade in removing the uncertainties that spring from fragmented, intensely competitve operations.

The lesson is one that deserves attention. In an age of consolidation and diminishing competition, in an age of monopolies and oligopolies, not all facets of American industry trod the same path. The cargo mills made up a facet where technological advance and integration of operations were not unknown, but where consolidation was halting, limited, and largely unsuccessful. The story of this industry, and others like it, is as much a part of the history of the age of enterprise as are the operations of the John D. Rockefellers of the age. And, though the builders of the cargo trade failed to arrive at the same final destination as other industrialists of the period, their influence was, within the region at least, just as lasting. Lumbermen who followed them built, and continue to build, on a foundation of knowledge and experience these pioneers provided. From British Columbia to the redwoods, their influence continues to be felt, both by the new generation of lumbermen and by the larger citizenry which continues to look, as did the owners of the cargo mills, out to sea as well as inland for opportunity and profits.

1896 George Emerson was reporting that the Northwestern Lumber Co.'s cargo mill at Hoquiam was rapidly approaching the point where it would be the second or third most important of those mills shipping by rail. See Northwestern Lumber Co. to A. Simpson, 13 Mar. 1896, EL.

APPENDIX I

LUMBER PRICES IN SAN FRANCISCO, 1849–99
(In dollars per MBF)

DATE	REDWOOD				OREGON PINE			
	Rough			Planed	Rough			Planed
	At Mill	Cargo Lots	Retail		At Mill	Cargo Lots	Retail	
1849 July	...	300–350	...	...	...	300–350	...	...
Oct.	...	250–275	...	...	100	250–275	...	...
1850 Jan.	...	100–150	...	...	150	100–150	...	...
Apr.	...	60–100	...	...	75–80	60–100	...	...
July	...	...	...	60	50–60	...	...	60
1851 Jan.	...	...|	...	50	40–50	...	...	50
July	...	...	...	50–60	35–45	...	...	50–60
1852 Jan.	...	...	...	75	...	...	...	75
Aug.	...	45–55	...	...	25	45–47	...	...
1853 Jan.	...	80–90	...	...	...	55–60	...	...
July	...	50–55	...	...	20	35	...	...
1854 Jan.	...	40	...	...	...	...	...	...
Mar.	...	45	...	...	...	40	...	...
July	...	25	...	...	10–15	30–32	...	...
1855 Jan.	...	18–20	...	...	...	20	...	...
July	...	...	...	...	...	...	...	...
1856 Jan.	...	24.50	...	...	...	...	...	...
July	...	23	...	...	...	...	...	...
1857 Jan.	...	22	...	...	...	22.50–26	...	...
July	...	18–20	...	...	...	23	...	...
1858 Jan.	14	23	...	25	...	22	...	...
July	...	...	...	...	...	...	...	...
1859 Jan.	...	...	...	...	...	...	...	...
July	...	25	...	...	14	...	...	...
1860 Jan.	...	...	...	...	12	...	...	...
July	...	...	...	...	...	...	...	...
1861 Jan.	...	...	...	...	...	...	...	...
July	...	23	...	...	...	...	...	...
1862 Jan.	...	20	...	...	...	...	...	...
July	...	16–18	...	...	...	...	...	...
1863 Jan.	...	16	20	26	...	...	...	...
July	...	16	...	26	...	...	...	...

DATE	REDWOOD				OREGON PINE			
	Rough			Planed	Rough			Planed
	At Mill	*Cargo Lots*	*Retail*		*At Mill*	*Cargo Lots*	*Retail*	
1864 Jan.	...	18	...	30	...	...	...	...
July	...	18	...	30	11	...	...	...
1865 Jan.	...	16–18	...	26	...	...	...	...
July	...	16	...	28	...	18	...	...
1866 Jan.	...	18	20	28	10	22.50	...	...
July	...	18	...	28	...	...	...	...
1867 Jan.	...	20	...	30	...	22.50	...	...
July	...	20	...	30	...	18	...	...
1868 Jan.	...	20	...	30	...	...	...	...
July	...	20	...	32	...	20	...	...
1869 Jan.	...	20–22.50	...	32	...	20	...	...
July	14	20	40–42 [?]	35	...	...	...	...
1870 Jan.	...	16–18	...	24	...	14–16	...	...
July	...	15–18	...	24	...	14–16	...	...
1871 Jan.	...	16–18	...	24	10–11	17	...	...
July	...	15	...	26	8.50–10	14–15	...	...
1872 Jan.	...	16	...	30	9–10	16	...	...
July	16	16–17	...	30	10–12	17–18	...	...
1873 Jan.	...	20	20	32	...	...	15–16	...
July	...	20	20	32	...	12.50–15	15	...
1874 Jan.	...	17	20	27	...	...	16–17	...
July	...	14–16	20	25	9–10	...	20	...
1875 Jan.	...	15–20	20	30	9–12	...	20	...
July	...	18	22.50	30	...	17–18	22.50	...
1876 Jan.	...	18	20	30	9–12	17–18	22.50	...
July	...	18	20	30	11	18	20	...
1877 Jan.	...	15	20	25	11	18	20	...
July	...	14	20	24	11	18	20	...
1878 Jan.	...	13	20	23	9–12	16–18	20	...
July	...	10–14	...	24	10	16–18	18	...
1879 Jan.	...	14	...	22	...	...	...	...
July	...	12	...	18	...	...	...	...
1880 Jan.	...	15	18	20	...	14	18	...
July	...	16	18	23	...	14	18	...
1881 Jan.	...	16	20	26	...	16	20	...
July	...	17	21	26	...	17	21	...

DATE	REDWOOD				OREGON PINE			
	Rough			Planed	Rough			Planed
	At Mill	*Cargo Lots*	*Retail*		*At Mill*	*Cargo Lots*	*Retail*	
1882 Jan.	...	18	22	28	...	18	22	...
July	...	18	22	30	...	18	22	...
1883 Jan.	...	18	22.50	...	...	18	22.50	...
July	...	18	22.50	...	...	18	22.50	...
1884 Jan.	...	18	22.50	...	...	18	22.50	...
July	...	18	22.50	...	...	18	22.50	...
1885 Jan.	...	...	...	...	...	...	...	...
July	...	13	...	...	...	...	...	...
1886 Jan.	...	14	17	...	...	...	15	...
July	...	...	18	...	...	...	15	...
1887 Jan.	...	16	20	...	9	...	18	...
July	...	17	21	...	...	...	19	...
1888 Jan.	...	19	...	...	11	...	20	...
July	...	18	...	...	11	...	19	...
1889 Jan.	...	15	...	...	11	14	...	...
July	...	14	19	...	11	...	19	...
1890 Jan.	...	14	...	...	...	14	20	...
July	...	16	20.50	...	10	14	20.50	24
1891 Jan.	...	16	21	...	10	15	21	24.50
July	...	14	19	...	9	13	19	22.50
1892 Jan.	...	14	19	...	9	13	18	21.50
July	...	14	19	...	9	12.50	18	...
1893 Jan.	...	13–14	19	...	9.50	12	17	...
July	...	15	19	...	9.50	12	17	...
1894 Jan.	...	15	19	...	9.50	12	17	...
July	...	14	17	...	9.50	12	13	...
1895 Jan.	...	14	17	...	9.50	10–12	13	16.50
July	...	14	17	...	...	10–12	13	...
1896 Jan.	...	14	16	...	9.50	10–12	13	...
July	...	13	16	...	8	11	13	...
1897 Jan.	...	13	16.50	...	10	12.25	13	...
July	...	13	15	...	...	12.50	13–15	...
1898 Jan.	...	12	19	...	...	12.50	17	...
July	...	13	16	...	...	12.50	15–16	...
1899 Jan.	...	13	14	...	...	13.50	15	...
July	...	14	17	31	...	13.50	16	...

Sources: Prices current published in various newspapers, especially San Francisco *Alta California;* Redwood Manufacturers Association, *Redwood Lumber* (San Francisco: Redwood Manufacturers Association, 1882); J. R. Hanify & Co., *Wholesale Price List Established March 28, 1899* (San Francisco: J. R. Hanify & Co., 1899); *Pacific Coast Wood and Iron,* 1884–1900.

Note: The prices listed in this appendix should be used with caution. As is made clear in Chapter 11, there was often a difference between the published prices of firms and/or associations and what buyers were in fact obliged to pay. Moreover, grading standards and terminology changed from time to time, thus prices might change either more or less than the figures supplied seem to indicate. There are insufficient data available to make adjustments in the figures quoted here. Finally, the available sources often fail to make it clear just what category of lumber the prices cited are for. To reduce confusion, the author has omitted data that appear to reflect clerical or typographical errors and has placed a question mark in brackets behind other figures that seem especially open to question. Newspapers listed lumber prices at irregular intervals. Prices derived from newspapers are for the earliest date on which lumber prices are given following 1 January and 1 July.

In spite of the problems associated with these data, this appendix can be useful in indicating approximate short-range fluctuations in price, certain long-range trends, and the relative prices of redwood and Oregon pine (Douglas fir). It should be used, of course, in conjunction with the related portions of the text.

APPENDIX 2

LUMBER PRODUCTION, 1849–1900

TABLE 1: LUMBER PRODUCTION BY MAJOR POLITICAL SUBDIVISION
(in MBF)

Year	California	Oregon	Washington	British Columbia
1849[a]	5,000	16,853	4,080	. . .
1859	196,000	41,169	77,125	1,750[b]
1869	353,842	75,193	128,743	25,000[b]
1879	326,340	177,171	160,176	50,000[b]
1889	528,554	444,565	1,061,560	67,612
1899	737,760	734,181	1,428,205	252,580

TABLE 2: ESTIMATED LUMBER PRODUCTION DOUGLAS FIR/COAST
REDWOOD REGION (in MBF)

Year	California	Oregon	Washington	British Columbia
1849[a]	. . .	16,852	4,080	. . .
1859	. . .	41,169	77,125	2,000[b]
1869	196,899	59,000	109,000	22,000[b]
1879	238,121	140,000	134,000	40,000[b]
1889	286,000	360,000	882,000	45,000[b]
1899	398,859	596,151	1,320,067	200,000[b]

SOURCES: F. L. Moravets, *Production of Lumber in Oregon and Washington, 1869–1948*, Forest Survey Report no. 100 (Portland: U.S. Forest Service, Pacific Northwest Forest and Range Experiment Station, 1949); Richard H. May, *A Century of Lumber Production in California and Nevada*, Forest Survey Report no. 20 (Berkeley: U.S. Forest Service, California Forest and Range Experiment Station, 1953); H. N. Whitford and Roland D. Craig, *Forests of British Columbia* (Ottawa: Canada, Commission of Conservation, Committee on Forests, 1918); Edmond S. Meany, Jr., "History of the Lumber Industry of the Pacific Northwest to 1917."

[a] Lumber production for Oregon in 1849 includes only those counties of Oregon Territory located south of the Columbia River; Washington production figures for 1849 are for those counties of Oregon Territory north of the Columbia River.

[b] Author's estimates based on scattered, fragmentary sources, primarily reports published in *Canada Lumberman*.

TABLE 3: REDWOOD REGION LUMBER PRODUCTION BY COUNTIES, 1858–1900 (in MBF)

Year	Santa Cruz	Sonoma	Mendocino	Humboldt	Del Norte	Total
1858	10,000					
1859				9,575		
1860	10,000		35,000 [?]			
1861				14,968		
1862						
1863						
1864						
1865	9,200			12,693		
1866	10,500		40,114	18,942		
1867	12,346		46,000	20,375		
1868	9,000 [?]			30,250		
1869	19,000				2,000	196,899
1870					4,000	
1871						
1872			50,000	27,648		
1873	27,600			33,943		
1874				48,635		
1875	21,000		50,000	77,000	7,000	
1876	15,000 [?]		53,000	64,000	10,000	
1877						
1878						
1879				32,349		238,121
1880	50,000 [?]			36,969	4,000	
1881	40,000	3,918	54,037	71,068	1,440	130,505
1882						152,517
1883		18,948	39,324	61,200	4,262	250,000
1884			53,000			208,455
1885		4,400	74,050	82,300	4,050	215,000
1886	42,002	7,140	88,393	93,147	14,515	248,439
1887				104,519		
1888						
1889			96,000	126,957		286,000
1890			116,035	160,000		
1891			99,438	152,517		
1892			98,754	166,855		
1893			116,753	160,000		
1894		9,901	82,011	110,898		
1895		8,289	110,062	127,037	4,008	
1896	11,000		87,744	99,772		
1897			103,621	99,973		
1898			93,054	95,817		
1899			99,712	117,993		398,859
1900			94,841	114,159		

SOURCES: U.S. Forest Service, California Forest and Range Experiment Station, *Bibliography of Early California Forestry* (Berkeley: U.S. Forest Service, 1939–41); California State Board of Forestry, *Biennial Report, 1885–1886*; Owen C. Coy, *The Humboldt Bay Region, 1850–1875*; *Pacific Coast Wood and Iron*, 1884–1900.

APPENDIX 3

LUMBER CONSUMPTION IN SAN FRANCISCO, 1860–1900

TABLE 1: LUMBER CARGOES RECEIVED IN SAN FRANCISCO
BY POINT OF ORIGIN, 1860–67 (in MBF)

Year	From Puget Sound	From Columbia River	From California[a]	Total
1860	36,227	6,488	32,808	75,523
1861	50,208	6,380	50,655	107,243
1866[b]	77,049	2,856	60,174	140,079
1867[b]	80,400	566	66,665	147,631

SOURCES: San Francisco *Alta California*, 7 Jan. 1862, 9 July 1867.

[a] Undoubtedly largely redwood, but with some Oregon pine and other species in cargo shipments; probably no pine from the interior reaching San Francisco this early.

[b] Estimates for the year based on data for the first six months, as data for the entire year not available.

Year	Redwood	Oregon Pine	Total
1863	43,022		
1864	41,591	66,000	107,591
1865	53,097	78,000	131,097
1866	60,174		
1867	66,665		
1868	84,754		
1869	81,899		
1870	87,706		
1871	75,295		
1872	94,268	126,754	232,214
1873	70,622	117,420	198,586
1874	92,326	139,856	248,147
1875	116,164	163,695	300,099
1876	115,951	161,338	304,624
1877	104,470	159,742	279,892
1878	93,546	148,128	258,711
1879	87,493	124,293	227,085
1880	80,348	114,390	214,385
1881	95,414	130,553	251,739
1882	94,606	138,738	233,344
1883	103,195	145,374	248,569
1884	103,941	177,305	281,246
1885	115,253	175,858	291,111
1886	98,246	182,998	281,244
1887	118,695	209,909	328,604
1888	135,782	266,449	402,231
1889	161,910	313,694	475,609
1890	166,274	237,196	403,470
1891	171,721	268,223	439,944
1892	186,809	253,135	439,944
1893	162,074	214,844	376,918
1894	104,565	214,354	318,919
1895	135,578	256,523	392,101
1896	109,613	258,431	368,044
1897	132,599	230,922	363,521
1898	118,148	215,036	333,184
1899	141,465	253,222	394,687
1900	136,760	236,934	373,694

SOURCES: Redwood Manufacturers Association, *Redwood Lumber: Statistics for the Port of San Francisco from 1862* (San Francisco: Redwood Manufacturers Association, 1882); California State Board of Forestry, *Biennial Report, 1885–1886*; San Francisco *Alta California*; California Lumber Dealers Exchange, *Annual Report* (San Francisco, 1883); *Pacific Coast Wood and Iron*, 1884–1900.

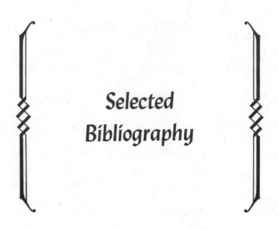

Selected
Bibliography

I. PRIMARY SOURCES

A. UNPUBLISHED MATERIALS

1. Personal Papers

Ainsworth, John C. Papers. University of Oregon library, Eugene.
Ames, E. G. Papers. University of Washington library, Seattle.
Emerson, George H. Letter books (microfilm). University of Washington library, Seattle.
Leidesdorff, William A. Papers. California Historical Society Library, San Francisco.
———. Papers. Henry E. Huntington Library, San Marino, Calif.
McLoughlin, John. Private papers (microfilm). University of Washington library, Seattle.
Minto, John. Papers. Oregon Historical Society, Portland.
Semple, Eugene. Papers. University of Washington library, Seattle.
Stearns, Abel. Papers. Henry E. Huntington Library, San Marino, Calif.
Walker, Elkanah. Papers. Henry E. Huntington Library, San Marino, Calif.
Yesler, Henry, Papers. University of Washington library, Seattle.

2. Company Records

Adams and Company. Papers. Henry E. Huntington Library, San Marino, Calif.
Astoria Mill Company. Account book. Oregon Historical Society, Portland.
Gardiner Mill Company. Papers. University of Oregon library, Eugene.
Hoquiam Mill Company. "Shipyard Lumber Book." University of Washington library, Seattle.
Hudson's Bay Company. Fort Nisqually papers. Henry E. Huntington Library, San Marino, Calif.

John Kentfield and Company. Papers. Bancroft Library, Berkeley, Calif.

Oregon Improvement Company. Papers (microfilm). University of Washington library, Seattle.

Pacific Export Lumber Company. Records of charters. Oregon Historical Society, Portland.

Pope and Talbot, Incorporated. Papers. Pope and Talbot, Inc., San Francisco, Calif.

Port Blakely Mill Company. Papers. University of Washington library, Seattle.

Washington Mill Company. Papers (microfilm). University of Washington library, Seattle.

William Kyle and Sons. Papers. University of Oregon library, Eugene.

3. Diaries, Memoirs, and Reminiscences

Abrams, William Penn. Diary. Bancroft Library, Berkeley, Calif.

Ainsworth, John C. "Autobiography of John C. Ainsworth, Oregon Capitalist." University of Oregon library, Eugene.

Brown, Charles. "Statement of Recollections of Early Events in California." Bancroft Library, Berkeley, Calif.

Douglas, George. "History of the Pacific Lumber Company . . . as told to D[erby] Bendorf" (typed transcript). Bancroft Library, Berkeley, Calif.

Pond, Ananias Rogers. "Journal, 1852–1862." Henry E. Huntington Library, San Marino, Calif.

Stimson, Thomas D. "A Record of the Family, Life and Activities of Charles Douglas Stimson" (photocopy). University of Washington library, Seattle.

Wilde, W. H. "Chronology of the Pacific Lumber Company, 1869 to 1945" (typescript). Bancroft Library, Berkeley, Calif.

4. Government Documents

Oregon Board of Pilot Commissioners. "Record of Arrivals and Departures, June 1, 1898 to May 31, 1906." Oregon Historical Society, Portland.

United States Customs House (Astoria, Ore.). "Marine List and Memoranda." Oregon Historical Society, Portland.

———. Papers (microfilm). Oregon Historical Society, Portland.

United States Customs House (Puget Sound). Papers. Federal Records Center, Seattle.

United States Department of State. Consular despatches, Hong Kong. Microfilm copy M-108. National Archives, Washington, D.C.

———. Consular despatches, Shanghai. Microfilm copy M-112. National Archives, Washington, D.C.

———. Consular despatches, Sydney, New South Wales. Microfilm copy M-173. National Archives, Washington, D.C.

———. Consular despatches, Tientsin. Microfilm copy M-114. National Archives, Washington, D.C.

B. PUBLISHED MATERIALS

1. Letters, Journals, and Diaries

Colvile, Eden. *London Correspondence Inward from Eden Colvile, 1849–1852.* Edited by E. E. Rich, assisted by A. M. Johnson. Publications of the Hudson's Bay Record Society, 19. London: Hudson's Bay Record Society, 1956.

Duflot de Mofras, Eugène. *Travels on the Pacific Coast.* Translated by Marguerite Eyer Wilbur. 2 vols. Santa Ana, Calif.: Fine Arts Press, 1937.

Hargrave, James. *The Hargrave Correspondence, 1821–1843.* Edited by G. P. de T. Glazebrook. Publications of the Champlain Society, 24. Toronto: The Champlain Society, 1938.

Larkin, Thomas Oliver. *The Larkin Papers: Personal, Business, and Official Correspondence of Thomas Oliver Larkin, Merchant and United States Consul in California.* Edited by George P. Hammond. 10 vols. Berkeley: University of California Press for the Bancroft Library, 1951–64.

McLoughlin, John. *Letters of Dr. John McLoughlin Written at Fort Vancouver, 1829–1832.* Edited by Dr. Burt Brown Barker. Portland: Binfords and Mort for the Oregon Historical Society, 1948.

———. *The Letters of John McLoughlin from Fort Vancouver to the Governor and Committee.* Edited by E. E. Rich. 3 vols. Publications of the Champlain Society, 4, 6, 7. Toronto: The Champlain Society, 1941–44.

Norton, Mrs. Zachariah. "Voyage of the Sequin, 1849." *OHQ* 34 (1933):255–58.

Simpson, George. *Fur Trade and Empire: George Simpson's Journal, Entitled Remarks Connected with the Fur Trade in the Course of a Voyage from York Factory to Fort George and Back to York Factory, 1824–25, with Related Documents.* Edited with a new introduction by Frederick Merk. Rev. ed. Cambridge, Mass.: The Belknap Press of Harvard University Press, 1968.

———. *Part of Dispatch from George Simpson, Esqr., Governor of Ruperts Land, to the Governor & Committee of the Hudson's Bay Company, London, March 1, 1829. . . .* Edited by E. E. Rich. Publications of the Champlain Society, Hudson's Bay Company Series, 10. Toronto: The Champlain Society, 1947.

Sutter, John A. *The Diary of Johann August Sutter.* San Francisco, Calif.: Grabhorn Press, 1932.

———. *New Helvetia Diary: A Record of Events Kept by John A. Sutter and His Clerks at New Helvetia from September 9, 1845, to May 25, 1848.* San Francisco, Calif.: Grabhorn Press, 1939.

Sylvester, Avery. "Voyages of the Pallas and Chenamus, 1843–45." *OHQ* 34 (1933):259–72, 359–71.

Wood, Tallmadge B. "Letters of Tallmadge B. Wood." Edited by Florence E. Baker. *OHQ* 3 (1902):395–98; 4 (1903):80–85.

2. Recollections, Reminiscences, and Memoirs

Ames, E. G. "Port Gamble, Washington." *WHQ* 16 (1925):17–19.

Brown, John Henry. *Reminiscences and Incidents of Early Days of San Francisco (1845–50)*. San Francisco, Calif.: Grabhorn Press, 1933.

Clayson, Edward, Sr. *Historical Narratives of Puget Sound: Hood's Canal, 1865–1885: The Experience of an Only Free Man in a Penal Colony.* Seattle, Wash.: R. L. Davis Printing Co., 1911.

Denny, Arthur A. *Pioneer Days on Puget Sound.* Edited by Alice Harriman. Seattle, Wash.: A. Harriman Co., 1908.

Dollar, Robert. *Memoirs of Robert Dollar.* 4 vols. San Francisco, Calif.: for the author by W. S. Van Cott and Co., 1917–25.

Duhaut-Cilly, Auguste. "Account of California in the Years 1827 and 1828." Translated by Charles Franklin Carter. *CHSQ* 8 (1929): 131–66, 214–50, 306–36.

Hopkins, Caspar T. "The California Recollections of Caspar T. Hopkins." *CHSQ* 25 (1946):97–120, 255–66, 325–46; 26 (1947):63–75, 175–83, 253–66, 351–63; 27 (1948):65–73, 165–74, 267–74, 339–51.

Judd, Laura Fish. *Honolulu: Sketches of Life in the Hawaiian Islands from 1828 to 1861.* Honolulu, Hawaii: Star-Bulletin Publishing Co., 1928.

Meares, John. *Voyages Made in the Years 1788 and 1789, from China to the North West Coast of America. . . .* London: Logographic Press, 1790.

Newmark, Harris. *Sixty Years in Southern California, 1853–1913, Containing the Reminiscences of Harris Newmark.* Edited by Maurice H. and Marco R. Newmark. 4th ed., rev. and augm. Los Angeles: Zeitlin and Ver Brugge, 1970.

Ormsby, Warren. "Peeling the Tanoak." *Forest History* 15 (1972):6–10.

Rutledge, Peter J., and Richard H. Tooker. "Two Views of the Dolbeer Donkey." *Forest History* 14 (1970):18–29.

Seymore, W. B. "Port Orchard Fifty Years Ago." *WHQ* 8 (1941):257–60.

Whiting, Perry. *Autobiography of Perry Whiting, Pioneer Building Material Merchant of Los Angeles.* Los Angeles: Smith-Barnes Corp., 1930.

Williams, Edward C. "The First Redwood Operations in California." *Pioneer Western Lumberman* 58 (1912):9–13.

3. Trade and Manufacturers' Association Reports and Government Documents

California State Agricultural Society. *Transactions* (1864–65, 1878, 1879). Sacramento: State Printer, 1866, 1878, 1880.

California State Board of Forestry. *Third Biennial Report* (1889–90). Sacramento: State Printer, 1890.

———. *Fourth Biennial Report* (1891–92). Sacramento: State Printer, 1892.

California State Mineralogist. *Second Biennial Report* (1880–82). Sacramento: State Printer, 1882.

China, Inspectorate General of Customs. *Decennial Reports on the Trade, Navigation, Industries, Etc., of the Ports Open to Foreign Commerce in China and Corea. . . .* Shanghai: Imperial Maritime Customs, 1904.

China, Inspectorate General of Customs. *Returns of Trade and Trade Reports* (1882–1900). Shanghai: Imperial Maritime Customs, 1883–1901.

Endacott, G. B. *An Eastern Entrepôt: A Collection of Documents Illustrating the History of Hong Kong.* London: Her Majesty's Stationery Office, 1964.

Hall, Henry. *Report on the Shipbuilding Industry of the United States.* United States Census Office, *Report on the Tenth Census, 1880*, vol. 8. Washington, D.C.: U.S. Government Printing Office, 1884.

Portland Board of Trade. *Annual Report* (various years) (microfilm). University of Oregon library, Eugene.

Redwood Manufacturers' Association. *Redwood Lumber: Statistics for the Port of San Francisco from 1862.* San Francisco, Calif.: Redwood Manufacturers' Association, n.d.

Selfridge, E. A. *American Lumber in Japan.* Washington, D.C.: U.S. Government Printing Office, 1928.

United States Bureau of Foreign Commerce. *American Lumber in Foreign Markets.* United States Bureau of Foreign Commerce, Special Consular Reports, vol. 11. Washington, D.C.: U.S. Government Printing Office, 1894.

4. Journals

Canada Lumberman and Millers', Manufacturers', and Miners' Gazette, 1880–1902.

Hunt's Merchants' Magazine and Commercial Review, 1846–70.

Pacific Coast Wood and Iron, 1884–1900.

Pacific Lumber Trade Journal, 1895–1900.

West Shore, 1875–91.

5. Newspapers

Arcata (California) *Humboldt Times*, 1854–1900 (after 1858 published in Eureka).

Florence (Oregon) *The West*, 1893–1904.

Hong Kong *China Mail*, 1850–60.

Hong Kong *Daily Advertiser*, 1871–73.

Hong Kong *Times, Daily Advertiser and Shipping Gazette*, 1873–76.

Honolulu *Pacific Commercial Advertiser*, 1856–84.

Honolulu *Polynesian*, 1840–63.

Olympia (Washington) *Columbian*, 1852–53.

Olympia *Pioneer and Democrat*, 1852–61.

Oregon City *Oregon Spectator*, 1846–55.

Portland *Oregonian*, 1861–1900.

San Francisco *Alta California*, 1849–91.

Scottsburg (Oregon) *Umpqua Weekly Gazette*, 1854–55.

Seattle *Weekly Intelligencer*, 1867–80.

Shanghai *North China Herald*, 1850–1900.

Tientsin *Chinese Times*, 1886–91.

Tientsin *Peking and Tientsin Times*, 1894–1902.

Vancouver (Washington) *Clarke County Register*, 1884–87.

Vancouver (Washington) *Independent*, 1880–93.
Victoria (British Columbia) *Colonist*, 1858–1900.

II. SECONDARY SOURCES

A. THESES AND DISSERTATIONS

Bauer, Patricia M. "The History of Lumbering and Tanning in Sonoma County, California, since 1812." Master's thesis, University of California, Berkeley, 1950.

Buchanan, Iva L. "An Economic History of Kitsap County, Washington, to 1889." Ph.D. dissertation, University of Washington, 1930.

Clark, Donald Hathway. "An Analysis of Forest Utilization as a Factor in Colonizing the Pacific Northwest and in Subsequent Population Transitions." Ph.D. dissertation, University of Washington, 1952.

Cox, John H. "Organizations of the Lumber Industry of the Pacific Northwest, 1889–1914." Ph.D. dissertation, University of California, Berkeley, 1937.

Finger, John Robert. "Henry L. Yesler's Seattle Years, 1852–1892." Ph.D. dissertation, University of Washington, 1968.

Gedosch, Thomas. "Seabeck: The Story of a Company Town, 1856–1886." Master's thesis, University of Washington, 1967.

Huff, Boyd Francis. "The Maritime History of San Francisco Bay." Ph.D. dissertation, University of California, Berkeley, 1956.

Hynding, Alan Andrew. "A Biography of Governor Eugene Semple." Ph.D. dissertation, University of Washington, 1965.

Johnson, Robert E. "Schooners Out of Coos Bay." Master's thesis, University of Oregon, 1953.

Lawrence, Joseph Collins. "Markets of Capital: A History of the Lumber Industry of British Columbia." Master's thesis, University of British Columbia, 1960.

MacDonald, Alexander Norbert. "Seattle's Economic Development, 1880–1910." Ph.D. dissertation, University of Washington, 1959.

Meany, Edmond S., Jr. "History of the Lumber Industry of the Pacific Northwest to 1917." Ph.D. dissertation, Harvard University, 1935.

Melendy, Howard Brett. "One Hundred Years of the Redwood Lumber Industry, 1850–1950." Ph.D. dissertation, Stanford University, 1952.

Sanford, Everett R. "A Short History of California Lumbering Including a Descriptive Bibliography of Material on Lumbering and Forestry in California." Master's thesis, University of California, Berkeley, 1931.

Tattersall, James N. "The Economic Development of the Pacific Northwest to 1920." Ph.D. dissertation, University of Washington, 1960.

B. BOOKS AND PAMPHLETS

Allen, G. C., and Audrey G. Donnithorne. *Western Enterprise in Far Eastern Economic Development: China and Japan.* London: Allen and Unwin, 1954.

[Alley, B. F.] *History of San Mateo County, California, including its geography, topography, geology, climatography, and description* . . . San Francisco, Calif.: B. F. Alley, 1883.

Bagley, Clarence B. *History of Seattle from the Earliest Settlement to the Present Time.* 3 vols. Chicago: S. J. Clarke Publishing Co., 1916.

Bancroft, Hubert Howe. *Chronicles of the Builders of the Commonwealth: Historical Character Study.* 7 vols. San Francisco, Calif.: The History Co., 1891–92.

————. *History of British Columbia, 1792–1887.* San Francisco, Calif.: The History Co., 1887.

————. *History of California.* 7 vols. San Francisco, Calif.: A. L. Bancroft and Co., 1884–90.

————. *History of Oregon.* 2 vols. San Francisco, Calif.: The History Co., 1886–88.

A Brief Statement of the Political Value of the Hawaiian Treaty to the United States. N.p. [1882?].

Brown, Alan K. *Sawpits in the Spanish Redwoods.* San Mateo, Calif.: San Mateo County Historical Association, 1966.

Childs, John M. *Del Norte County As It Is.* Crescent City, Calif.: John M. Childs, 1894.

Clar, C. Raymond. *California Government and Forestry: From Spanish Times to the Creation of the Department of Natural Resources in 1927.* Sacramento, Calif.: State of California, Department of Natural Resources, Division of Forestry, 1959.

————. *Harvesting and Use of Lumber in Hispanic California.* Sacramento, Calif.: Sacramento Corral of Westerners, 1971.

Clark, Donald H. *Eighteen Men and a Horse.* Seattle, Wash.: Metropolitan Press, 1949.

Clark, Norman H. *Mill Town: A Social History of Everett, Washington, from Its Earliest Beginnings on the Shores of Puget Sound to the Tragic and Infamous Event Known as the Everett Massacre.* Seattle: University of Washington Press, 1970.

Coman, Edwin T., Jr., and Helen M. Gibbs. *Time, Tide and Timber: A Century of Pope & Talbot.* Stanford, Calif.: Stanford University Press, 1949.

Coy, Owen C. *The Humboldt Bay Region, 1850–1875: A Study in the American Colonization of California.* Los Angeles: California State Historical Association, 1929.

Defebaugh, James Elliott. *History of the Lumber Industry of America.* 2 vols. 2nd ed. Chicago: The American Lumberman, 1906–7.

Dixon, Les B. *The Birth of the Lumber Industry in British Columbia.* Vancouver: British Columbia Lumberman, 1956.

[Edgar Cherry and Company]. *Redwood and Lumbering in California Forests.* . . . San Francisco: Edgar Cherry and Co., 1884.

Elchibegoff, Ivan M. *United States International Timber Trade in the Pacific Area.* Stanford, Calif.: Stanford University Press, 1949.

Fairburn, William Armstrong. *Merchant Sail.* 6 vols. Center Lovell, Me.: Fairburn Marine Educational Foundation, 1945–55.

Genzoli, Andrew M., and Wallace E. Martin. *Redwood Bonanza, a Frontier's Reward: Lively Incidents in the Life of a New Empire.* Eureka, Calif.: Schooner Features, 1967.

Grattan, C. Hartley. *The Southwest Pacific to 1900, A Modern History: Australia, New Zealand, the Islands, Antarctica.* Ann Arbor: University of Michigan Press, 1963.

————. *The United States and the Southwest Pacific.* Cambridge, Mass.: Harvard University Press, 1961.

Greenwood, Gordon. *Early American-Australian Relations from the Arrival of the Spaniards in America to the Close of 1830.* Melbourne, Australia: Melbourne University Press, 1944.

Griffin, Eldon. *Clippers and Consuls: American Consular and Commercial Relations with Eastern Asia, 1845–1860.* Ann Arbor, Mich.: Edwards Brothers, 1938.

Haden-Guest, Stephen, et al., eds. *A World Geography of Forest Resources.* New York: Ronald Press, 1956.

Harrison, E. S. *History of Santa Cruz County, California.* San Francisco, Calif.: Pacific Press Publishing Co., 1892.

Hawley, A. T. *The Climate, Resources and Advantages of Humboldt County, California, Described in a Series of Letters to the San Francisco Daily Evening Bulletin.* . . . Eureka, Calif.: J. E. Wyman and Son, 1879.

Hittell, John S. *The Commerce and Industries of the Pacific Coast of North America;* . . . San Francisco, Calif.: A. L. Bancroft and Co., 1882.

————. *A History of the City of San Francisco and Incidentally of the State of California.* San Francisco, Calif.: A. L. Bancroft and Co., 1878.

Holbrook, Stewart H. *The American Lumberjack.* Rev. ed. of *Holy Old Mackinaw.* New York: Collier Books, 1962.

————. *Green Commonwealth: A Narrative of the Past and a Look at the Future of One Forest Products Community.* 2nd ed., enl. Seattle, Wash.: for Simpson Logging Co. by Frank McCaffrey, 1945.

Hoopes, Chad L. *Lure of Humboldt Bay Region.* Dubuque, Iowa: W. C. Brown Book Co., 1966.

Hopkins, Caspar T. *Ship Building on the Pacific Coast (No. 2).* San Francisco, Calif.: Board of Marine Underwriters, 1874.

Hutchins, John G. B. *The American Maritime Industries and Public Policy, 1789–1914: An Economic History.* Cambridge, Mass.: Harvard University Press, 1941.

Jarves, James J. *History of the Hawaiian or Sandwich Islands.* Boston: Tappan and Bennet, 1843.

Johansen, Dorothy O., and Charles M. Gates. *Empire of the Columbia: A History of the Pacific Northwest.* New York: Harper and Brothers, 1957. 2nd ed., 1967.

John Kentfield and Company et al. *Petition to the U.S. Senate and House of Representatives by Ship Builders, Ship Owners and Lumber Merchants of*

Pacific Coast States, Relating to the Treaty of Reciprocity between the United States and the Hawaiian Islands. San Francisco, Calif.: Cunningham, Curtis, and Welch, 1886.

Levin, Jonathan V. *The Export Economies: Their Pattern of Development in Historical Perspective.* Cambridge, Mass.: Harvard University Press, 1960.

Lockley, Fred. *History of the Columbia River Valley from the Dalles to the Sea.* 3 vols. Chicago: S. J. Clarke Publishing Co., 1928.

Lower, A. R. M. *The North American Assault on the Canadian Forest: A History of the Lumber Trade between Canada and the United States.* Toronto: Ryerson Press, 1938.

Lucia, Ellis. *Head Rig: The Story of the West Coast Lumber Industry.* Portland, Ore.: Overland West Press, 1965.

Mears, Eliot Grinnell. *Maritime Trade of the Western United States.* Stanford, Calif.: Stanford University Press, 1935.

Menefee, C. A. *Historical and Descriptive Sketch Book of Napa, Sonoma, Lake and Mendocino Counties comprising sketches of their topography, productions, history, scenery, and peculiar attraction.* Napa City, Calif.: Reporter Publishing House, 1873.

Moravets, F. L. *Lumber Production in Oregon and Washington, 1869–1948.* Forest Survey Report, no. 100. Portland: Pacific Northwest Forest and Range Experiment Station, 1949.

Morgan, Theodore. *Hawaii, a Century of Economic Change, 1778–1876.* Cambridge, Mass.: Harvard University Press, 1948.

[Munro-Fraser, J. P.] *History of Sonoma County with a Full and Particular Record of Spanish Grants. . . .* San Francisco, Calif.: Alley, Bowen and Co., 1880.

[Murray, W. H.] *The Builders of a Great City: San Francisco's Representative Men: The City, Its History and Commerce. . . .* San Francisco, Calif.: San Francisco Journal of Commerce, 1891.

Newell, Gordon, ed. *The H. W. McCurdy Marine History of the Pacific Northwest: An Illustrated Review of the Growth and Development of the Maritime Industry from 1895. . . .* Seattle, Wash.: Superior Publishing Co., 1966.

Palmer, Lyman L. *History of Mendocino County, California.* San Francisco, Calif.: Alley, Bowen and Co., 1880.

Perloff, Harvey S., et al. *Regions, Resources, and Economic Growth.* Baltimore, Md.: Johns Hopkins Press, for Resources for the Future, 1960.

Pomeroy, Earl. *The Pacific Slope: A History of California, Oregon, Washington, Idaho, Utah, and Nevada.* New York: Alfred A. Knopf, 1965.

Porter, Glenn, and Harold C. Livesay. *Merchants and Manufacturers: Studies in the Changing Structure of Nineteenth-Century Marketing.* Baltimore, Md.: Johns Hopkins Press, 1971.

Radius, Walter A. *United States Shipping in Transpacific Trade, 1922–1928.* Stanford, Calif.: Stanford University Press, 1944.

Rich, E. E. *Hudson's Bay Company, 1670–1870.* 3 vols. New York: Macmillan, 1960.

Ryder, David Warren. *Memories of the Mendocino Coast: being a brief account of the discovery, settlement and development of the Mendocino coast, together with the correlated history of the Union lumber company and how the coast and the company grew up together.* San Francisco, Calif.: Union Lumber Co., 1948.

Soulé, Frank, John H. Gihon, and James Nisbet. *The Annals of San Francisco: Containing a Summary of the History of the First Discovery, Settlement, Progress, and Present Condition of California, and a Complete History of All the Important Events Connected with Its Great City: To Which Are Added Biographical Memoirs of Some Prominent Citizens.* New York: D. Appleton and Co., 1855.

Stanger, Frank M. *Sawmills in the Redwoods: Logging on the San Francisco Peninsula, 1849–1967.* San Mateo, Calif.: San Mateo County Historical Association, 1967.

Tate, Merze. *Hawaii: Reciprocity or Annexation.* East Lansing: Michigan State University Press, 1968.

Taylor, Paul S. *The Sailors' Union of the Pacific.* New York: Ronald Press, 1923.

Throckmorton, Arthur L. *Oregon Argonauts: Merchant Adventurers on the Western Frontier.* Portland: Oregon Historical Society, 1961.

Underhill, Reuben L. *From Cowhides to Golden Fleece: A Narrative of California, 1832–1858, Based upon Unpublished Correspondence of Thomas Oliver Larkin of Monterey, Trader, Developer, Promoter, and Only American Consul.* Stanford, Calif.: Stanford University Press, 1939.

Vinnedge, Robert W. *The Pacific Northwest Lumber Industry and Its Development.* New Haven, Conn.: Yale University, School of Forestry, 1923.

Wright, E. W., ed. *Lewis and Dryden's Marine History of the Pacific Northwest: An Illustrated Review of the Growth and Development of the Maritime Industry, from the Advent of the Earliest Navigators to the Present Time, with Sketches and Portraits of a Number of Well Known Marine Men.* Portland, Ore.: Lewis and Dryden Printing Co., 1895.

[W. W. Elliott and Company]. *History of Humboldt County, California.* San Francisco, Calif.: W. W. Elliott and Co., 1882.

———. *Santa Cruz County, California: Illustrations descriptive of its scenery, fine residences, public buildings, manufactories, hotels, farm scenes, business houses, schools, churches, mines, mills, etc. . . .* San Francisco, Calif.: W. W. Elliott and Co., 1879.

Zon, Raphael, and William N. Sparhawk. *Forest Resources of the World.* 2 vols. New York: McGraw-Hill Book Co., 1923.

C. ARTICLES, ESSAYS, AND PAPERS

Baker, Abner. "Economic Growth in Portland in the 1880s." *OHQ* 67 (1966):105–23.

Beckham, Stephen Dow. "Asa Mead Simpson, Lumberman and Shipbuilder." *OHQ* 68 (1967):259–73.

Berner, Richard C. "The Port Blakely Mill Company, 1876–1889." *PNQ* 57 (1966):158–71.

Bradley, Harold Whitman. "California and the Hawaiian Islands, 1846–1852." *PHR* 16 (1947):18–29.

Buchanan, Iva L. "Lumbering and Logging in the Puget Sound Area in Territorial Days." *PNQ* 27 (1936):34–53.

Burgess, Sherwood. "Lumbering in Hispanic California." *CHSQ* 41 (1962):237–48.

———. "The Forgotten Redwoods of the East Bay." *CHSQ* 30 (1951):1–15.

Carranco, Lynwood, and Mrs. Eugene Fountain. "California's First Railroad: The Union Plank Walk, Rail Track, and Wharf Company Railroad." *Journal of the West* 3 (1964):243–56.

Clar, C. Raymond. "John Sutter, Lumberman." *Journal of Forestry* 56 (1958):259–65.

Clark, Donald H. "Sawmill on the Columbia." *The Beaver* 281 (1950):42–44.

Cox, John H. "Trade Associations in the Lumber Industry of the Pacific Northwest, 1899–1914." *PNQ* 41 (1950):285–311.

Cox, Thomas R. "The Lower Columbia Lumber Industry, 1880–1893." *OHQ* 67 (1966):160–78.

———. "Lumber and Ships: The Business Empire of Asa Mead Simpson." *Forest History* 14 (1970):16–26.

———. "The Passage to India Revisited: Asian Trade and the Development of the Far West, 1850–1900." In *Reflections of Western Historians*, edited by John A. Carroll, pp. 85–103. Tucson: University of Arizona Press, 1969.

Dickie, David W. "The Pacific Coast Steam Schooner." In *Historical Transactions, 1893–1943*, pp. 39–48. New York: Society of Naval Architects and Marine Engineers, 1945.

Doig, Ivan. "John J. McGilvra & Timber Trespass: Seeking a Puget Sound Timber Policy, 1861–1865." *Forest History* 13 (1970):6–17.

Drury, Aubrey. "John Albert Hooper." *CHSQ* 31 (1952):289–305.

Gibbs, Helen M. "Pope and Talbot's Tugboat Fleet." *PNQ* 42 (1951):302–23.

Hutchinson, William H. "California's Economic Imperialism: An Historical Iceberg." In *Reflections of Western Historians*, edited by John A. Carroll, pp. 67–83. Tucson: University of Arizona Press, 1969.

Ingersoll, Ernest. "In a Redwood Logging Camp." *Harper's New Monthly Magazine* 66 (1883):193–210.

Lamb, W. Kaye. "Early Lumbering on Vancouver Island." *British Columbia Historical Quarterly* 2 (1938):31–53, 95–121.

Lawrence, Joseph Collins. "California's Influence on the Industrial and Commercial Development of British Columbia, 1858–1885." Paper presented at the 61st annual meeting of the Pacific Coast Branch, American Historical Association, Santa Clara, Calif., 29 Aug. 1968.

Loehr, Rodney C. "Saving the Kerf: The Introduction of the Band Saw Mill." *Agricultural History* 23 (1949):168–72.

Lomax, Alfred L. "Dr McLoughlin's Tropical Trade Route." *The Beaver* 295 (1964):10–15.

——. "Hawaii–Columbia River Trade in Early Days." *OHQ* 43 (1942):328–38.

Manning, William R. "The Nootka Sound Controversy." In American Historical Association *Annual Report* (1904), vol. 1, pp. 279–478. Washington, D.C.: U.S. Government Printing Office, 1905.

Maunder, Elwood R. "Building on Sawdust." *PNQ* 51 (1960):57–62.

Meany, Edmond S. "First American Settlement on Puget Sound." *WHQ* 7 (1916):136–43.

Melendy, H. Brett. "Two Men and a Mill: John Dolbeer, William Carson, and the Redwood Lumber Industry in California." *CHSQ* 38 (1959):59–71.

Palais, Hyman, and Earl Roberts. "The History of the Lumber Industry in Humboldt County." *PHR* 19 (1950):1–16.

Parker, Robert J. "Larkin, Anglo-American Businessman in Mexican California." In *Greater America: Essays in Honor of Herbert Eugene Bolton*, edited by Adele Ogden and Engel Sluiter, pp. 415–29. Berkeley: University of California Press, 1945.

——. "Larkin's Monterey Business: Articles of Trade, 1833–1839." *Historical Society of Southern California Quarterly* 24 (1942):54–60.

Scammon, C. M. "Lumbering in Washington Territory." *Overland Monthly* 5 (1870):55–60.

Smalley, Brian H. "Some Aspects of the Maine to San Francisco Trade, 1849–1852." *Journal of the West* 6 (1967):593–603.

Stokes, Evelyn. "Kauri and White Pine: A Comparison of New Zealand and American Lumbering." *Annals of the Association of American Geographers* 56 (1966):440–50.

Throckmorton, Arthur L. "George Abernethy, Pioneer Merchant." *PNQ* 48 (1957):76–88.

Thrum, Thomas G. "History of the Hudson's Bay Company Agency in Honolulu." Hawaiian Historical Society *Eighteenth Annual Report* (1911), pp. 35–49.

Vassault, F. I. "Lumbering in Washington." *Overland Monthly*, new series 20 (1892):23–32.

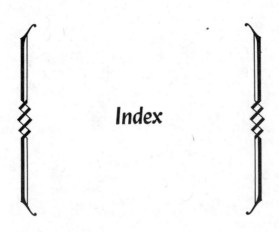

Index